ENCYCLOPEDIA OF FIRE PROTECTION

Second Edition

DENNIS P. NOLAN, P.E., PHD

Australia • Canada • Mexico • Singapore • Spain • United Kingdom • United States

Encyclopedia of Fire Protection, Second Edition

Dennis P. Nolan, P.E., PhD

Vice President, Technology and Trades ABU:
David Garza

Director of Learning Solutions:
Sandy Clark

Acquisitions Editor:
Alison Weintraub

Product Manager:
Jennifer A. Thompson

Marketing Director:
Deborah S. Yarnell

Channel Manager:
William Lawrensen

Marketing Coordinator:
Penelope Crosby

Production Editor:
Toni Hansen

Editorial Assistant:
Maria Conto

Printed in the United States of America
1 2 3 4 5 XX 08 07 06

For more information contact
Thomson Delmar Learning
Executive Woods
5 Maxwell Drive, PO Box 8007,
Clifton Park, NY 12065-8007
Or find us on the World Wide Web at
www.delmarlearning.com

Library of Congress Cataloging-in-Publication Data

Nolan, Dennis P.
Encyclopedia of fire protection / Dennis P. Nolan.—2nd ed.
p. cm.
Includes bibliographical references and index.
ISBN 1-4180-2014-1 (alk. paper)
1. Fire protection engineering—Encyclopedias. I. Title.

TH9116.N65 2006
628.9′2203—dc22

2006005171

NOTICE TO THE READER

Dedicated to

Kushal, Nicholas, and Zebulon

and also

Mom and Dad

Words are, of course, the most powerful drug used by mankind.

Rudyard Kipling

(1923)

The world, an entity out of everything, was created by neither gods nor men, but was, is and will be eternally living fire, regularly becoming ignited and regularly becoming extinguished.

Heraclitus

The Cosmic Fragments, no. 20 (c. 480 B.C.)

Contents

Preface

How to Use this Book

It is intended that this book will greatly aid those individuals just entering the professional fire service, experienced individuals and those individuals from other organizations and industries seeking fire protection knowledge.

Therefore, students in fire protection or fire protection engineering programs, as well as professional firefighters and fire protection engineers will find this a handy reference.

Much information on fire protection topics is available from various sources, but most sources do not define their terminology or topics, nor do they provide a history or source of their meaning. Having worked in the fire protection profession for over 30 years, I still find terminology that is unfamiliar. Additionally, with the advancement of fire technology and scientific investigations, new fire protection terms and methods are constantly being developed, and obsolete or archaic terms are being retired. Older dictionaries and references have also unwittingly omitted numerous terms. The purpose of this book is to provide a reference source for the background, meaning, and description of fire protection terms being used in government, industry, research, and education today. Additionally pictures, diagrams, and tables or graphs further enhance the reader's knowledge of the subject.

The field of fire protection encompasses various unrelated organizations and industries, such as the insurance field, research entities, and educational organizations. Many of these organizations may not realize that their individual terminology may not be understood by individuals or even be compatible with the nomenclature used outside their own sphere of influence. It is therefore prudent to have a basic understanding of these individual terms.

This encyclopedia is based mainly on the terminology used in United States fire codes and regulations. When terminology has been used that is significantly known internationally, this has also been noted.

New to this Edition

The second edition has been expanded to include additional terminology, more photos and diagrams and further cross-references:

- ***New and Updated Terms*** help students and professionals keep current with changes in the industry.
- ***Extensive Art*** provides a visual for various terms and concepts to aid learning and ensure accurate references.
- ***Cross References*** relate common terms and concepts to help further understanding of topic.

Also Available

This book serves as a handy companion to the following Thomson Delmar Learning titles:

Firefighter's Handbook: Essentials of Firefighting and Emergency Response, 2E

Provides coverage of Firefighter I and II training to prepare students to take their certification exam. (Order #: 1-4018-3575-9)

Introduction to Fire Protection, 3E by Robert Klinoff

Introduces students to the fire service, including coverage of the process, as well as the necessary skills to be a successful firefighter. (Order #: 1-4180-0177-5)

Visit us at www.firescience.com

Acknowledgments

The following individuals or organizations have graciously provided information, advice, or assistance in the preparation of this book.

Kathleen H. Almand, Society of Fire Protection Engineers (SFPE); Dennis J. Berry, National Fire Protection Association (NFPA); Dr. John L. Bryan, Professor Emeritus, University of Maryland; Dr. Douglas Drysdale, University of Edinburgh; Federal Emergency Management Administration (FEMA); Thomas Gallacher, Consultant; Larry Jackman, Greenheck Fan Corporation; Trevor Kletz, Loughborough University; Arthur C. Stevens (posthumously), Consultant; Roger S. Wilkins, Grinnel Corporation; Kristin Young, Cambridge Consulting Corporation; and the staff of Thomson Delmar Learning.

The author and Publisher would like to thank the following reviewers for the comments and suggestions they offered during the development of each edition. Our gratitude is extended to:

Richard Beckman
Rio Hondo College
LaHabra Heights Fire Dept.
LaHabra Heights, CA

Terrence Malone
District of Columbia Fire Dept.
Washington, DC

David Einspahr
Aims Community College
Greeley, CO

Lee Cooper
Wisconsin Indianhead Technical College
New Richmond, WI

Theodore Cashel
Township of Princeton Fire Dept.
Princeton, NJ

Gary Coley
Fox Valley Technical College
Neenah, WI

Rudy Horist
Elgin Community College
Elgin Fire Dept.
Elgin, IL

Lonnie Inzer
Pikes Peak Community College
Colorado Springs, CO

Bill Lowe
Clayton County Fire Department
Fayetteville, GA

Lee Silvi
Lakeland Community College
Mentor, OH

Eddie Smith
Crafton Hills College
Yucaipa, CA

About the Author

Dr. Dennis P. Nolan has had a long career devoted to fire protection engineering, risk engineering, loss prevention engineering, and system safety engineering. He holds a degree of Doctor of Philosophy (PhD) in business administration from Berne University, a Master of Science degree in Systems Management from Florida Institute of Technology and a Bachelor of Science degree in Fire Protection Engineering from the University of Maryland. Dr. Nolan is a US-registered Fire Protection Engineering Professional Engineer in the State of California.

He is currently on the executive staff of the Saudi Arabian Oil Company (Saudi Aramco), located in Abqaiq, Saudi Arabia, site of some of the largest oil and gas fields in the world. This operation also contains the largest oil and gas production and separation facilities in the world. The magnitude of the risks, worldwide sensitivity, and foreign location make this one of the most highly critical fire-risk operations in the world. He has also been associated with Boeing, Lockheed, Marathon Oil Company, and Occidental Petroleum Corporation in various fire protection engineering, risk analysis, and safety roles in several locations in the United States and overseas. As part of his career, he has examined oil production, refining, and marketing facilities under severe conditions and in various unique worldwide locations, including Africa, Asia-Pacific, Europe, the Middle East, Russia, and North and South America. His activity in the aerospace field has included engineering support for the NASA Space Shuttle launch facilities at the Kennedy Space Center (and for those facilities undertaken at Vandenburg Air Force Base, California) and for the "Star Wars" defense systems.

Dr. Nolan has received numerous safety awards and is a member of the American Society of Safety Engineers. He was a member of the National Fire Protection Association, Society of Petroleum Engineers, the Society of Fire Protection Engineers, and the Fire Protection Working Group of the UK Offshore Operators Association (UKOOA). He is the author of many technical papers and professional articles in various international fire safety publications. He has written three previous books, *Application of HAZOP and What-If Safety Reviews to the Petroleum, Petrochemical, and Chemical Industries; Handbook of Fire and Explosion Protection Engineering Principles for Oil, Gas, Chemical, and Related Facilities;* and *Fire Fighting Pumping Systems at Industrial Facilities.* For many years, Dr. Nolan has also been listed in *Who's Who in California* and has been included in the sixteenth edition of *Who's Who in the World.*

Ablation—Removal of material by erosion, evaporation, or reaction for short-term protection against high temperatures. Ablative materials form a thermal layer to protect the underlying structure. Spacecraft typically utilize an ablative material (heat shield) for reentry into Earth's atmosphere.

Ablative Water—A firefighting agent made by mixing water with additives to form a dense, heat-absorbing blanket. It is usually used to fight forest fires and is commonly dropped by aircraft.

Ablaze—Term used to describe a material vigorously on fire or burning. *See also* **Burn; Combustion.**

Abort Switch—A manually activated switch provided for fixed gaseous fire-suppression systems to cancel the signal to release and discharge the system agent. The use of an abort switch is preferred over other manual means, such as portable fire extinguishers or when a false alarm condition is immediately known, to avoid the unnecessary release of large quantities of fire suppression agent. An abort switch is only practical where individuals may be present to immediately investigate the cause of the system activation and have time to activate the switch.

Absolute Pressure—A pressure measurement that includes atmospheric pressure in the amount indicated. *See also* **Atmospheric Pressure.**

Absorption—A method of controlling a spilled material by the application of a substance that is able to absorb (take in or suck up) the subject material that has spilled.

Accelerant—A material (gas, liquid, or solid) used to initiate or promote the spread of a fire incident. Most accelerants are highly flammable. An arsonist often uses an accelerant such as gasoline to initiate his or her actions, which can be discovered during a fire investigation or reconstruction. The rate of accelerant flame-spread can be compared to flame spreads for the combustible contents of the overall fire spread during a building fire to determine if an accelerant were involved and the actual accelerant used.

Accelerator—A device provided on a dry pipe sprinkler system, at the alarm check valve, to promote the rapid operation of the valve. It directs the trapped system air pressure above the valve through a small diameter line to pass directly below the valve clapper once an initial drop in system pressure is sensed by the accelerator device. Once activated it equalizes the pressure above and below the valve and allows the alarm check valve to open rapidly, while an "exhauster" discharges the air above the alarm check valve to the atmosphere. Normally the pressure downstream of the alarm check valve (which is higher than the upstream side) is released by the activation of a sprinkler head on the sprinkler system due to the fire event.

A

This decrease has to be transmitted back to the alarm check valve before it will open, which necessarily delays its operation. An accelerator is an optional feature of the dry pipe alarm check valve that must be specified at the time of design and purchase. Due to the high water velocity at the time of initial alarm check valve operation, an accelerator may be susceptible to an inrush of water that may cause foreign matter and corrosion concerns. As a result, accelerators are fitted with antiflooding devices to prevent water from entering them. They may also be called quick opening devices (QODs). *See also* **Exhauster; Quick Opening Device (QOD).**

Accident—An unplanned event, sometimes but not necessarily injurious or damaging, that interrupts an activity; a chance occurrence arising from unknown causes, an unexpected happening due to carelessness, ignorance, and the like. It may lead to ill health, injury, property damage, environmental impact, loss of income, increased liabilities, or loss of prestige or self-esteem. All accidents are considered preventable if suitable precautions are used, and therefore current practice used the term incident rather than accident to describe the events.

Accidental Alarm—A fire alarm set off and transmitted through accidental activation of an automatic or manual fire alarm device, frequently caused by low air pressure on automatic sprinkler dry pipe valves, excessive or unexpected heat from industrial processes, cold weather, or human error. *See also* **False Alarm.**

Accordion Horseshoe Load or **Fold**—A method of storing fire hose. The hose is nested in a square-cornered, U-shaped fashion, similar to a horseshoe shape packed on Edge. Fire hoses on fire trucks are arranged in a particular fashion for ease of use and distribution at a fire incident. May also be called horseshoe load. See Figure A-1. *See also* **Accordion (Hose) Load** or **Fold; Donut Roll; Flat Load.**

Accordion (Hose) Load or **Fold**—A method of storing fire hose. The hose is configured in an accordion style (back and forth) and held in vertical stacks or laid in horizontal layers. See Figure A-2. *See also* **Accordion Horseshoe Load** or **Fold; Flat Load Hose Load; Hose Rack.**

Act of God—Terminology used to describe events that are beyond human control and generally of natural occurrence (earthquake, lightning, tornado) that may cause a fire to develop. Historical and predictive frequencies of occurrence are used to determine the risk from these events. This term is commonly used in the fire insurance industry.

Active Fire Protection (AFP)—A fire protection method that requires manual, mechanical, or other means of initiation, replenishment, or sustenance for its performance during a detected hazard or fire incident. Typical activations include switching on, directing, injecting, or expelling in order to combat smoke, flame, or

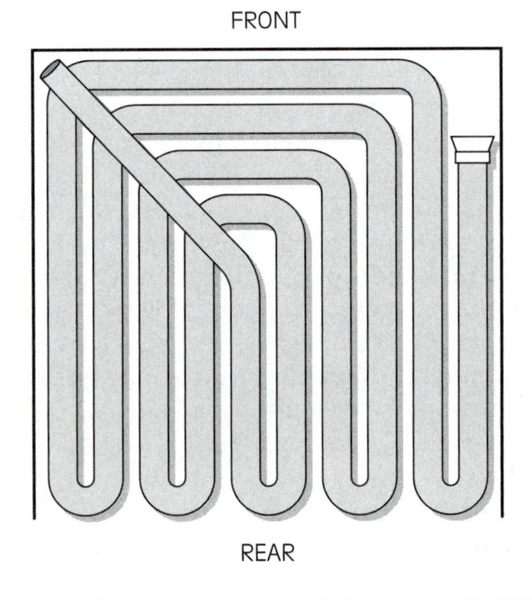

Figure A-1 Accordion horseshoe load (simplified for clarity).

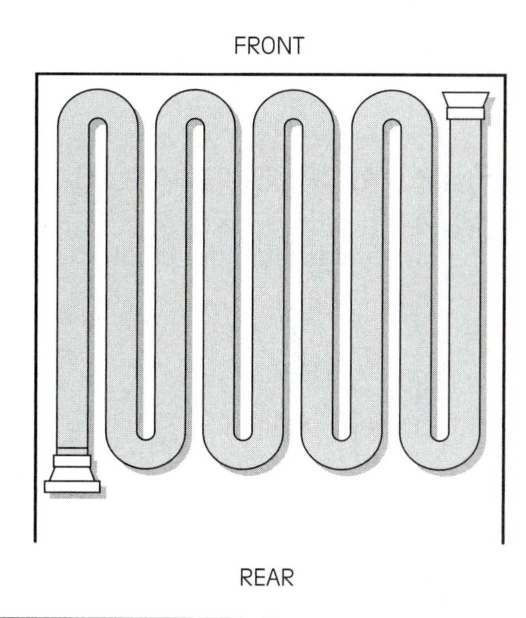

Figure A-2 Accordion (hose) load (simplified for clarity).

thermal loadings. Fire sprinkler systems and manual firefighting efforts are examples. Active systems are commonly composed of an integrated detection, signaling, and automated fire control system. Active fire protection systems are contrasted with passive fire protection systems that are in place and do not require additional actions at the immediate time of the fire for their protective qualities to be achieved. In general, passive fire protection systems are preferred over active systems due to inherent reliability because of their in situ nature. Active systems may be automatically or manually applied. *See also* **Automatic Fire Protection; Passive Fire Protection (PFP).**

Active Smoke Detection System—A fire detection system where smoke is transported to and into a sampling port to aspirate smoke into the detector-sensing chamber rather than relying totally on outside forces such as fire plume strength or environmental air flows. These systems actively draw smoke into the sensing chamber through the use of suction fans. Smoke in the immediate vicinity of the sampling ports is drawn into the detector-sensing chamber. See Figure A-3. *See also* **Aspirating Smoke Detection (ASD); Passive Smoke Detection System; Very Early Smoke Detection and Alarm (VESDA) System.**

Actual Delivered Density (ADD)—The rate at which water is actually deposited from operating sprinklers onto the top horizontal surface of a burning combustible array. ADD is a measure of the sprinkler distribution pattern and penetration capability in the presence of a fire plume. As the delay in sprinkler activation lengthens, the fire size increases and the ADD decreases. ADD is a function of the fire plume velocity, the water drop momentum and size, and the distance the drops must travel. The sooner water is applied to a growing fire, the lower the required delivered density (RDD) and the greater the actual delivered density. When the ADD is less than the RDD, the sprinkler discharge will no longer be effective to achieve early fire suppression. This terminology is mainly used in the evaluation of sprinklers in test environments. *See also* **Required Delivered Density (RDD).**

Actuate—The action of setting a fire alarm signal device or fire protection device into operation by either automatic or manual means.

Adapter—A hose connection device that allows two hoses to be linked together when they each contain different threads. Adapters have one end with a receiving end and the other with an inserting end. *See also* **Coupling.**

Adiabatic Flame Temperature—The maximum possible flame temperature that can be achieved by a particular combustion process (with no heat loss from the combustion). For example, the adiabatic flame temperatures for hydrocarbon fuels

A

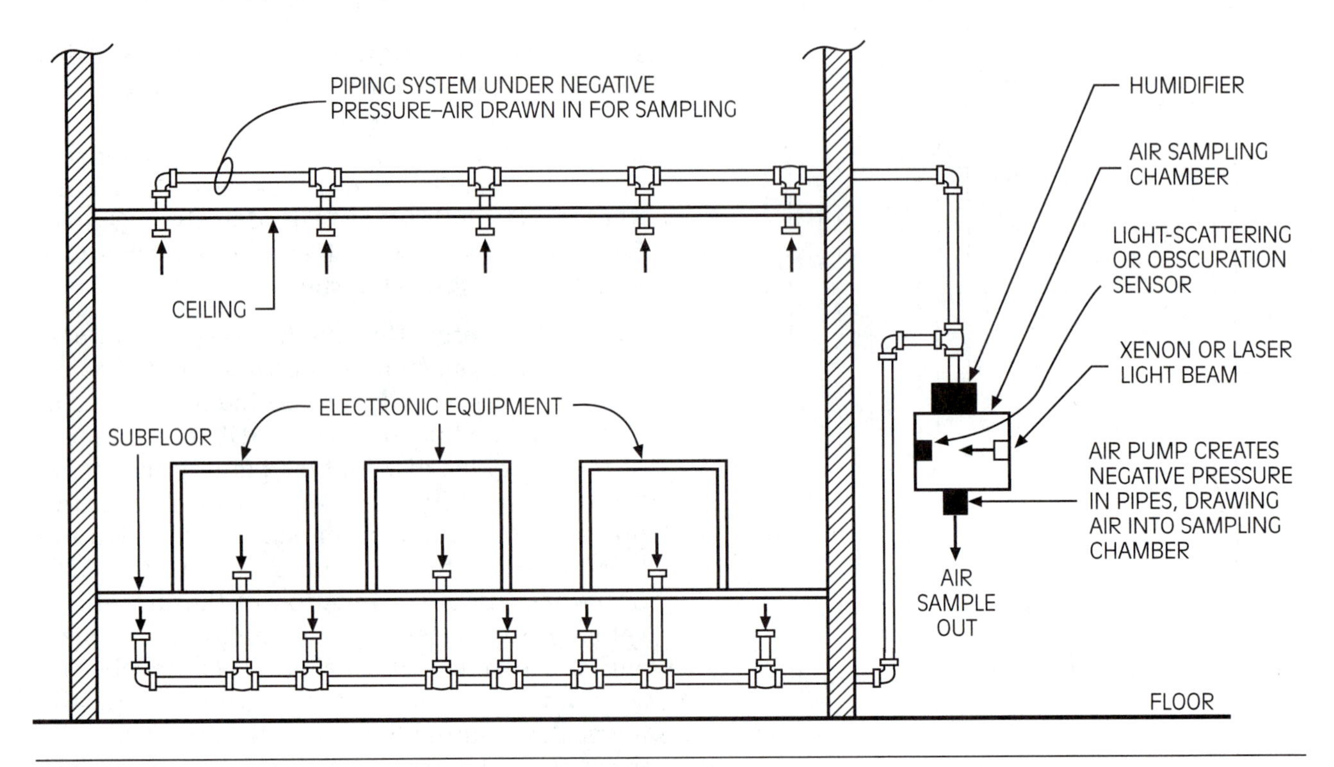

Figure A-3 Active smoke detection system.

burning in air range from 3,632°F to 4,172°F (2,000°C to 2,300°C) with actual measured laminar flame temperatures of 3,272°F to 3,632°F (1,800°C to 2,000°C). *See also* **Flame Temperature.**

Adsorption—A method of decontamination for individuals, equipment, and the environment. An adsorptive material, which acts like a sponge, will collect a spilled material when it is applied to it. The resulting material is to be treated a waste, and it will not change the characteristics of the material that has been collected. Additionary compatibility between the spilled material and the adsorptive material should be investigated to avoid adverse reactions.

Advanced Exterior Firefighting—Offensive firefighting activities performed on the outside of an enclosed structure that are begun after the fire incipient stage. They are usually related to major site fire hazards (flammable liquids, gases, or complicated hardware). *See also* **Offensive Firefighting.**

Advancing Line—A portion of a fire hose that is moved forward to the area of fire control and suppression.

Aerial Attack—The provision of firefighting materials from aircraft (helicopters or airplanes). Water, equipment, supplies, and personnel may be transported by aircraft and provided to the fire scene. Aerial attack operations are commonly employed for forest, brush, grassland, or wildland fires. *See also* **Fire Bombers.**

Aerial Device—An aerial ladder, elevating platform, aerial ladder platform, or water tower designed to position personnel, handle materials, provide egress, and discharge water at elevated locations. An aerial device may also refer to a pyrotechnical device.

Aerial (Fire) Apparatus—A fire service vehicle equipped with an aerial ladder, elevating platform, aerial ladder platform, or water tower. It is designed and equipped to support firefighting and rescue operations by positioning personnel, handling materials, providing continuous egress, or discharging water at elevated locations, which are beyond the length of ground ladders.

Aerial Ignition—As used in wildland fire fighting, the ignition of fuels by dropping incendiary devices or materials by aircraft to create a firebreak.

Aerial Ladder—A fire truck with an extendable ladder on a power-operated turntable used in fire rescue operations at elevated locations. The ladder is provided in several sections and extends to various heights. Extension of the ladder may be by electrical, hydraulic, or mechanical power devices. Early aerial ladders were made of wood and called "the big stick." Modern ladders are made of aluminum or steel. Before the advent of modern aerial ladders, water towers were used to apply water to the upper floors of buildings. The first successful aerial ladder in America was patented by firefighter Daniel Hayes of the San Francisco Fire Department in 1868. It was raised by firefighters using a series of gears and pulleys. Later models had spring assistance. See Figure A-4. *See also* **Aerial Ladder Platform; Water Tower.**

Aerial Ladder Platform—A type of elevating platform that includes the continuous escape capabilities of an aerial ladder. *See also* **Aerial Ladder; Aerial Platform.**

Aerial Platform—The snorkel aerial platform truck was invented in 1958 by fire commissioner Robert Quinn, who was inspired by watching workers moving an aerial basket. The aerial platform allows for stability and maneuverability that the ladder truck does not provide. It also provides a more stable rescue platform since personnel will not slip on water-soaked ladders. The platforms can typically support 900 lbs. (408 kg) of weight and provide a stream of 1,000 gpm (63 l/sec). It may also be called an elevating platform. *See also* **Aerial Ladder Platform; Elevating Platform; Water Tower.**

Afterdamp—The gas left after an explosion of firedamp in a mine, usually carbon monoxide. *See also* **Firedamp.**

Afterflame—Term used for the measurement of flame resistance of a specimen undergoing fire testing that measures the time of flaming of a specimen after the applied simulated fire exposure is discontinued. One of the tests for the flame resistance of materials, particularly textiles, is the length of time an afterflame exists (afterflame time). Afterflame periods are typically measured in seconds.

Figure A-4 Aerial ladder truck.

Afterglow—After initial combustion, some substances continue to emit light char, and smolder without flame. During this period, there can be a luminous glowing called afterglow. One of the ratings for the fire hazard of materials, particularly textiles, is the amount of afterglow that occurs.

Air (Aspirated) Sampling Smoke Detector—*See* **Active Smoke Detection System.**

Air Tanker—*See* **Tanker.**

Aircraft Rescue and Firefighting (ARFF)—Fire fighting operations that involve fixed or rotary wing aircraft, usually associated with airports or landing strips.

Airport Crash Truck—A fire pumper that sprays foam or dry chemicals on aircraft that are on fire at airports or have the potential for a fire incident. See Figure A-5. It may also be called an Aircraft Rescue Firefighting (ARFF) vehicle. *See also* **Crash Truck.**

Alarm—Signal indicating an emergency requiring an immediate action, such as an alarm for fire from a manual box, a water flow alarm, or an alarm from an automatic detection system. Alarms can be visual (flashing or strobe light/beacon), audible (bells, horns, buzzers), or both.

Alarm Check Valve—A valve provided in a sprinkler system that prevents the backflow of water in the system. It also initiates a fire alarm once water flow through the valve is achieved. Trapped water pressure in the system maintains the (clapper) valve in a closed position until the water pressure is released by the activation of a sprinkler head(s). This causes the valve to open or "trip." A tap on the valve is connected to a pressure switch. The pressure switch can then activate a fire alarm signal. See Figure A-6. *See also* **Clapper Valve; Dry Pipe Valve; Retard Chamber; Sprinkler System; Waterflow Alarm; Waterflow Detector.**

American Fire Sprinkler Association (AFSA)—A nonprofit international association organized in 1981 that represents open shop fire sprinkler contractors. AFSA was formed partly as a result of discussions by the National Fire Sprinkler Association (NFSA) to provide a sprinkler fitter training curriculum for open shop contractors. It is dedicated to the educational advancement

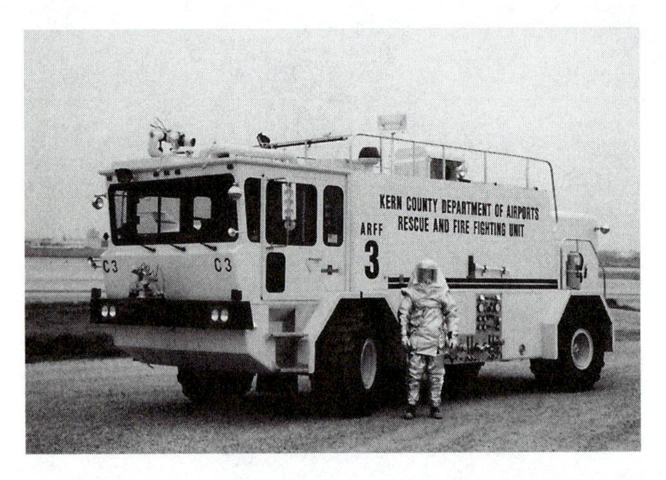

Figure A-5 Aircraft rescue firefighting (ARFF) vehicle.

Figure A-6 Alarm check valve in sprinkler riser, cutaway view showing internal clapper.

of its members and the promotion of the use of automatic fire sprinkler systems. It also provides training and consulting services, and acts as a communication forum and a liaison with other interested organizations. It provides representation on automatic fire sprinkler issues of concern to its members. *See also* **National Fire Sprinkler Association (NFSA).**

American Insurance Association (AIA)—A leading property and casualty insurance trade association representing over 435 companies that underwrite more than $120 billion in premiums annually. It was previously known as the National Board of Fire Underwriters (NBFU), which was established in 1866 and published the National Building Code in 1905. The code was used as a model for adoption by cities, as well as a basis to evaluate the building regulations of cities for insurance grading purposes. The AIA took its current form in 1964 when the old American Insurance Association merged with the National Board of Fire Underwriters (NBFU) and the Association of Casualty and Surety Companies. The AIA represents its members at all levels of government concerning legislative, regulatory, and legal issues. The insurance industry is primarily regulated at the state level, and the AIA is represented in every state. It is the only insurance trade organization that has provided this representation. *See also* **National Board of Fire Underwriters (NBFU).**

American National Fire Hose Connection Screw Thread—A standard cut of grooves for screwed fire appliance couplings and fixed connections adopted in the United States. The number and angles of grooves are based on a standardized pattern and have been specified by fire codes. The development and specifications for hose couplings were prepared by the National Fire Protection Association (NFPA) as early as 1898. Before 1905, most American fire departments adopted whatever fire hose coupling threads were convenient. This led to incompatibility problems for mutual aid assistance to locations that had a different coupling in use. Consequently additional fire damage could occur. The disastrous fire in Baltimore, Maryland in 1904 revealed the need for standardized fire hose couplings. As a result, the NFPA organized a committee with the assistance of the American Water Works Association (AWWA) and the International Association of Fire Chiefs (IAFC) to resolve the concern. An NFPA committee was appointed in 1905 and established a national standard thread of 2.5 in. (6.35 cm) and larger fire hose connections. Work on smaller fire hose threads was started in 1916. NFPA 1963, *Standard for Fire Hose Connections,* is now used to specify fire hose connections and has been applied throughout most of the United States. Fire hose threaded connections in the United States are referred to as American National Fire Hose Connection Screw Thread. They are commonly abbreviated as NH (also known as NST or NS). *See also* **National Standard Thread (NST);** the section on **Notable Fires Throughout History,** Entry on 1904 Baltimore, Maryland, USA.

American National Standards Institute (ANSI)—A voluntary private, nonprofit organization founded in 1918, composed of over 1,300 members that creates, adopts, and sets guidelines and standards for a wide range of technical areas. It is the umbrella body for standardization in the United States. Many National Fire Protection Association (NFPA) standards for fire safety have been incorporated as ANSI standards.

American Society of Testing Materials (ASTM)—A source of voluntary consensus standards on materials, products, systems, and services. Many fire testing standards have been developed by ASTM.

A

Analog Detector—A fire detector that senses data code. The central processor of the fire alarm panel determines if there is a fire present by interpreting the signal.

Anchor Point—A location determined to be safe from fire effects in which to begin line construction on a wildland fire. It is used to reduce the chance of fire fighters being flanked by a fire.

Annunciator Panel (AP)—A central control device for receiving and processing electronic input from the activation of fire detection devices, displaying or reporting these signals in recognizable manner (e.g., visual and audible) for emergency services and if specified, then acting on occurrences of fires within a building, for example, activating alarms. There are two types of annunciator panels: conventional panels and analogue addressable panels. Conventional panels usually have a small number of circuits, each circuit covering a zone within the building. A small map of the building is often placed near the main entrance with the defined zones drawn up, and light emitting diodes (LEDs) indicating whether a particular circuit/zone has been activated. Another common method is to have the different zones listed in a column, with an LED next to each zone name. The main drawback with conventional panels is that you cannot determine which device has been activated within a circuit. The fire may be in one small room, but as far as the brigade can tell, a fire could exist anywhere within a zone. Analogue panels are usually much more advanced than their conventional counterparts, with a higher degree of programming flexibility and single point detection. See Figure A-7. *See also* **Fire Alarm Control Panel (FACP).**

Apparatus—A vehicle or group of vehicles of any type that are utilized by the fire services.

Application Rate—The amount of foam concentrate or foam solution required to suppress a fire incident.

Applicator Pipe—A pipe attached to a nozzle that is curved or shaped to allow direct application of water spray from the nozzle over a burning object.

Approved—Term used by national or local fire codes to denote equipment, devices, tools, or systems meeting acceptable standards of fire safety performance for a particular purpose as verified by a recognized testing or professional agency. *See also* **Classified; Factory Mutual (FM); Labeled; Listed; Underwriter's Laboratories (UL).**

Aqueous Film Forming Foam (AFFF)—*See* **Foam.**

Aquifer—A geological formation, group of formations, or part of a formation that is capable of yielding water in a significant amount that would be useful for supplying fire protection applications. The capacity of the aquifer is dependent on its permeability, its proximity to sources of replenishment (such as rainwater and runoff) and other users of the aquifer. Underground sources of water can also contain many types of contaminants that can cause corrosion of fire protection systems or be harmful to personnel. These include, but are not limited to, pH level, salts such as chlorides, and entrained gases such as hydrogen gases, carbon dioxide (CO_2), and hydrogen sulfide (H_2S). Where the quality of water is questionable, it should be analyzed to determine its suitability.

Aramid—Generic name for high strength, flame resistant synthetic fabric used in fire fighters clothing. *See also* **Nomex.**

Arc—A luminous, high intensity electrical discharge in a gas or a vapor. Arcing may be a source of ignition for a fire if it is of sufficient energy. It may also be called spark discharge. *See also* **Ignitable Liquid; Incendive Arc; Incendive Spark; Non-Incendive; Spark; Spark Ignition.**

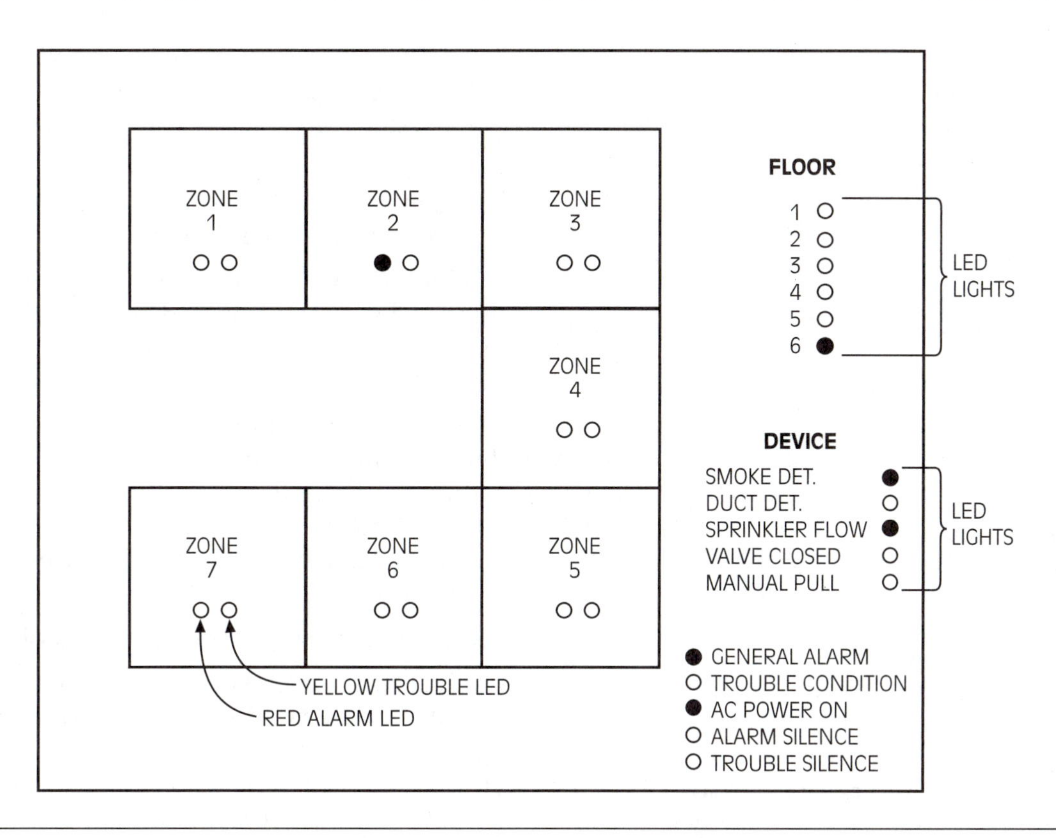

Figure A-7 Typical annunciator panel display.

Area Protection—The provision of water spray protection for a specified surface area for the purpose of fire protection.

Argonite™—A gaseous fire suppressant agent based on a naturally occurring substance that prevents harm to the environment if the agent is discharged into the atmosphere. It extinguishes fires based on the principle of oxygen depletion.

Around-the-Pump Proportioner—A foam proportioning arrangement whereby the pressure drop between the discharge and suction sides of a water pump is used to induce foam concentrate into the firewater stream by a suitable variable or fixed orifice connected to a venturi inductor in a bypass between the pump suction and the pump discharge. It may also be called a pump proportioning system. *See also* **Eductor; Proportioning.**

Arson—The crime of maliciously and intentionally, or recklessly, starting a fire or causing an explosion. An arsonist may utilize highly flammable liquids or explosives to spread the fire quickly and deliberately place obstructions to impede the actions of firefighters. Precise legal definitions vary among jurisdictions, wherein it is defined by statutes and judicial decisions. Arson is defined in the legal profession as the malicious and voluntary burning of the property of others without their consent.

If an act of arson causes death, the arsonist is guilty of murder even if he did not intend to kill. The definition of arson in legal terms has been so expanded that most jurisdictions have divided arson statutes into two or more degrees, reserving the heavier punishments for those occurrences that pose a danger to human life. Usually, those include the burning of habitable dwellings such as houses, stores, or factories, as well as vehicles, bridges, or trees. Some states also impose a higher penalty for arson committed for the purpose of concealing or destroying evidence of another crime. Lighter penalties are assigned to the new categories of arson endangering primarily property. Thus, it is arson to burn personal as well as real property or for a person to burn his own house to defraud an insurance company.

A fire caused by accident or ordinary carelessness is not arson, because maliciousness is lacking. A person may be guilty of arson, however, if he or she acts recklessly and disregards the consequences. An arsonist's motive is often irrelevant if he or she acts voluntarily and without the consent of the owner of the property. Under some legal standards for insanity, it may be a defense that the arsonist suffers from pyromania, an irresistible urge to set fires. *See also* **Fire Bomb; Firebug; Fire Investigation; Incendiarist; Pyromania.**

Arson Hotline—A special telephone answering operation for the purpose of receiving information from the public on arson incidents, to assist in the investigation and prosecution of those responsible for the arson incident.

Arsonist—*See* **Incendiarist.**

Articulating Boom Ladder—A fire service vehicle containing a series of booms and platform on the end. The platform is positioned by the movement of the booms.

Ash—Powdery residue from the burning of a combustible material, usually consisting of metal oxides or silicates. *See also* **Ash, Fly; Soot.**

Ash, Fly—Fine solid particles of ashes, dust, and soot carried out from a burning fuel. *See also* **Ash; Ember; Firebrand.**

Asphyxiation—A condition that causes death due to a lack of oxygen and an excessive amount of carbon monoxide or other toxic gases in the bloodstream.

Aspirating Discharge Device—A foam water delivery device that is specifically designed to aspirate and mix air into a liquid foam solution (concentrate and water) to generate foam with a specific discharge pattern for fire protection applications.

Aspirating Nozzle—*See* **Nozzle, Aspirating.**

Aspirating Smoke Detection (ASD)—The continuous sampling, through induction and transportation of a representative sample of air, from an area to a single smoke detector mounted nearby. An aspirating smoke detection system consists of a number of small-diameter sampling tubes distributed across a ceiling and/or across air conditioning supply points or other specified locations with holes in the tubes at various intervals to sample the ambient air. An aspirator (air suction pump) draws air into the tubes' holes to a central smoke detector. The sampled air is filtered to remove large particles and then passed to a sensing chamber to detect smoke particles. One sampling detector can cover 2,153 sq. ft. (200 m^2), equivalent to twenty point detectors. Aspirating smoke detection systems were originally developed for the Australian Post Office to prevent interruption in telecommunication facilities from fire hazards. *See also* **Active Smoke Detection System; Smoke Detector; Very Early Smoke Detection and Alarm (VESDA) System.**

Assistance Agreements—*See* **Automatic Aid.**

Astragal—A horizontal or vertical molding (overlapping or wraparound) attached to the meeting edge of one leaf of a pair of doors to protect against weather conditions; to minimize the passage of light between the doors; or to retard the passage of smoke, flame, or gases during a fire, and in the case of a half (Dutch) door, to ensure that the lower leaf of the door closes in conjunction with the upper leaf.

Atmospheric Pressure—The pressure exerted by the weight of the atmosphere. At sea level it is approximately 14.7 psi (101 kPa or 760 mm of mercury) under standard conditions. Atmospheric pressure decreases with an increase of altitude above sea level and increases below sea level.

Atmospheric Transmissivity—The ratio of the incident radiation to the emitted radiation from a fire, taking into account the loss of radiation due to scatter and absorption from the atmosphere. *See also* **Absolute Pressure.**

Atomization—The separation of atoms and molecules into an unconnected state where they are in suspension rather than in liquid form. *See also* **Water Mist.**

Attack, Fire—Activities undertaken by the fire service in combating a fire incident. *See also* **Initial Attack; Defenisve Attack; Direct Attack; Offensive Attack.**

Attack (Hose) Line—A firewater hose attached to a nozzle used in manual firefighting efforts to attack a fire incident. *See also* **Hose, Fire; Pre-connected.**

Authority Having Jurisdiction (AHJ)—The responsible organization (private or governmental) that requires, accepts, and approves fire protection measures for a location within its domain based on the regulations adopted for that domain. Most US states have a Fire Marshal's Office that is assigned the task of implementing and enforcing fire safety regulations. Private organizations usually have a fire protection engineer or loss prevention engineer or an insurance engineer who independently reviews project designs and existing facilities for compliance with governmental and company requirements.

Autoextended—The fire scenario in which the fire extends out of an opening of a building, for example, window, flares up the side of the building and penetrates into an opening directly above, for example, another window.

Autoignition—The initiation of the combustion process by external heat without the application of a spark or flame. *See also* **Autoignition Temperature (AIT); Spontaneous Ignition.**

Autoignition Temperature (AIT)—The minimum temperature of a vapor and air mixture at which it is marginally self-igniting in the absence of an independent ignition source such as a spark or flame. For example, carbon disulfide has an autoignition temperature of 212°F (100°C), and if it contacts a hot surface of this temperature, it will ignite. Synonymous with ignition temperature, it has also been defined as the spontaneous-ignition temperature. The autoignition temperature depends on many factors such as ignition delay, concentration of vapors, environmental effects (volume, pressure, oxygen content), catalytic material, and flow conditions. *See also* **Ignition Delay; Ignition Temperature; Kindling Temperature** or **Kindling Point.**

Automatic Aid—A prearranged commitment used among adjacent (but under separate governmental jurisdiction) fire departments to assist each other in fighting fires (or other emergencies) without regard to jurisdictional boundaries. It is used in areas where there are county islands within city limits or where a fire in one area directly impacts another agency's area may also be called **Assistance Agreements.** *See also* **Mutual Aid.**

Automatic Ball Drip—A drain valve mechanism that automatically drains collected water from a low point in piping when pressure is removed. It is commonly provided to remove water from points that are subject to freezing to prevent pipe failure or blockages during system use, such as fire department connections, hose headers, fire pump test connections, and trimming in fire protection valves.

Automatic Door Release—A device on a self-closing door that holds the door open during normal operation but causes it to close by releasing electromagnetic holders when activated by a signal, such as from a fire alarm. The closed door prevents the spread of fire and smoke. For personnel exit applications and other locations where smoke spread is a concern, fusible links or other similar heat-activated door-closing devices are not recommended because smoke may pass through the door opening before there is sufficient heat to melt the fusible device. *See also* **Automatic Fire Door.**

Automatic Fire Alarm Association (AFAA)—An organization founded in 1953 that represents the automatic fire detection and fire alarm industry. Its membership consists of manufacturers, distributors, regional associations, users, and local authorities. It evaluates codes and standards for the proper application of fire alarm and detection devices and also provides training and seminars on fire alarm systems.

Automatic Fire Alarm System—A system of controls, initiating devices, and alarm signals in which all or some of the initiating circuits are activated by automatic devices, such as smoke detectors. *See also* **Fire Alarm; Manual Pull Station (MPS).**

Automatic Fire Detection System—An arrangement of detectors that ascertains the presence of a combustion process by sensing heat, smoke, or flames. The detectors can be self-annunciating or connected to a fire alarm control panel (FACP) to initiate other alarm devices or signaling systems. Various codes and standards are used to design and provide a fire alarm system, notably NFPA 72, *National Fire Alarm Code. See also* **Fire Detection; Fire Detector.**

Automatic Fire Door—A fire door that closes immediately following the detection of a fire incident. Most fire doors that automatically close are held open by electromagnetic holders. When the fire is detected by a fire detection system, power is removed from the magnetic holders, and the doors swing closed from their self-closers. It may also be called an automatic-closing fire door. *See also* **Automatic Door Release; Fire Door; Fire Door, Power Operated; Fire Door, Self-Closing.**

Automatic Fire Extinguishing System—A fire suppression system that automatically senses the occurrence of a fire and signals a control device for the application of extinguishing agents. It is designed to distribute and apply the extinguishing agents in sufficient quantities and densities to effect fire control and extinguishment. No human intervention is required. Almost all automatic fire extinguishing systems are required to be designed and installed according to prescribed rules and regulations that ensure they are reliable and effective for fire protection applications.

Automatic Fire Protection—Active fire protection measures that are activated immediately following a fire incident without human intervention. Automatic fire protection systems provide for both fire detection and extinguishment. *See also* **Active Fire Protection (AFP).**

Automatic Hydrant Valve—A valve when fitted to a hydrant outlet that automatically opens to allow water to flow into the hose line.

Automatic (or Constant) Pressure Nozzle—A nozzle with a spring mechanism built-in that reacts to pressure changes and adjusts the flow and resultant reach of the nozzle.

Automatic Sprinkler—A water spray device, commonly a nozzle with a perpendicular spray deflector attached, used to deflect the water spray to a predetermined area at a predetermined density for the purpose of fire protection. Automatic sprinklers for fire protection applications are available in a variety of configurations and decorative features. The sprinklers for fire protection are normally required to be tested or approved by an independent testing agency (UL or FM). Sprinklers may be installed with an open orifice (dry sprinklers) or provided with an element that contains the system water supply (wet sprinklers). Water is sprayed from the sprinkler once the heat from a fire causes a fusible element in the sprinkler head to melt. Once the element has melted it releases a tension mounted cap on the outlet of the sprinkler. The sprinkler outlet is connected to a water distribution pipe network. Water pressure in the pipe network directs water onto the sprinkler deflector, providing the water spray. See Figure A-8. *See also* **Sprinkler.**

Automatic Sprinkler System—*See* **Sprinkler System.**

Automatic Transfer Switch (ATS)—A self-actuating electrical switch for transferring one or more load conductor connection(s) from one source of power to another. They are used for electric firewater pump controllers to ensure a source of power is available to start the pump if a single power source fails.

Auxiliary Appliances—A fire service term for protective devices which usually refers to sprinkler and standpipe systems.

Auxiliary Drain—An additional drain provided on the piping of a sprinkler system and located where the system forms a low point that cannot be drained by the system main drain. It is used to prevent freezing in the system, perform corrosion control measures, or for normal maintenance operations.

Auxiliary Equipment—Auxiliary fire vehicles are equipped with specialized equipment for effecting rescue, ventilating buildings, and performing salvage operations. Aerial ladders that typically extend to 100 ft. (30.5 m) are carried on hook-and-ladder vehicles that also hold various kinds of tools and equipment, including heavy-duty jacks and air bags, extrication tools, oxyacetylene torches, self-contained breathing apparatus, and resuscitators. Other more basic equipment includes axes, shovels, picks, battering rams, power saws, hooks, and wrenches. Elevating platform trucks can raise firefighters and equipment, including the water delivery system, as high as 100 ft. (30.5 m). Rescue trucks carry a wide assortment of specialized emergency equipment, including the type that might be used in building collapses and cave-ins. Field communications units carry sophisticated electronic equipment to manage fire and emergency operations. Salvage trucks carry implements for reducing water damage, including large waterproof covers, de-watering devices, and tools for shutting off water flow from sprinkler heads.

Figure A-8 Assorted sprinklers.

Hazardous materials response units are staffed with specially trained personnel equipped with protective clothing and monitoring devices for use at chemical spills and similar incidents.

Available Flow—The amount of water that can be moved to extinguish a fire. The amount is dependent on water supplies, pumps, and hose for pipe size and length.

Back Burn—The process of burning land areas to establish a firebreak in vegetation in order to stop the spread of a wildland fire. *See also* **Backfire** or **Backfiring; Firebreak; Fireline.**

Backdraft or **Smoke Explosion**—An explosion resulting from the sudden introduction of air (oxygen) into a confined space containing oxygen deficient superheated products of incomplete combustion. Backdraft explosive velocities are subsonic (deflagration). Therefore they preclude the production of a pronounced centralized area of damage or "seat" of an explosion. A backdraft may occur during firefighting operations because of inadequate or improper ventilation procedures. The term backdraft is thought to have originated from the fluttering, fluctuating behavior of smoke immediately before the explosion. *See also* **Updraft.**

Backfire or **Backfiring**—Another fire started to extinguish or control an oncoming fire, as in a forest, by burning an area in the path of the oncoming flames. Backfiring is accomplished by instituting a carefully controlled burn of a strip of forest or vegetation on the leeward side of the blaze. When the fire reaches the burned area it can go no further and will not endanger firefighters. *See also* **Back Burn; Backfire Preventer; Backing Fire.**

Backfire Preventer (Arrester)—A device to prevent flame propagation (flame arrester), shut off the fuel supply, and relieve pressure in a piping network as a result of a backfire. *See also* **Backfire** or **Backfiring; Safety Blowout (Backfire Preventer).**

Backflow Preventer—An arrangement or device provided in a public potable water system to prevent the flow of water from a fire protection system back into the distribution main. The backflow preventer is provided to prevent contamination of public potable water supply from end source users. Break tanks, double check valves (American Water Works Association [AWWA] Standard C510, *Double Check Valve Backflow-Prevention Assembly*), and reduced pressure assemblies are commonly employed. A reduced pressure assembly is similar to a double check valve but keeps the pressure between the two check valves at a lower pressure than the supply side. The AWWA classifies backflow preventers from Class 1 to 6 (least to most) depending on the amount of contamination hazards in the fire protection system to the potable water system. *See also* **Break Tank.**

Backing Fire—Terminology used in U.S. Forest Service firefighting to describe a wildland fire traveling against the prevailing wind direction. *See also* **Backfire** or **Backfiring.**

Backpack Pump—A pump and small tank carried on an individual's back, primarily for use in wildland fires by forest firefighters. See Figure B-1.

Back Stretch or **Flying Stretch**—An attach hose line lay or arrangement with the fire apparatus

Figure B-1 Backpack pump.

at the hydrant and the hose line is stretched back from the apparatus to the fire. The flying stretch is a version of the back stretch where the apparatus is located at the fire, the hose attack portion is removed and the apparatus proceeds to the hydrant.

Backup Line—A secondary firefighting hose provided in reserve of the primary firefighting hose. It is used in case the primary is inadequate and to protect personnel using the primary hose for close firefighting attack operations.

Baker-Strehlow Vapor Cloud Explosion Model—A mathematical and graphical model of a vapor cloud explosion blast. The model is based on the premise that a vapor cloud explosion can only occur within that portion of a flammable vapor cloud that is congested or partially confined. It provides a plot of peak pressure wave duration as well as a plot of peak overpressure with a series of curves to account for various deflagration strengths as well as detonation. The model provides fairly broadly defined categories to choose the parameter values and a degree of congestion and fuel reactivity. It uses numerical and experimental data relating to the structure of blast waves. Application of the Baker-Strehlow method requires the estimation of maximum flame speed attained and equivalent energy of the explosion. *See also* **TNT Equivalence Vapor Cloud Explosion Model.**

Balanced Pressure Pump Proportioning—A foam proportioning system that uses positive displacement foam concentrate pumps to correctly proportion a foam concentrate. It operates over a wide range of flows and pressures. See Figure B-2. *See also* **In-Line Balanced Proportioner (ILBP); Proportioning.**

Balloon Frame Construction—A type of structural framing used in building construction. It features wall studs that span the full height of the building. Floor joists are secured to ribbon boards at the side of the wall studs. This results in interconnecting passages from the foundation to the roof. If these passages are not firestopped, they act as wooden chimneys capable of conducting and spreading flame, heat, and smoke throughout the structure during a fire incident.

Bambi Bucket—Slang term used by wildland fire fighters for the collapsible bucket slung

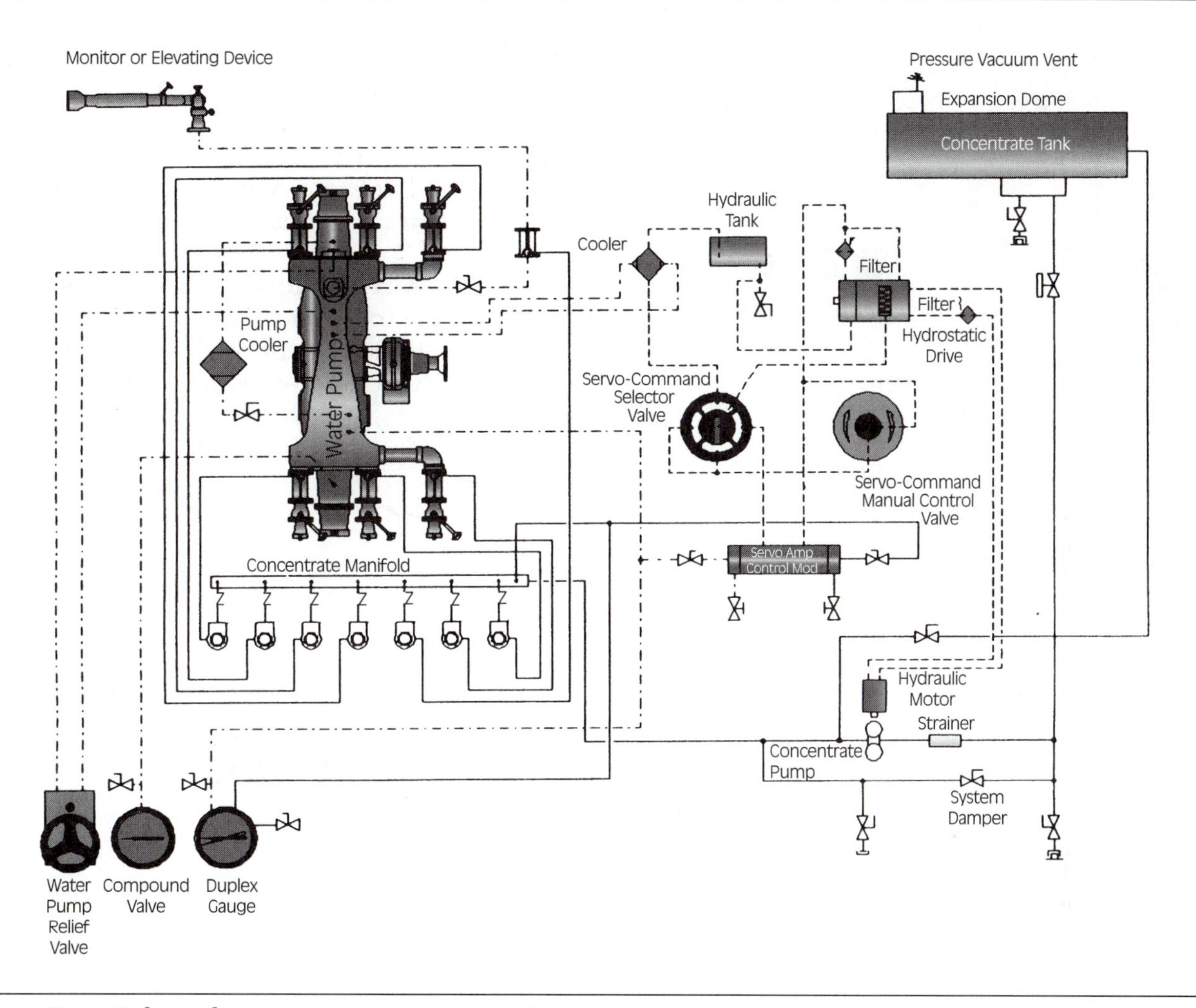

Figure B-2 Balanced pressure pump proportioning system.

below a helicopter which is used for collecting water from a variety of sources for fire suppression activities.

Bangor Ladder—A large extension ladder. Because of its size, it requires stabilizing poles (called tormentors or staypoles) to secure it when it is raised or lowered.

Beam Smoke Detection (BSD)—*See* **Linear Beam Smoke Detection.**

Becket Bend (Knot)—A type of knot used in the fire services that is used for tying ropes together that are of equal diameter. The double becket bend is used to tie ropes together of unequal diameters. See Figure B-3.

Bimetallic Strip—A fire detection method that uses heat input into a metal strip to cause an alarm. The device consists of two strips of metal that have different coefficients of

B

Figure B-3 Becket bend knot.

thermal expansion. Heat input to the metal strips causes it to bend. This action is arranged so that an electrical contact can be opened upon bending of the strip, activating a signal that can be converted into an alarm for fire detection from heat detection. See Figure B-4.

Bimetallic Wire—A method of fire detection that uses a pair of wires that sense the generation of heat and cause a signal to be sent to a monitoring device. Heat causes an electromagnetic

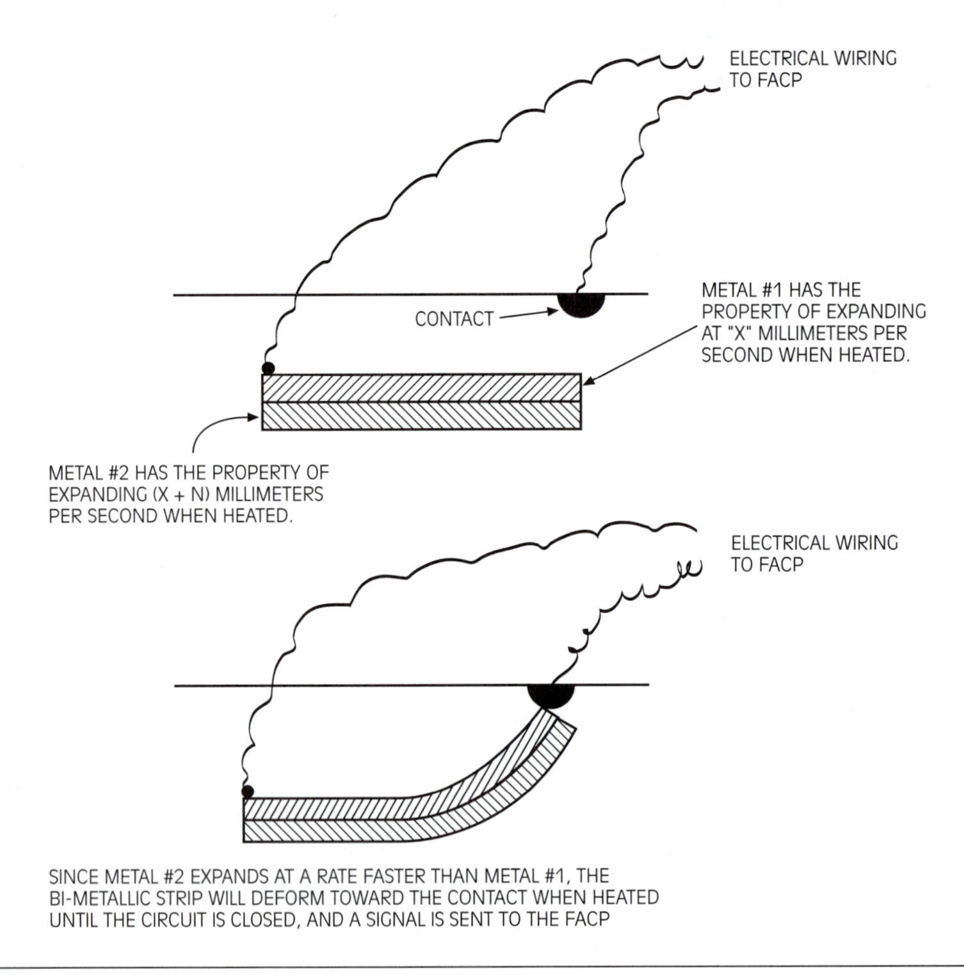

Figure B-4 Bimetallic strip.

force to occur in a pair of wires composed of two pieces of different metals along any point in its length. The electromotive force appears at the free end of the wire and is translated into a fire alarm. The wires do not have to be run in a straight fashion and can be routed in any direction as required to protect the hazard. *See also* **Linear** or **Line-Type Heat Detection (LHD).**

Blacken—To extinguish flames, quench burning embers, and wet charred materials so that they become black. It is similar to the term damp down. *See also* **Damp Down; Quench.**

Bladder Tank—A storage tank for firefighting foam concentrate that uses an internal flexible membrane called a bladder or diaphragm made of a reinforced rubber (elastomeric) material. It makes use of available water pressure as the force to press against or squeeze (collapse) the membrane of the bladder tank to force out the foam concentrate to a proportioner device (orifice or venturi), similar to squeezing a balloon. Bladder tanks may be in a horizontal or vertical orientation and are constructed to the requirements of the American Society of Mechanical Engineers (ASME), Boiler and Pressure Vessel Code, Section VII for unfired pressure vessels. See Figure B-5. *See also* **Foam Proportioning; Pressure Proportioning Tank.**

Figure B-5 Bladder tank.

Blanketing (or **Padding), Gas**—A method of fire protection for a combustible liquid storage tank or vessel. The vapor space of the container is maintained in an inert or fuel-enriched condition (that is, not within the flammable range) to prevent ignition. Common methods employed in the petrochemical industries include a continual application of fuel gas to the vapor space or the funneling of combustion exhaust gases from internal combustion engines (commonly employed on large crude-oil tanker cargo hold spaces) at a slightly higher pressure than ambient conditions. This prevents the entrance of free air into the container and keeps the atmosphere at an enriched state, preventing the formation of a combustible mixture. *See also* **Flammable Limit** or **Flammability Limits; Inerting.**

Blast—A transient change in gas density, pressure (both positive and negative), and velocity of the air surrounding an explosive point. The most common sources of blasts are from the ignition of semi- or unconfined vapor cloud explosions, the detonation of high explosives, or the rupture of high-pressure vessels. The initial change can be either gradual or discontinuous. A discontinuous change is commonly referred to as a shock wave, and a gradual change as a pressure wave. *See also* **Explosion; Overpressure; Shock Wave.**

Blast Pressure Front—The expanding leading edge of an explosion reaction. It separates a major difference in pressure between normal ambient pressure ahead of the blast pressure front and potentially damaging high pressure at and behind the front. *See also* **Blast Wave.**

Blast Relief Panel—Parts of an enclosure wall, ceiling, or roof designed to increase the area of venting in an explosion. This is accomplished by opening or removing a specific panel by the force of the explosion and venting the blast pressure to the outside atmosphere, while leaving the remaining portion of the enclosure intact. The size of the panels or vents are dependent on the plant strength and plant volume potential explosion intensity. The need of post-blast relief fire suppression capability also has to be considered when blast relief panels are provided. *See also* **Explosion Vent.**

Blast Resistant—A generic term for the ability of a location or equipment to withstand the effects of an explosion pressure wave. The blast-resistant requirements of a particular location are dependent on the amount of explosive material that may be present and, generally, the distance from the explosion ignition point. Blast resistant construction techniques are also used to enhance the resistance to blast effects for buildings, such as monolithic construction, deflective arrangements, absence of openings, etc. *See also* **Explosion Resistant; Fire Resistant.**

Blast Wall—A structural division designed expressly for the purpose of resisting specific blast loads. A blast wall may require additional mechanisms, to actually retain it in place and resist overpressure loads. A blast wall may or may not be rated as a firewall. Blast walls that harmlessly deflect overpressures rather than absorb them are more efficient and preferable.

Blast Wave—An overpressure front traveling outward from an explosion point. *See also* **Blast Pressure Front.**

Blaze—Terminology for a free-burning fire characterized as spectacular in flame evolution. *See also* **Free-Burning.**

Blitz Attack—A method of firefighting that aggressively applies water to a fire using a large-size fire stream, 2.5 in. (65 mm) or larger.

Blow-Off Caps or **Blowout Plugs**—Protective caps or plugs fitted to the outlets of deluge system nozzles or open sprinkler heads to protect them from the accumulation of foreign matter (dirt, insects, nests, etc.) that may plug or affect the water discharge. They may also be fitted to nozzles to allow pre-filling or priming of the pipe network. The caps are usually tightly fitted and are pushed away (blown off) by the water pressure of the system during activation. A small wire attached to the cap and pipework retains the cap after it has been released. It is usually fitted to systems in which foreign material may be normally present in the atmosphere (outside or for a process the system is protecting). A faster water application to the hazard may also be achieved if caps are provided on the nozzles and the pipe network is prefilled with water up to the nozzles. Where it is applied to prevent particulate ingress, it may also be called a dust cap. *See also* **Preprimed System.**

Blowout—Commonly a form of jet fire, but it normally refers to a high pressure release from a well that may or may not be ignited. Blowouts are primarily combustible liquids or gases that are not initially ignited but may be deliberately ignited to avoid further hazards from the accumulation of gas cloud and a possible explosion. A blowout occurs from the release of a high pressure flow of oil or gas from a high pressure supply source, usually an uncontrolled well with a substantial fallout of heavy liquids that may also ignite on or around the wellsite as pool fires. Blowouts may occur during the initial exploration drilling activities for oil and gas reservoirs when unexpected high pressure is rapidly and unexpectedly encountered during the drilling operations. Reservoir pressure is commonly counterbalanced with drilling mud, the density of which is changed to meet the ongoing drilling conditions (backpressure in the well). A *kick* warns of the potential of an impending blowout. Control of blowouts requires specially trained firefighters and well control experts experienced in the suppression of well-head fires and who are able to rapidly cap the wellheads emitting dangerous liquids and gases after extinguishing the flames. The usual method is to detonate an explosive at the wellhead to blow out the fire while simultaneously keeping the area cool to prevent re-ignition. Other methods include drilling a relief well to intersect the existing well and divert the oil and gas away from the uncontrolled well on fire. *See also* **Blowout Preventer (BOP).**

Blowout Preventer (BOP)—A mechanism to rapidly close and seal off a well borehole to prevent a blowout. It consists of rams and shear rams, usually hydraulically operated, fitted at the top of the well being drilled. It is activated if well pressures are encountered that cannot be controlled by the drilling process systems (drilling mud injection) and could lead to a blowout of the well. *See also* **Blowout.**

Blowup Fire—A wildland fire that increases in intensity very suddenly or has a very high rate of spread that is beyond control. Blowup fires are usually accompanied by high convective currents. A blowup fire may occur because of favorable atmospheric conditions, increased wind velocities, or low humidity. *See also* **Conflagration; Firestorm; Firewhirl Flare-up.**

Blunt Start—The removal of the incomplete thread at the end of a hose coupling to avoid cross threading and damage. *See also* **Higbee Indicators** or **Cut.**

Boiling Liquid Expanding Vapor Explosion (BLEVE)—A catastrophic rupture of a pressurized vessel containing a liquid at a temperature above its normal boiling point with the simultaneous ignition of the vaporizing fluid. A short-duration, intense fireball occurs if the liquid is flammable. During the rupture of the vessel, a pressure wave may be produced and fragments of the containment vessel will be thrown considerable distances.

The conditions that allow a BLEVE to occur are:

a) The material is initially in a liquid form

b) It must be confined in a container and is above its boiling point

c) There must be a failure of the container due to weakness by flame impingement, structural weakness, failure or poor design of the relief valve, or impact, such as due to rail or vehicle collision.

A BLEVE usually develops when the metal of a vessel is heated to over 1,000°F (538°C), which is of interest to fire protection professionals. Most BLEVE conditions occur when the vessel is slightly less than $^1/_2$ to $^3/_4$ full of liquid. The liquid vaporization energy is such that heavy pieces of vessels may be propelled very far distances—up to 0.75 miles (1.2 km)—and deaths from burns have been recorded to individuals who have been as much as 250 ft. (76 m) from a vessel. As protection against BLEVE development, pressure vessels are provided with a cooling spray to limit heat input and temperature of the vessel shell. *See also* **Depressurization; Emergency Venting; Flame Impingement.**

Boilover—A boiling liquid eruption in a hazardous material storage tank. It is usually described as an event in the burning of certain materials in an open tank when, after a long period of quiescent burning, there is a sudden increase in fire intensity associated with the expulsion of burning material from the tank. Boilovers are caused by heating water trapped at the bottom of a tank to vaporization, resulting in a boiling expulsion of the resultant steam

B

through the layer of fluid above it. A heat wave is created at the top of the tank due to burning at the surface of the fuel, which travels to the bottom of the tank. Oils subject to boilovers have a wide range of boiling points. These characteristics are present in most crude oils and can be produced in synthetic mixtures. Boilovers will not occur to any significant degree with water-soluble liquids or light hydrocarbon products such as gasoline. *See also* **Heat Wave; Latent Heat.**

Bonfire—A large fire purposely built in an open area, primarily for cultural celebration purposes after a harvest of crops. Fuel for the fire usually consists of cut trees, vegetation, or lumber.

Booster Hose or **Line**—A small-diameter (0.75 or 1.0 in. [1.9 or 2.54 cm]) firewater hose provided on a mobile fire apparatus. It is usually stored on a reel. It can be quickly charged and handled by one person for immediate use on small fires. The hose is fabric-reinforced, and rubber-lined and covered. It may also be called a hand line or red line. *See also* **Firewater Pump, Booster Handline.**

Booster Pump—*See* **Firewater Pump, Booster.**

Booster Tank—The fixed water tank on a pumper that supplies hand lines at a fire until a connection to a water source can be made. Most tanks are between 500 and 1,000 gallons (1,893 to 3,785 liters).

Bowline Knot—A type of knot used in the fire service, usually to form a loop in natural fiber ropes. It is not typically used due to the advent of synthetic fiber ropes. See Figure B-6.

Figure B-6 Bowline knot.

Branch Line (BL)—The piping of a firewater sprinkler system on which the sprinkler heads are directly mounted.

Branchpipe—A British term for a section of a fire hose.

Brands, Flying—*See* **Ember; Fire Screen.**

Breach—Interruption of the continuity of a fire resistance rated assembly or barrier. It can also refer to a purposely made opening in a barrier to accomodate rescue, hose line operations, or other actions.

Break Glass Unit—Term for a manual fire alarm activation station that is fitted with a breakable glass cover or other tamper-resistant device made of glass. Use of a glass cover prevents accidental or malicious activation. *See also* **False Alarm; Manual Activation Callpoint (MAC); Manual Pull Station (MPS).**

Break Tank—A relatively small intermediate holding tank provided at the supply for a firewater protection system to prevent contamination to the public potable water supply system. It receives water from the public potable supply system through an air gap (or break) in the line, which feeds into the break tank, thus preventing reverse flow of liquid back into the public supply. *See also* **Backflow Preventer.**

Bresnan Distributor—A fire-fighting nozzle used for basements or cellars when firefighters cannot make a direct attack on the fire. It is inserted through an opening in the floor and usually has six or four tips or openings designed to rotate in circular pattern to provide adequate water application in the enclosure. *See also* **Cellar (Pipe) Nozzle.** See Figure B-7.

British Thermal Unit (Btu)—The quantity of heat required to raise the temperature of one pound of

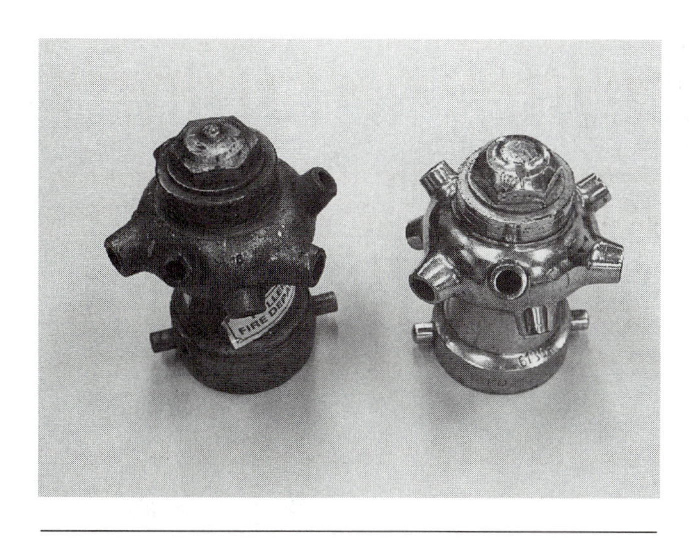

Figure B-7 Bresnan distributor nozzles.

water 1°F (–17°C) at the pressure of one atmosphere and the temperature of 60°F (15.5°C).

Broadcast Burning—Intentional burning of a wildland area. It is usually started for the disposal of unwanted timber cuttings rather than as a prescribed burning for the removal of materials constituting fire hazards. *See also* **Prescribed Burning** or **Fire; Spot Burning.**

Brushfire—A classification of a wildfire to describe a fire in vegetation that is less than 6 ft. (1.8 m) tall, such as grasses, grains, bushes, and saplings. Brushfires mainly occur during dry conditions. The spread of a brushfire is prevented by using permanent or temporary firebreaks and clearing brush that is adjacent to buildings to a significant distance. *See also* **Wildland Fire.**

Bucket Brigade—A chain of persons acting to put out a fire by passing buckets of water from hand to hand from the source of water supply to the point of water application onto the fire. The buckets are returned to the source of the water and the process is repeated. They are used where other mechanical methods (pumps, piping, and hydrants) are not available. Bucket brigades were commonly used before the advent of modern fire pumping apparatus and date from ancient times. The buckets were normally made of leather and some were highly decorated. Fire brigade buckets are now considered antiques and are highly prized by collectors. See Figure B-8. *See also* **Fire Bucket.**

Building Construction Types—There are five general types of building construction classifications defined for fire protection purposes. They are classified according to their fire-resistive properties. They include the following:

Fire Resistive—A broad range of structural systems capable of withstanding fires of specified intensity and duration without failure. Common fire-resistive components include masonry load-bearing walls, reinforced concrete or protective steel columns, and poured or precast concrete floors and roofs (Ref. NFPA 220, Type 1).

Heavy Timber—Characterized by masonry walls, heavy-timber columns and beams, and heavy plank floors. Although not immune to fire, the large mass of the wooden members slows the rate of combustion. Heavy timber construction can be used where the smallest dimension of the members exceeds 5.5 in. (14 cm). When timbers are this large, they are charred but not consumed in a fire and are generally considered akin to a fire-resistant type of construction (Ref. NFPA 220, Type 4).

Noncombustible—Type of structure made of noncombustible materials in lieu of fire-resistant materials. Steel beams, columns, and masonry or metal walls are used (Ref. NFPA 220, Type 2).

Ordinary—Consists of masonry exterior load-bearing walls that are of noncombustible construction. Interior framing, floors, and roofs

B

Figure B-8 Bucket brigade.

are made of wood or other combustible materials, whose bulk is less than that needed to qualify as heavy-timber construction. If the floor and roof construction and their supports have a one-hour fire resistance rating and all openings through the floors (stairwells) are enclosed with partitions having a one-hour fire resistance rating, then the construction is classified as "protected ordinary construction" (Ref. NFPA 220, Type 3).

Wood Frame—Building construction characterized by use of wood exterior walls, partitions, floors, and roofs. Exterior walls may be sheathed with brick veneer, stucco, or metal-clad or asphalt siding (Ref. NFPA 220, Type 5).

See also **Combustible Construction.**

Building Occupancy—The primary activity for which a building is designed and built. Fire code requirements are based on the risk a building occupancy represents, and, therefore, various building occupancies are normally defined by a fire code. Common building occupancies include assembly, business, educational, factory or industrial, hazardous or high hazard, institutional, mercantile, residential, storage and utility, special, or miscellaneous.

Bunker Gear—*See* **Turnout Clothing** or **Gear.**

Burn—A generic term meaning to involve in a fire or to ignite into combustion. Very rapid burning is termed an explosion. Burning may be controlled or uncontrolled. Controlled burning is used in heating, cooking, and other domestic or industrial uses. Uncontrolled burning results from unconfined combustion, which may cause injuries or property damage. *See also* **Ablaze; Combustion; Explosion; Fire; Free-Burning.**

Burnable—Something that is capable of undergoing combustion or being consumed by fire.

Burnback—Flames traveling back to a combustible liquid surface area and re-igniting it, although it had previously been extinguished by foam application. It is caused by inadequate foam application, the use of outdated foam, foam used over a long period (aging), or the heat degradation of foam. *See also* **Burnback Resistance, Foam.**

Burnback Resistance, Foam—The ability of a firefighting foam to resist direct flame impingement such as would occur with a partially extinguished petroleum fire or with Class A foam in exposure protection and pretreatment. It is a feature of a foam blanket to retain aerated moisture. Burnback resistance is also a function of the amount of foam that has been applied to a fire exposure. *See also* **Burnback.**

Burn Building—Term for a permanent building used to train firefighters under real conditions and effects of fire and smoke. See Figure B-9. *See also* **Smokehouse.**

Burned Area—Part of a damaged area of a material or location that has been destroyed by combustion or pyrolysis effects.

Burner Management System (BMS)—A control system dedicated to boiler-furnace safety, operator assistance in the starting and stopping of fuel preparation and burning equipment, and prevention of misoperation of and damage to fuel preparation and burning equipment.

Figure B-9 Burn building.

Burning Ban—A specified limitation on open-air burning within a defined area, usually due to sustained high fire danger.

Burning Behavior—All the physical or chemical changes that occur when an item is exposed to an ignition source.

Burning Point—*See* **Fire Point.**

Burning Rate—The mass of fuel consumed in a fire per unit of time also known as "mass loss rate."

Burning Velocity—The rate at which the flame consumes the unburned gas. Also the velocity of the flame relative to the unburned gas. *See also* **Flame Speed; Fundamental Burning Velocity.**

Burn Injury—An injury to the skin and deeper tissues caused by hot liquids, flames, radiant heat, and direct contact with hot solids, caustic chemicals, electricity, or electromagnetic (nuclear) radiation. A first-degree burn injury occurs with a skin temperature of about 118°F (48°C), and a second-degree burn injury occurs with a skin temperature of 131°F (55°C). Instantaneous skin destruction occurs at 162°F (72°C). Inhaling hot air or gases can also burn the upper respiratory tract. Approximately two million persons suffer serious burns in the United States each year; of these approximately 115,000 are hospitalized and 12,000 die. The severity of a burn depends on its depth, its extent, and the age of the victim. Burns are classified by depth as first, second, and third degree. First-degree burns cause redness and pain (sunburn) and affect only the outer skin layer. Second-degree burns penetrate beneath the superficial skin layer and are marked by edema and blisters (scalding by hot liquid). In third-degree burns, both the epidermis and dermis are destroyed, and underlying tissue may also be damaged. It has a charred or white leathery appearance and initially there may be a loss of sensation to the area. The extent of a burn is expressed as the percent of total skin surface that is injured. Individuals less than 1 year old and over 40 years old have a higher mortality rate than those between 2 and 39 years old for burns of similar depth and extent. Inhalation of smoke from a fire also significantly increases mortality.

Thermal destruction of the skin permits infection, which is the most common cause of death for extensively burned individuals. Body fluids and minerals are lost through the wound. The lungs, heart, liver, and kidneys may be affected by infection and fluid loss.

First aid for most burns involves the application of cool water as soon as possible after the burn. Burns of 15 percent of the body surface or less are usually treated in hospital emergency rooms by removing dead tissue (debridement), dressing with antibiotic cream (often silver sulfadiazine), and administering oral pain medication. Burns of 15 to 25 percent of the body surface often require hospitalization to provide intravenous fluids and avoid complications. Burns of more than 25 percent of the body surface are usually treated in specialized burn centers. Aggressive surgical management is directed toward early skin grafting and avoidance of such complications as dehydration, pneumonia, kidney failure, and infection. Pain control with intravenous narcotics is frequently required. The markedly increased metabolic rate of severely burned patients requires high-protein nutritional supplements given intravenously and by mouth. Extensive scarring of deep burns may cause disfigurement and limitation of joint motion. Plastic surgery is often required to reduce the effects of the scars. Psychological problems often result from scarring.

Investigations are underway to improve burn victims' nutritional support, enhance the immune response to infection, and grow skin from small donor sites in tissue culture to cover large wounds.

Since over 50 percent of all burns are preventable (separation, barriers, protective clothing, etc.), safety programs can significantly reduce the incidence of burn injuries. See Figure B-10. *See also* **Fire Injury; Thermal Burn.**

Burnout—(1) The point in time at which flame ceases, (2) the deliberate burning of an area in front of a forest fire to stop its advance, (3) the unintended escape of flame through the wall of a pyrotechnic chamber during functioning of a fireworks device, (4) a building that has undergone complete destruction of all combustible material from a fire incident, or (5) stress to an individual firefighter due to the number of calls responded to in a given period of time. *See also* **Fuels Management.**

Burn Pattern—The evidence of a fire's origin and growth. It is based on the size, shape, and

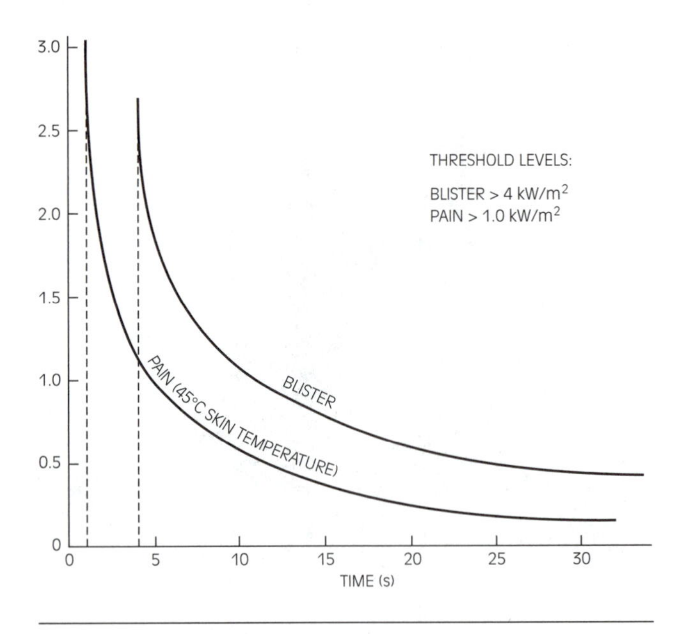

Figure B-10 Incident radiant heat flux on bare skin.

configuration of markings or indications left after a fire incident from the combustion process particles, flame impingement, materials consumed, etc. Examination of a burn pattern is useful for arson investigators and others interested in a fire's origin and growth. *See also* **Depth of Char; V Pattern.**

Burn Pit—A means of safely disposing of combustible liquid wastes in the process industries through combustion. Combustible waste liquids are directed to a pit or bermed area through a collection header. The pit is provided with an ignitor system (pilot flame) to initiate combustion of the liquids delivered to the pit. Burn pits that are poorly designed, maintained, or if inadequately lined, can pose an environmental threat to groundwater supplies. *See also* **Flare.**

Burst Test Pressure—A destructive test performed on a fire hose or piping to determine its maximum water pressure tolerance. The burst test pressure is reached by pumping water into the appliance until the specified water pressure is reached. The requirements of NFPA 1961, *Standard on Fire Hose,* requires the burst-test pressure rating of a fire hose to be not less than three times the service test rating of the fire hose. *See also* **Kink (Pressure) Test; Proof Test Pressure.**

Bushfire—An Australian term for a forest fire. It usually occurs in dry and windy conditions. *See also* **Forest Fire.**

C

Caking—The tendency of fire protection dry powder to form into lumps due to the presence of moisture. Modern fire protection dry chemical powders are provided with additives to prevent caking, which prohibits the free flow of dry chemical agent application for fire protection purposes. *See also* **Dry Powder.**

Calorimetric Bomb Test—A test procedure to determine the heat of combustion of a material using a bomb calorimeter. The material is burned in a high-pressure bomb calorimeter in one atmosphere of oxygen. The temperature rise produced by the combustion process is measured and the heat of combustion is calculated. *See also* **Heat of Combustion.**

Carbon Dioxide (CO_2) Fire Extinguisher—A steel cylinder filled with liquid carbon dioxide (CO_2), which when released, expands suddenly into a gas that is applied for fire control and suppression. The rapid expansion may cause so great a lowering of temperature that the CO_2 may solidify into powdery "snow." This snow volatilizes (vaporizes) on contact with the burning substance, producing a blanket of gas. Carbon dioxide gas that is applied to a fire, cools and smothers the flame causing fire extinguishment to occur. The fire extinguisher is made so that it is generally easily portable by hand or mounted on wheels for large cylinders. It is fitted with a nonconductive hose and discharge horn, which is manually directed to the fire location by the operator of the extinguisher. See Figures C-1 and C-2.

Carbon Dioxide (CO_2) Fire Suppression Agent—A naturally occurring gas used for fire protection applications. Carbon dioxide is a noncombustible gas that can penetrate and spread to all parts of a fire, diluting the available oxygen concentration so that it will not support combustion. It also reduces the gasified fuel in the fire environment and cools the fire zone to aid in fire control and extinguishment. A CO_2 system extinguishes fires in practically all combustibles except those that have their own oxygen supply or contain certain metals that cause decomposition of the carbon dioxide. Carbon dioxide is primarily employed when water cannot be used and a fire must be controlled or suppressed by suffocation. Carbon dioxide does not conduct electricity and therefore it can be used on electrical equipment that is energized. It will not freeze or deteriorate with age. Carbon dioxide is a dangerous gas to human life since it displaces oxygen. CO_2 concentrations above 9 percent are considered hazardous, though 30 percent or more is needed to effect fire extinguishment. NFPA 12, *Carbon Dioxide Extinguishing Systems,* provides a table specifying the exact concentration requirements by volume for specific hazards. As a guide, 1 lb. (0.45 kg) of CO_2 liquid may produces 8 cu. ft. (0.23 m^3) of free gas at atmospheric pressure. Carbon dioxide extinguishes a fire almost entirely by smothering. It does have a

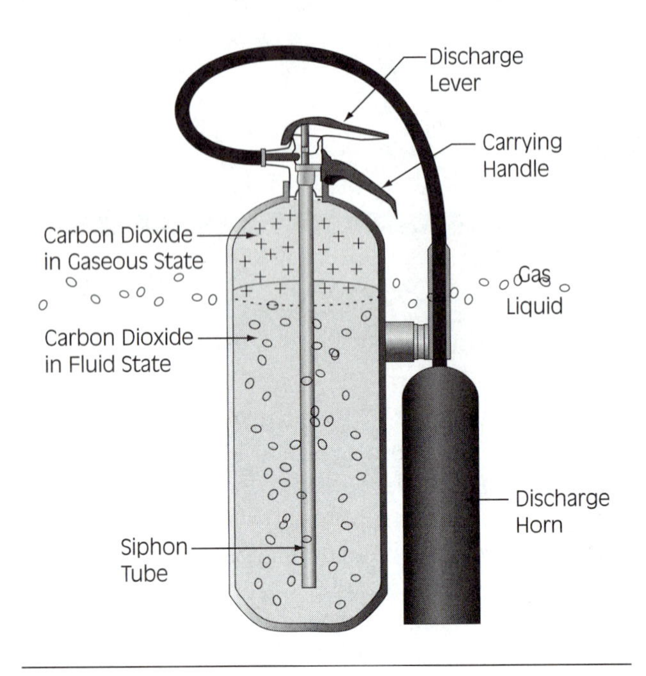

Figure C-1 CO_2 fire extinguisher's internal arrangement.

Figure C-2 Typical CO_2 fire extinguishers.

limited cooling effect of about 100 Btu per pound. Where rotating equipment such as turbines or compressors is involved, a primary and supplemental discharge is used. The supplemental discharge is used to account for leakages during "run-down" of the equipment. Concentrations are to be achieved within one minute and are to be maintained for 20 minutes within the protected enclosure. Carbon dioxide systems may be ineffective if used outdoors, since wind may dissipate the vapors rapidly. Therefore where required for outside applications, larger storage cylinders are used. It has a vapor density of 1.5 and will settle to the lower portions of an enclosure where ventilation effects do not occur. Carbon dioxide is the most commonly utilized liquefied gas fire extinguisher. Carbon dioxide for fire extinguishment is applied through portable fire extinguishers, local application systems, or total flooding systems. *See also* **Carbon Dioxide (CO_2) Fire Suppression System; Fire Suppressant Agent.**

Carbon Dioxide (CO_2) Fire Suppression System— A fire suppression system that consists of carbon dioxide contained in a liquid condition in a pressurized storage container, a valve for releasing the material, a distribution system, discharge devices for applying carbon dioxide onto a fire, and a means for actuating the systems (automatic detection or manual release devices). Systems are usually specified as either high pressure (individual cylinders) or low pressure (bulk storage), depending on the pressure of the storage container of carbon dioxide. Low-pressure systems are used where large quantities of carbon dioxide are needed for extinguishment, usually above 2,000 lbs. (907 kg). High-pressure systems contain CO_2 in upright steel cylinders, which are at ambient temperatures of 32°F to 120°F (0°C to 49°C), at about 850 psi (5,860 kPa). Low-pressure systems contain CO_2 in a storage container that is kept in a large, insulated, refrigerated tank at 0°F (–18°C) and approximately 300 psi (2,068 kPa). Fixed CO_2 fire suppression systems were first available in the 1920s. Guidance for the design and operation of fixed CO_2 systems is provided in NFPA 12, *Standard for Carbon Dioxide Extinguishing Systems.* NFPA began work on its standard for carbon dioxide extinguishing systems in 1928 and adopted it as an official standard in 1929. See Figure C-3. *See also* **Carbon Dioxide (CO_2) Fire Suppression Agent; Fire Extinguisher, Portable; Total Flooding.**

Figure C-3 Carbon dioxide low pressure fire suppresion system.

Carbon Tetrachloride (CCl_4)—One of the earliest chemicals to be used as a firefighting agent for portable fire extinguishers (circa 1907), is now considered obsolete. It is a colorless, nonflammable, toxic liquid having an odor resembling chloroform. When carbon tetrachloride liquid is discharged on a fire, its thermal decomposition causes the formation of gaseous phosgene. Phosgene has a density about three and one-half times that of air. This causes the fire to extinguish by oxygen exclusion or smothering by the phosgene gas. Carbon tetrachloride is now considered an obsolete fire extinguishing agent due to the noxious and toxic decomposition products associated with CCl_4 when it is applied to a fire (inhalation of phosgene causes severe lung injury, the full effects appearing several hours after exposure). In the 1930s a number of deaths and injuries were attributed to the agent or its breakdown products and it was generally outlawed in the 1950s for fire protection applications. It is no longer referred to by NFPA for such applications. It is currently used as a solvent and as a refrigerant. *See also* **Fire Extinguisher, Portable.**

Cartridge Operated Fire Extinguisher—A portable manually operated fire extinguisher which uses a separate cartridge, of stored expelling gas, attached to side of extinguisher agent storage container, usually for regular and multipurpose dry chemical and most dry powder class D fires. A puncturing lever is depressed at the time of use, which ruptures a disk and the expelling gas is released in the agent tank which pressurizes it. The dispension of agent by the unit is controlled by a nozzle at the end of hose that is connected to the agent storage tank. Cartridge type extinguishers are used when the suppression agent may be susceptible to excessive "caking" and needs to be agitated before it is used. *See also* **Dry Chemical Fire Extinguisher.**

Cavitation—Formation of a partial vacuum (creating gas bubbles), in a liquid by a swiftly moving solid body (a propeller). Cavitation may occur in a firewater pumping system due to improper design, arrangement, or installation. The generation and collapse of the gas bubbles produce a vibration and sometimes severe mechanical strain on the pumping system, reducing performance and causing accelerated deterioration of the pumping components (especially the impeller). Specific design and installation requirements are set forth in NFPA 20, *Installation of Centrifugal Fire Pumps,* to prevent cavitation from occurring in fixed firewater pump installations. *See also* **Net Positive Suction Head (NPSH); Net Positive Suction Head Required (NPSHr).**

Ceiling Layer—The buoyant layer of hot combustion gases and smoke produced by a fire in an enclosed area. Since they rise to the ceiling of the enclosure due to their buoyancy, they are referred to as the ceiling layer.

Cellar (Pipe) Nozzle—A special firefighting nozzle normally used to attack fires in the lower levels of a structure such as basements and cellars. It consists of a nozzle mounted on the end of a pipe that is placed or pushed through the floor from the area above the intended level of application. The nozzle normally rotates or sprays water completely around the area where it has been inserted. It is primarily used where fire conditions or accessibility hinder the use of flexible firefighting hoses to the intended area of water application. Due to its usefulness, it is sometimes applied to dock, attic, and transportation container fires; to fires in other concealed spaces; and for internal exposure protection. Cellar nozzles provide a pattern of water spray with a radius from 12 ft. to 30 ft. (3.7 m to 9.1 m). Some nozzles are directional and can be remote controlled. It may also be called a distributor nozzle or revolving cellar nozzle. See Figure C-4. *See also* **Bresnan Distributor; Distributor Nozzle.**

Figure C-4 Cellar nozzle.

Cellulosic Fire—A fire with a fuel source composition predominantly of cellulose (wood, paper, cotton, etc.). A fire involving these materials is relatively slow growing, although its intensity may ultimately reach or exceed that of a hydrocarbon fire. Standard building fire barriers are based on a cellulosic fire exposure as defined by ASTM-E-119, *Standard Test Methods for Fire Tests of Building Construction and Materials;* ISO Standard No. 834. Cellulosic fires reach a maximum temperature of just over 1,652°F (900°C). *See also* **Fire Barrier; Fire Resistance Rating.**

Central Station System (or **Central Station Fire Alarm System**)—A fire alarm system controlled and operated by a designated business for fire alarm system operation and maintenance. All signals generated by the system report to a central station (office) and are acted upon as required. *See also* **Proprietary Supervising Station Fire Alarm System.**

Centrifugal Pump—A type of pump that utilizes centrifugal force to impact pressure to a fluid.

It consists of an impeller that is encased and rotated at high speed to force water movement. The force of rotation imparts a velocity pressure to the fluid, causing it to flow *See also* **Fire Pump; Net Positive Suction Head (NPSH); Pump, End Suction and Vertical In-Line; Pump, Horizontal Split Case; Pump, Vertical Shaft, Turbine Type.**

C Factor—A relative roughness coefficient used in mathematical calculations of friction losses for water flow in pipes for fire protection systems when using the Hazen-Williams friction loss formula. C factors are dependent on the smoothness of the internal surfaces of pipes and are features of the pipe material and system age. A high C factor (120) represents a smooth internal pipe and a low C factor (80 or 90) is a rough internal pipe surface. The C factor decreases as the level of friction within the pipe interior surface increases. When computerized hydraulic programs are used to determine water pressures and flow conditions within a water distribution system, the friction coefficients to use are specified as part of the input data.

Common piping C factors used for fire protection applications include the following:

Unlined Cast or Ductile Iron	100
Asbestos Cement, Cement-Lined Cast or Ductile Iron, Cement-Lined Steel and Concrete	140
Polyethylene, Polyvinyl Chloride (PVC), and Fiberglass Epoxy	150
Copper	150

See also **Hazen-Williams Formula.**

Char—Carbonaceous material that has been burned and has a blackened appearance. The formation of char or charring is the production of carbonaceous deposits rather than complete combustion to carbon dioxide (CO_2). Many natural and synthetic polymers exhibit this feature. The extent of char depends on many factors, such as physical form, molecular structure, and the availability of oxygen (air). Charring produces an insulating layer in a material and can be effective in reducing the burning rate by restricting the heat flow to the unpyrolyzed layer. Intumescent paints simulate a char layer by swelling and thereby produce an insulating layer when heated. *See also* **Depth of Char; Isochar.**

Char Blisters—Convex areas of carbonized material separated by cracks or crevasses that form on the surface of a char. They form on materials such as wood as the result of pyrolysis or burning.

Chemical (Fire) Engine—An early firefighting apparatus or rig that used a chemical reaction to generate carbon dioxide (CO_2) in order to expel water very rapidly from a onboard storage tank. They were introduced to overcome the delay encountered from the operation of manual or steam pumpers (circa 1860s). They used the reaction of sulfuric acid with a bicarbonate solution in an onboard tank to form carbon dioxide gas to expel water under pressure ($H_2SO_4 + 2NaHCO_3 \rightarrow Na_2SO_4 + 2H_2O + 2CO_2$). Generally some of the undissolved sodium bicarbonate was also expelled in the process, giving the water a milky appearance and creating a mess whenever it was applied. They were extremely popular and useful, but of limited capacity. Chemical apparatus became obsolete (circa 1930) with the introduction and popularity of centrifugal fire pumps mounted on fire vehicles (pumpers) and improvements in public water supplies. At the time of their use, many fire service individuals believed that the mixture of water and undissolved bicarbonate along with carbon dioxide possessed fire extinguishing properties of up to 40 times that of water. There is no scientific proof that this was true, however this idea persisted. It may also be called a chemical fire apparatus. See Figure C-5. *See also* **Chemical Extinguisher.**

C

Figure C-5 Chemical (fire) engine.

Chemical Extinguisher—A portable fire extinguisher that utilized a chemical reaction at the time of use to form a pressurizing medium for expelling the contents. Most chemical extinguishers used bicarbonate of soda, water, and sulfuric acid. The soda and acid were kept separate inside in the extinguisher prior to use. The extinguisher was turned upside down to mix the contents prior to use. The first practical chemical extinguishers appeared about 1865. Chemical extinguishers are now considered obsolete due to the advances in pre-pressurized containers or *cartridge* operated portable fire extinguishers and the possible delay caused by waiting for the completion of the chemical reaction to occur. It is also commonly called a soda-acid extinguisher. *See also* **Chemical (Fire) Engine.**

Chemical Flash Fire—The ignition of a combustible gas or vapor that produces an outward expanding flame front or fireball. Chemical flash fires produce heat of 1,000°F to 1,900°F (538°C to 1,038°C). Room flashovers commonly produce temperatures of 1,200°F to 1,500°F (649°C to 816°C). A chemical flash fire depends on the size of the gas cloud.

Chemical Foam—A foam produced for fire protection applications but now considered obsolete due to the advancement and superior features of mechanical foams. Chemical foam is produced by the reaction of an alkaline salt solution (usually bicarbonate of soda) with an acid salt solution (generally aluminum sulphate), which is mixed with about 90 percent water. This produces carbon dioxide (CO_2) gas that becomes trapped in the bubbles of aluminum hydrate with a foaming agent, usually protein hydrolysates. This forms a fire resistant foam solution. Chemical foams can only be mixed at the time of the incident. There is a very narrow temperature range 60°F to 85°F (15.6°C to 29.4°C) over which chemical foam reactions are adequate to produce acceptable foam. They require large amounts of powder, have extensive maintenance requirements because of the corrosive nature of the chemical powders, must be used in close proximity to the incident, and frequently clog and jam. The foam is very stable but has a tendency to bake and crack. This allows fissures to form, which allow combustible vapors to escape and combustion to occur. These concerns have caused chemical foams to be largely replaced by mechanical foams where no mixing of powders or chemical reaction is required. Chemical foam was originally patented in 1887. *See also* **Foam Concentrate; Mechanical Foam.**

Chemical Smoke— Nontoxic smoke that is artificially created from chemical compounds for training and testing purposes (firefighter training, integrity testing, etc.). *See also* **Cold Smoke; Smoke Bomb.**

Chemical Wagon or **Appliance**—*See* **Chemical (Fire) Engine.**

Chimney—A noncombustible structure that contains one or more internal passageways provided for the removal or conveying of fire or flue gases to the outside atmosphere where they may be safely dispersed. *See also* **Fireplace.**

Chimney Effect—The updraft caused by hot air or gases rising in a confined area. Chimney effects can occur in the stairwells of tall buildings that do not prevent the ingress of smoke and hot gases in the lower levels of the building from entering the stairwells. It may also be called stack effect. *See also* **Trench Effect.**

Chimney Fire—A fire that occurs in the flue of a chimney. Over a period of time the interior of a chimney stack collects or becomes coated with soot. If this soot layer becomes significantly thick, a fire may occur in the collected soot of the chimney flue instead of just in the hearth of the fireplace.

Circuit Breaker—A safety device for electrical circuits designed to open the circuit when abnormal conditions occur (overcurrent, abnormal voltage, high temperature, grounding, etc.) to prevent damage or overheating to the system and the possible occurrence of a fire. They are usually designed to permit opening and closing of the circuits manually but will automatically open the circuit during an abnormal condition. The term started to be used circa 1872. *See also* **Fuse; Overload, Electrical.**

Cistern—An impounded water supply for fire service and emergency uses. It consists of a below-ground water storage tank with an open top or loose cover. It is at atmospheric pressure. Water for fire protection applications from a cistern is usually obtained by a Fire Service pumper by drafting from the cistern, which is relayed to area of concern.

Clapper Valve—A valve consisting of a hinged plate that permits the flow of water in only one direction. It is similar to a check valve or a one-way valve. Clapper valves are commonly found as part of an alarm check valve or deluge valve for fire protection systems. *See also* **Alarm Check Valve; Deluge Valve.**

C

Class A Fire—Fires in ordinary combustible material such as wood, cloth, paper, rubber, and many plastics. Some plastics behave like Class A combustibles up to a point, but then they develop many attributes of a Class B fire. *See also* **Fire Class.**

Class A Fire Pumper—*See* **Fire Pumper.**

Class A Foam—Foam intended for use on Class A fires. *See also* **Foam Concentrate.**

Class B Fire—Fires in flammable liquids or gases, such as oils, liquid fuels, lubrication oils, hydraulic fluids, greases, tars, oil base paints, lacquers, aerosols, cleaning compounds, and cutting or fuel gases. *See also* **Fire Class.**

Class B Fire Pumper—*See* **Fire Pumper.**

Class B Foam—Foam intended for use on Class B fires. *See also* **Foam Concentrate.**

Class C Fire—Fires involving energized electrical equipment where the electrical nonconductivity of the extinguishing media is of prime importance. (When electrical equipment is de-energized, the fire becomes a Class A or B fire, and Class A or B fire extinguishers, respectively, can be used safely on the fire.) *See also* **Electrical Fire; Fire Class.**

Class D Fire—Fires in combustible metals such as magnesium, titanium, zirconium, sodium, or potassium. Temperatures of metal fires are generally higher than those of flammable liquid fires. Metal fires that contain their own oxidizing agents may continue to burn in inert atmospheres. Dust clouds of most metals are explosive. *See also* **Fire Class.**

C

Class K Fire—Such as vegetable or animal oils and fats classification of cooking oil fires. Similar to Class B fuels but involving high temperature cooking oils. *See also* **Fire Class.**

Classified—Products or materials that are specified for use and that meet standard test requirements for fire safety concerns. Standard test conditions are usually set by national or local code requirements and verified through independent testing laboratories (UL or FM). *See also* **Approved; Factory Mutual (FM); Labeled; Listed; Underwriter's Laboratories (UL).**

Classified Area—An area or zone defined as a three-dimensional space in which a flammable atmosphere is or may be expected to be present in such frequencies as to require special precautions and restrictions for the construction and use of electrical apparatus and hot surface exposures (lights) that act as an ignition source. Classified areas have specific restrictions based on the equipment involved, gases encountered, and the probability of leakage. Several national and industry institutions have guidelines and codes for the classification of areas, defined in the United States by NFPA National Electrical Code (NFPA 70), Flammable and Combustible Liquids Code (NFPA 30); American Petroleum Institute (RP 500), Institute of Petroleum (IP), etc. *See also* **Explosionproof; Flameproof; Intrinsically Safe (IS).**

Claw Tool—A specialized tool used by firefighters for forcible entry. It consists of a hook and fulcrum at one end of a long bar and a prying blade at the other end. The claw tool is now considered obsolete, and the halligan tool is commonly used instead. *See also* **Forcible Entry; Halligan Tool; Kelly Tool.**

Clean Agent—A volatile or gaseous fire extinguishing agent that is not electrically conductive and does not leave any residue during or after its application following evaporation. Common clean agents include Argonite, Carbon dioxide, Halon, Inergen, and FM-200. Although Halon is considered a clean agent, it may contribute to the Earth's ozone depletion and therefore is considered environmentally harmful. *See also* **Argonite; FM-200; Inergen.**

Clean Agent Fire Suppression System (CAFSS)—A fire suppression application system that utilizes a volatile or gaseous fire extinguishing agent that is not electrically conductive and does not leave any residue during or after its application following evaporation. *See also* **Inergen.**

Clean Burn—A fire pattern left on a surface where the soot has been burned away. *See also* **Burn Pattern.**

Closed Cup Tester—A test apparatus used to measure the flash point of a combustible liquid. (Ref. ASTM D 56, *Standard Method of Test for Flash Point by the Tag Closed Cup Tester.*) *See also* **Flash Point (FP).**

Closed Head Foam-Water Sprinkler System—A system using automatic sprinkler system with fusible heads that is charged with air, water, and foam solution. Upon activation of the fusible head, foam-water is discharged. This system does not air aspirate the foam. *See also* **Open Head System.**

Cloud Fire—A transient fire resulting from the ignition of a cloud of gas or vapor and not subject to significant flame acceleration from the effects of confinement or turbulence. It therefore can only occur after a relatively slow release of hydrocarbon vapors and in an open free space.

Clove Hitch—A type of knot used in the fire service, consisting of two half hitches. Usually used to tie the rope to an object. See Figure C-6. *See also* **Half Hitch.**

Code—*See* **Fire Code.**

Coefficient of Discharge—A mathematical factor applied to flow calculations from an opening

Figure C-6 Clove hitch.

that varies according to the arrangement and smoothness of the particular opening.

The following table identifies typical discharge coefficients used in the fire protection profession:

Outlet Description	Coefficient
Standard sprinkler [nominal 0.5 in. (1.27 cm) diameter outlet]	0.75
Standard orifice (sharp edge)	0.62
Smooth bore nozzle	0.96 to 0.98
Underwriters' playpipe	0.97
Deluge or monitor nozzle	0.997
Open pipe, smooth bore, well-rounded edge	0.90
Open pipe, purred opening	0.80
Hydrant butt, smooth or well-rounded outlet, full flow	0.90
Hydrant butt, square and sharp at hydrant barrel	0.80
Hydrant butt, square outlet projecting into barrel	0.70

See also **Hydrant Coefficient; Orifice.**

Cold Flow—A failure phenomenon associated with fusible elements. If a fusible element is subject to heating and cooling cycles, such as a sprinkler next to a space heater, it will soften and harden in a cyclic fashion. The repeated cycle will eventually cause the fusible element to operate, and in the case of a sprinkler system, accidental water release causing water damage.

Cold Smoke—Smoke that is produced from a smoldering fire. The fire itself does not generate adequate quantities of heat to produce a flaming fire. Cold smoke therefore lacks the buoyancy of smoke from a flaming fire because its low heat content does not generate a strong convection current. Cold smoke may be more difficult to detect by ceiling mounted smoke detectors due to its lack of buoyancy. *See also* **Chemical Smoke; Passive Smoke Detection System; Smoke; Smoke Bomb; Smoldering.**

Combination Apparatus—A piece of fire apparatus required to perform more than one function, usually called triple combination quads or quints.

Combination Nozzle—*See* **Nozzle, Combination.**

Combined Agent System—*See* **Dual Agent System.**

Combustible—In a general sense, any material capable of burning, generally in air under normal conditions of ambient temperature and pressure, unless otherwise specified. This implies a lower degree of flammability. Although there is no distinction between a material that is flammable and one that is combustible, NFPA 30, *Flammable and Combustible Liquids Code,* defines the difference between the classification of combustible liquids and flammable liquids based on flash point temperatures and vapor pressures. Combustion can also occur in cases where an oxidizer other than the oxygen in air is present, such as chlorine, fluorine, or chemicals containing oxygen in their structure. Combustible is a relative term since many materials will burn under one set of conditions and not burn under others. The term combustible does not indicate the ease of ignition for a material, the burning intensity, or the rate of burning. Only when the term is modified by adjectives such as *highly* as in highly combustible is a distinction made in common terminology. Therefore, additional parameters such as flame spread and smoke developed are required to define the degree of combustibility and are selected for building construction to reduce the risk of fire. *See also* **Combustible Liquid; Flammable Gas; Flammable Liquid.**

Combustible Construction—Building construction primarily of unprotected wood or wood byproducts. Combustible construction is inherently more hazardous for fire conditions to develop in preference to fire resistive or noncombustible construction. *See also* **Building Construction Types.**

Combustible Dust—Normally referred to as dust with a diameter of 420 microns or smaller, it can be a fire or explosion hazard when dispersed and ignited in the air or with any other available gaseous oxidizer. Each particular dust has its own flammability limits. *See also* **Dust Explosion; Explosive Limits.**

Combustible Gas Detector—An instrument designed to detect the presence or concentration of combustible gases or vapors in the atmosphere. It is usually calibrated to indicate the concentration of a gas as a percentage of its lower explosive limit (LEL) so that a reading of 100 percent indicates that the LEL has been reached. They use either a solid-state circuit, infrared (IR) beam, electrochemical, or dual catalytic bead for the detection of gas in an area. Portable monitors are used for personnel protection and fixed installations are provided for property protection.

Combustible Liquid—As generally defined, it is any liquid that has a closed-cup flash point at or above 100°F (37.8°C). Combustible liquids are classified as Class II or Class III, and flammable liquids are classified as IA, IB or IC:

Class II Liquid—Any liquid tested with a flash point at or above 100°F (37.8°C) and below 140°F (60°C).

Class IIIA—Any liquid tested with a flash point at or above 140°F (60°C), but below 200°F (93°C).

Class IIIB—Any liquid tested with a flash point at or above 200°F (93°C).

See also **Combustible; Flammable Liquid; Flammable Liquid Storage Locker or Cabinet; Flash Point (FP).**

Combustible Vapor Dispersion (CVD) Analysis—A mathematical estimation of the probability, location, and distance a release of combustible gases or vapors will exist until dilution naturally disperses the combustible gas or vapor concentration to below its lower explosive limit (LEL), or until they are no longer considered ignitable (usually taken as 50 percent of the LEL) based on a defined release rate, release pressure, orientation, and assumed ambient conditions. Conditions of little or no wind cause little or negligible natural dispersion whereas moderate to high winds produce effective dispersion effects. Commonly used in the process industries to estimate the potential blast effects, if the vapor were to ignite.

Combustion—A rapid exothermic chemical process. It involves the evolution of radiation effects (heat and light) as a result of a reaction of an oxidizer (usually oxygen in air) with an oxidizing material. Other physical phenomena that sometimes occur during combustion reactions are explosion and detonation. Combustion is generally defined as oxidation accomplished by burning.

Hydrogen burns oxygen, causing heat, which produces water (steam). Molecules of oxygen and hydrogen generally occur in pairs of atoms O_2 and H_2. At a sufficiently high temperature, each oxygen molecule collides violently with two hydrogen molecules. The collision breaks the hydrogen molecules apart and they reform as two fast-moving water molecules ($2H_2 + O_2 = 2H_2O$). When a carbon-based material is burned, the molecules react to form carbon dioxide (CO_2), or carbon monoxide (CO) (if insufficient oxygen is available) and water (H_2O) molecules.

The rate or speed at which the reactants combine is high, because of the nature of the chemical reaction itself and in part because more energy is generated than can escape into the surrounding medium. The result is that the temperature of the reactants is raised to accelerate the reaction even more. A familiar example is a lighted match. When a match is struck, friction heats the head to a temperature at which the chemicals react and generate more heat than can escape into the air, and they burn with a flame. If a wind blows away the heat or the chemicals are moist and friction does not raise the temperature sufficiently, the match goes out. Properly ignited, the heat from the flame raises the temperature of a nearby layer of the matchstick and of oxygen in the air adjacent to it, and the wood and oxygen react in a combustion reaction. When equilibrium between the total heat energies of the reactants and the total heat energies of the products (including the actual heat and light emitted) is reached, combustion stops. Flames have a definable composition and a complex structure; they are said to be multiform and are capable of existing at quite low temperatures as well as at extremely high temperatures. The emission of light in the flame results from the presence of excited particles and charged atoms, molecules, and electrons.

Combustion of solids can occur by two mechanisms, either flaming or smoldering. Flaming combustion takes place in the gas or vapor phase of a fuel after it decomposes from the heat effects of the flame or smoldering process. Combustion is usually thought of in a practical sense as a controlled event, whereas fire (which is also a combustion process) is thought of as an uncontrolled event. *See also* **Ablaze; Burn; Combustion Gases; Fire; Fire Tetrahedron; Fire Triangle; Flame; Pyrolysis; Smoldering.**

C

Combustion Chamber—That portion of an appliance within which combustion occurs as part of the operation of the appliance (furnace, boiler tube firebox).

Combustion Detector—A part of a primary safety control for a combustion process that is responsive directly to flame properties. *See also* **Supervised Flame.**

Combustion Gas Detector—A fire detector that reacts to combustion gases. It can provide warning of a fire presence in the early stages. *See also* **Fire Detector.**

Combustion Gases—Gases produced from the process of combustion. They are usually colorless toxic gases such as carbon monoxide, carbon dioxide, hydrogen chloride, and hydrogen cyanide given off from a fire often before visible smoke. The exact type and amount of combustion gases depends on the material being consumed by the combustion process and the ambient conditions of the environment (availability of oxygen). Combustion gases may result in personnel fatalities through asphyxiation or poisoning. *See also* **Combustion; Smoke Inhalation.**

Combustion, Incomplete—Burning that occurs with an insufficient supply of air so that an inefficient combustion process occurs and the burning substance is only partially consumed. It could be burned further with an additional or supplemental air supply to the combustion process to allow a complete combustion reaction to occur. The production of smoke from a combustion process is generally considered a sign of incomplete combustion. *See also* **Perfect Combustion; Smoke.**

Combustion Products—Effluents such as heat, gases, solid particulates, inerts, and liquid aerosols produced by combustion of a fuel, but excluding excess air.

Command—*See* **Incident Commander.**

Command Staff—Include the positions of safety officer, public information officer, and liasion officer.

Compartmentation—A method of fire protection for structural facilities whereby structural components (walls, floors, roof, etc.) of adequate fire resistance are used to contain a fire event, limit the size of a fire in a building, and prevent it from spreading to other areas. It is a passive fire protection measure provided at the time of building design and construction. Failure of compartmentation is normally caused by openings in the structural elements that are not firestopped and allow a fire or its effects to pass through it. *See also* **Fire Blocking; Fire Compartment; Passive Fire Protection (PFP).**

Compound (Pressure) Gauge—An instrument or device used to measure pressures both above (positive) and below (negative) atmospheric pressure. It may also refer to the pressure gauge on the intake of a fire pump. *See also* **Pressure Gauge.**

Compressed Air Foam System (CAFS)—A fire-fighting foam system that uses compressed air with foam solution to create a foam in an application hose or mixing chamber. It has a foam expansion ration of 1 to 60 and is used to suppress fires in Class A fuels. It uses a minimum pressure for the compressed air source of 100 psi (689 kPa). Compressed air foam systems are an optional feature provided to mobile fire trucks and are primarily used for wildfires. See Figure C-7. *See also* **Foam.**

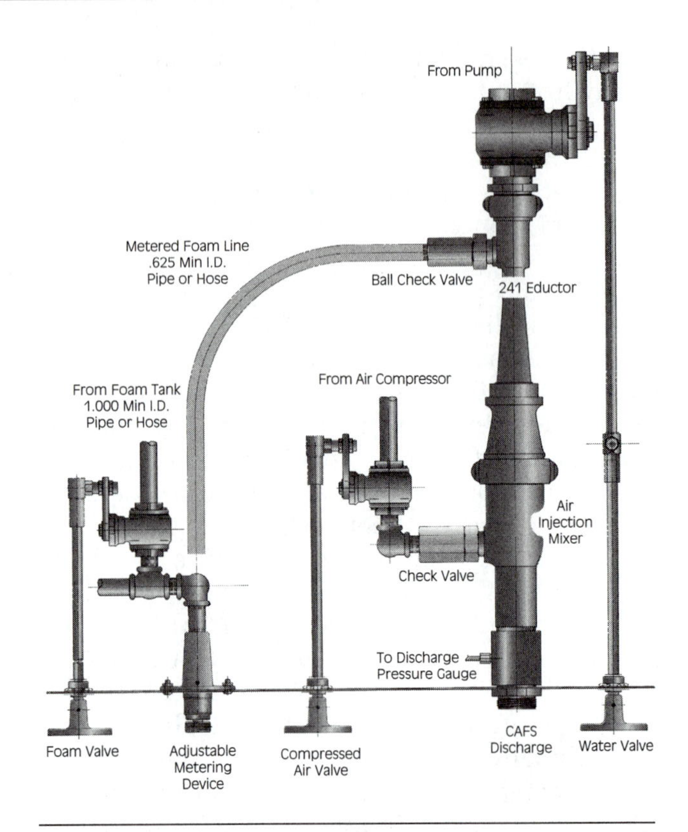

Figure C-7 Typical CAFS arrangement on mobile fire apparatus.

Computational Fluid Dynamics (CFD)—A type of mathematical modeling in which a region of a flow under study is subdivided by a grid into a large number of control volumes. Computational fluid dynamics is employed in the study of flame and fire growth.

Computer Aided Dispatch (CAD)—The assignment of fire services to an emergency incident with the aid of a computer system.

Concentration Limits of Flammability—Synonymous with flammable limits, explosive limits. *See also* **Flammable Limit** or **Flammability Limits.**

Conduction—The mode of heat transfer associated with solids in direct contact. Each solid has a temperature dependent K-factor, which is a measure of the rate of conduction. *See also* **Heat Conduction; Heat Transfer; K-Factor; Thermal Conductivity.**

Conductivity Method—A method to determine if a foam solution has been accurately proportioned. It is particularly useful for examining AFFF or alcohol-resistant AFFF produced foams and foams with a low proportioning percentage (1 percent or less). These foams exhibit a very low refractive index, making examination of their proportioning accuracy with a refractometer limited, and therefore the conductivity method is recommended. The conductivity method uses changes in electrical conductivity of water as foam concentrate is added to it. The conductivity method is very accurate if there are substantial changes in conductivity, since the foam concentrate is added to the water in small amounts. Because salty or brackish water is very conductive, the conductivity method may not be suitable due to small conductivity changes as foam concentrate is added to these liquids. As a test for these liquids, foam and water solutions can be made in advance in controlled conditions to determine if adequate changes in conductivity will be detected if the water source is salty or brackish. *See also* **Proportioning; Refractive Index; Refractometer.**

Conflagration—The spread of a fire over a considerable area destroying a large number of buildings or a substantial amount of property. May also include forest fires that involve structures and communities. A conflagration usually crosses natural firebreaks such as rivers, streets, etc. A large number of conflagrations occurred in the late 1800s and early 1900s in America because of the widespread use of wooden exterior building construction materials, poor building codes, inefficient water supplies, and equipment. Wooden roof shakes have been specifically mentioned as a contributing factor in most recent conflagrations in America. Conflagrations continue to occur in arid areas where exposed wooden construction is used to a large extent, such as California. It has been stated that almost every city in the world, both ancient and modern, has been involved in a conflagration of some extent. *See also* **Blowup Fire; Notable Fires Throughout History.**

Congressional Fire Services Caucus—A function of the US federal government. Founded in 1987, its purpose is to direct attention to the concerns of its fire safety constituents.

Congressional Fire Services Institute (CFSI)—A nonprofit, nonpartisan organization whose purpose is to educate members of the US Congress (House of Representatives and Senate) on issues concerning the fire and emergency services. It was founded in 1989, and is used by congressional members for advice on fire service viewpoints concerning legislative action.

Control Line—A wildland firefighting process that includes the removal of fuel, application of water, and use of natural resources to stop the advance or spread of a fire. It may also be applied to constructed or natural barriers that are used to control a fire. It may also be called a fireline. *See also* **Firebreak; Fireline.**

Control Panel, Fire Alarm (FACP)—*See* **Fire Alarm Control Panel (FACP).**

Controller—The cabinet, motor starter, circuit breaker, disconnect switch, or other device for starting and stopping an electric motor or internal combustion engine that drives a firewater pump. The controller is generally designed to receive external inputs (signals) that then cause automatic operation of the firewater pump. These signals may be the activation of an automatic sprinkler system, low water-system pressure, fire detection, etc. Guidance in the design and operation of controllers is provided in NFPA 20, *Standard for the Installation of Centrifugal Fire Pumps*.

C

Convection—The heat transfer associated with fluid movement around a heated body. Warmer, less dense fluid rises and is replaced by cooler, more dense fluid. Convection currents rise during a fire event due to the heat transfer to the surrounding air, causing it to rise and allow cooler air to enter the fire environment at the base of the fire. *See also* **Heat Convection; Heat Transfer; Thermal Convection.**

Convective Column—The air current that rises from a heat source or a fire. Air rises due to decreased density from being heated. *See also* **Plume.**

Cool Flame Ignition—*See* **Ignition Temperature.**

Cooling Spray—A water spray provided to lower the temperature of an object, to afford exposure protection to individuals or objects, or to cool a fire as a precursor to extinguishment. High-pressure vessels are commonly provided with a fixed cooling spray as protection against rupture or boiling liquid expanding vapor explosion (BLEVE) concerns (if they contain combustible materials), if they may be potentially exposed to heat input from a fire. *See also* **Boiling Liquid Expanding Vapor Explosion (BLEVE); Water Spray.**

Corps of Vigilēs—Paid Roman firefighters instituted by the Emperor Augustus in the year 6 CE. There was one cohort of 1,000 men for each of the seven areas of Rome. The Familia Publica were Roman firefighters, usually slaves that preceded the Corps of Vigilēs. The Corps of Vigilēs protected Rome for about 500 years. They had three basic duties as firefighters: an *Aquarius* or water man, an *Uncinarius* or hook man, and the *Siphonarius* or pump man. The *Aquarius* organized the water delivery through a bucket brigade, the *Siphonarius* took control of the water pump and was also the mechanic for it, and the *Uncinarius* was responsible for dragging burning material from roofs and walls. They also used wet blankets, centonēs, to beat out a fire or protect other exposures. Poles, axes, ladders, and brooms were also used.

Coupled Water-Motor Pump Proportioning—A proportioning method for the introduction of foam concentrate or wetting agent into a water delivery system for fire protection applications. It consists of a positive displacement pump in a water line coupled to a smaller positive displacement foam concentrate or wetting agent pump to provide proportioning. *See also* **Eductor; Proportioning.**

Coupling—A fitting provided at the end of firewater hose to allow it to be connected to an appliance, another hose, or a firewater source (pumper or hydrant). *See also* **Adapter.**

Crash Truck—A specialized firefighting vehicle provided at aircraft operations to attend to aircraft crashes. It is equipped to handle large combustible liquid spills, perform rescue operations, and operate on cross-country terrain. NFPA 414, *Standard for Aircraft Rescue and Firefighting Vehicles* provides guidance in the design and specification of these vehicles. *See also* **Airport Crash Truck.**

Creeping Fire—Terminology used by the forest service to describe a woodland fire that burns with a low flame and spreads slowly.

Crib—A common arrangement of wooden timber used in fire testing to simulate a given combustible fuel load. The wooden stack is arranged to allow adequate fire development within a prescribed area. Generally, a square or rectangular shape is used that alternately positions

Figure C-8 Fire test with crib.

wooden stringers, spaced apart from each other, on successive layers until the desired quantity and a configuration that is porous for air circulation is achieved. See Figure C-8. *See also* **Fire Test; Fuel Loading.**

Critical Function—An operation or activity essential to the continuing survival of a system. Any of those functions that are vital to the life of the system. Critical function evaluations are important when determining the survivability of fire protection and life safety systems during a fire incident.

Critical Incident Stress Debriefing (CISD)—Consolation provided to individuals to minimize the psychological stress that may occur after responding to a traumatic incident. It is a major part of the fire services, especially those fire departments provided with a station-based ambulance service. It may also be called post-traumatic incident debriefing. *See also* **Fire Trauma; Post Incident Trauma.**

Critical Radiant Flux—A test measurement recorded for heat input during testing of the fire resistance of floor coverings prescribed in NFPA 253, *Standard Method of Test for Critical Radiant Flux of Floor Covering Systems Using a Radiant Heat Energy Source.* It is the level of radiant heat energy at the most distant flameout point on the floor covering system.

Critical Temperature—The temperature at which a structural metal (such as steel) softens when heated and can no longer support a load. It is usually below its melting temperature.

Cross-Zoning—A method of fire detection whereby adjacent fire detectors are connected to different sensing circuits to the fire alarm control panel. Confirmed fire detection is only achieved if two detectors are activated, one from each of the separate alarm circuits. Cross-zoning is used primarily as a deterrent against false alarms and in particular where a fixed fire suppression system (such as a CO_2 system) is arranged to automatically discharge upon fire detection to avoid accidental release of the suppression gas. It may also be referred to as a voting system.

Crown Fire—*See* **Forest Fire.**

Curtain Boards—Curtain boards, also called draft stops, are provided for the containment of smoke and heat beneath a ceiling to the area of occurrence to prevent fire spread. Combined with roof vents, they allow the release of these combustion products to prevent the spread of a fire based on specific fire design conditions, and if sprinklers are provided, to promptly cause their activation. Curtain boards are generally

C

provided in aircraft hangars, on pier or lumber roofs, or similar structures. Preferably, curtain boards should be made of any substantial, noncombustible material that will resist the passage of smoke. They generally extend down from the ceiling for a minimum of 20 percent of ceiling height. Curtain boards are important for prompt and positive activation of the roof smoke and heat vents because they bank up combustion products (heat, gases, and smoke) in the curtained area. The distance between curtain boards should not exceed eight times the ceiling height to ensure that vents remote from the fire within the curtained compartment will be effective. NFPA 204M, *Guide for Smoke and Heat Venting*, provides guidance in the design and application of curtain boards. Curtain boards are not the same as curtain walls (also called panel walls), which are applied to the exterior of a building structure to enclose it. *See also* **Draft Curtain; Draft Stopping; Draft Wall; Smoke Stop.**

Damp Down—To wet down or "blacken" a fire. *See also* **Blacken.**

Damper, Fire—A device (damper) arranged to seal off airflow automatically through part of an air distribution system to resist the passage of heat and flame. It is usually an assembly of louvers arranged to close from the heat of a fire by melting a fusible link or through a remote activation signal. Fire dampers are required by all building codes to maintain the required level of fire resistance rating for walls, partitions, and floors when they are penetrated by air ducts or other ventilation openings. There are two significant ratings when applying a fire damper; the fire resistance rating and the airflow closure rating. The fire rating is dependent on meeting the fire resistant rating of the fire barrier being penetrated by the airflow duct and the airflow rating is either static or dynamic, depending on whether the air flow is automatically shut down upon fire detection. Ref. UL Standard 555, *Standard for Safety, Fire Dampers.* See Figure D-1. *See also* **Damper, Smoke; Fire Barrier.**

Damper, Smoke—A damper arranged to restrict the spread of smoke in a heating, ventilation and air conditioning (HVAC) air duct system. It is designed to automatically shut off air movement in the event of a fire. It is usually applied in a Passive Smoke Control System or as part of an Engineered Smoke Control System to control the movement of smoke within a building when the HVAC is operational in an engineered smoke control system. HVAC control fans are used to create pressure differences in conjunction with fixed barriers (walls and floors). Higher pressures surround the fire area and prevent the spread of smoke from the fire zone into other areas of the building. A smoke damper also can be a standard louvered damper serving other control functions, provided the location lends itself to the dual purpose. A smoke damper is not required to meet all the design functions of a fire damper. Smoke dampers are classified according to leakage rates: Class 1 (lowest), Classes 2, 3, and 4 (highest); elevated temperature 250°F (121°C), 350°F (177°C) or higher; and prescribed pressure and velocity differences at the damper (specific velocity of airflow when open and to close against a specific pressure differential). Ref. UL 555S, *Leakage Rated Dampers for Use in Smoke Control Systems.* See Figure D-2. *See also* **Damper, Fire; Smoke Barrier.**

Dead Load—The weight of a structure or building itself and any permanently installed or built-in equipment. *See also* **Impact Load; Live Load.**

Deck Gun (Deluge Set)—A portable heavy stream application device used for firefighting. It is set up on the ground at a fire scene and supplied by large diameter hoses for the application of large quantities of water onto a fire incident. *See also* **Deluge Set; Master Stream.**

Deep-Seated Fire—A fire that is embedded in the material being consumed by combustion.

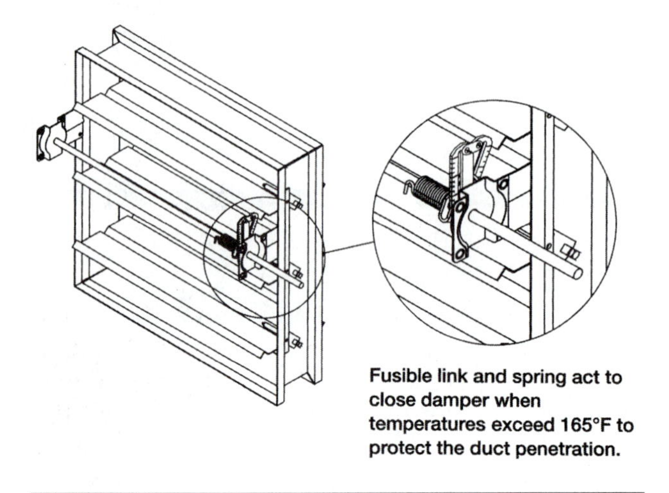

Figure D-1 Fire damper. *(Courtesy Greenheck Fan Corporation)*

Extinguishment of a surface fire does not guarantee a deep-seated fire may also be eliminated. Extinguishment of deep-seated fires requires an individual to investigate the interior of a material once the surface fire has been extinguished to determine if interior extinguishment has also been accomplished. If a deep-seated fire in an enclosed area is to be extinguished by a gaseous agent, the period of agent concentration has to be adequate to ensure suppression has been accomplished.

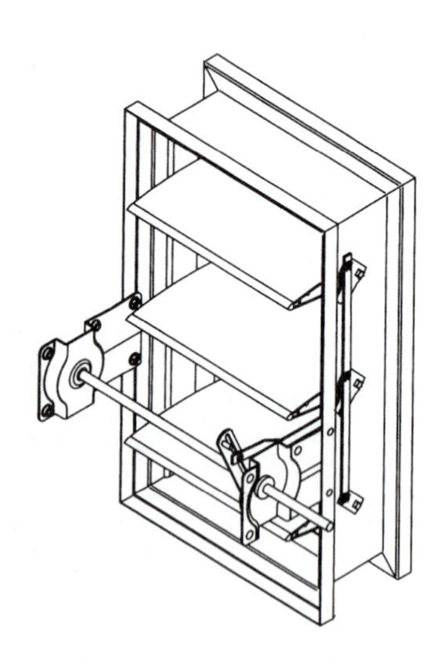

Figure D-2 Smoke damper. *(Courtesy Greenheck Fan Corporation)*

Defensive Firefighting or **Attack**—The mode of manual fire control in which the only fire suppression activities taken are limited to those required to keep a fire from extending from one area to another usually undertaken until sufficient resources arrive to undertake an offensive attack. Typically meant as an exterior attack for exposure protection. *See also* **Offensive Firefighting.**

Deflagration—Mechanism for the propagation of an explosion reaction through a flammable gas mixture that is thermal in nature. The velocity of the reaction is always less than the speed of sound in the mixture but is capable of causing damage. *See also* **Detonation; Explosive; Flame Speed.**

Deflagration Index—A measure of the explosibility of a combustible dust or gas that is computed from the maximum rate of pressure rise attained by combustion in a closed vessel. The purpose of deflagration index measurements is to predict the effect of the deflagration of a particular material (dust or gas) in a large enclosure without carrying out full-scale test work. *See also* **Explosibility Index; Index of Explosibility.**

Deflagration Isolation—A method of protection against a deflagration incident by employing equipment and procedures to interrupt the propagation of deflagration flame front at a specific point.

Deflagration Pressure Containment—A method of protection against deflagration overpressure exposure in a pressure vessel. Its design pressure is rated to withstand the maximum internal deflagration pressure that can be expected.

Deflagration Suppression—A method of detecting and preventing a combustion process in a confined space while it is in its incipient stage.

Deluge—The immediate release of a commodity, usually referring to a water release mechanism for fire protection purposes. Water deluge protection commonly consists of distribution piping and water spray nozzles that are connected to a supply valve. When the valve is activated (opened), water is immediately released through the system.

Deluge Set—A large capacity firewater nozzle classification. It generally consists of a short length of large diameter hose having a large nozzle with a tripod support at one end and a three- or four-way siamese or inlet connection at the supply end. See Figure D-3. *See also* **Deck Gun (Deluge Set).**

Deluge Sprinkler System—*See* **Sprinkler System.**

Deluge Valve—A fire protection valve that activates by the operation of a fire detection system or manually. It is designed to withhold the fire suppressant agent until the valve is activated. It consists of a clapper valve held by a latch within a valve housing. The latch is release by a fire detection or manual activating device, whereby the fire suppression agent discharge pressure forces open the clapper and disperses the agent into the distribution system. *See also* **Clapper Valve.**

Demand Curve—A graphical representation of the amount of water pressure and quantity needed by a protection system or facilities to adequately supply its devices based on hydraulic calculations of the system requirements. A demand curve is prepared against a supply curve to verify that the facility water supplies and requirements can be achieved. *See also* **Supply Curve.**

Deployment Plan—A predetermined set of instructions, procedures, or arrangements of fire service apparatus and personnel for specific types of incidents and specific locations. *See also* **Pre-Fire Plan; Pre-Incident Planning.**

Figure D-3 Deluge set.

Depth of Char—Term used by fire investigators to indicate the amount of time a material has burned. A deep char indicating a longer period of fire exposure. See Figure D-4.

Depressurization—A safety feature provided in the process industries to avoid the release of a vessel's contents due to fire exposure. It consists of the deliberate depressurizing and

D

D

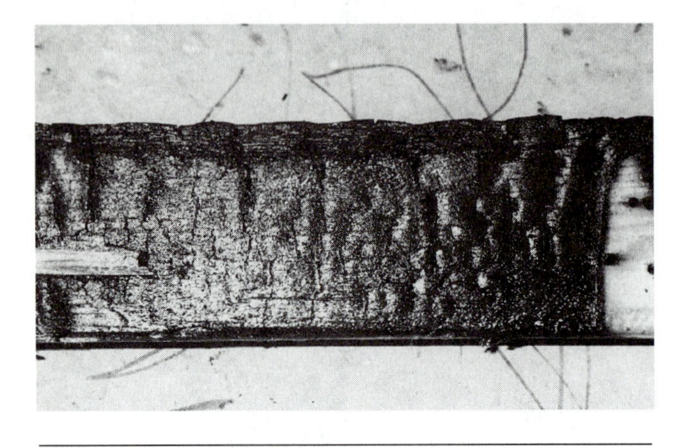

Figure D-4 Depth of char example.

removal of the contents of a process vessel to a disposal system due to weakening of the vessel shell from heat input. During a fire incident, a process vessel may be exposed to rapid heat input that reduces the strength of the vessel steel below its ability to contain the pressure level of the process. The vessel will then rupture, releasing its contents and escalating the incident. Guidelines for depressurization for the process industries are provided in American Petroleum Institute API RP 521, *Guide for Pressure Relieving and Depressuring Systems*. *See also* **Boiling Liquid Expanding Vapor Explosion (BLEVE); Emergency Venting.**

Design Density, Sprinkler—The quantity of water per square foot (m^2) that has been found by experience and testing to be effective in controlling and suppressing a fire based on occupancy.

Detector Check Valve—A check valve (one-way valve) used to detect leakage or the misuse of water intended only for fire protection applications. It consists of a weighted clapper on the main flow that is normally in a closed position. Small water flows are forced through a metered bypass. Metering of limited flow indicates a leak or misuse of the system. Increased demand required by the fire protection system causes the weighted clapper to fully open, allowing full flow for the fire emergency.

Detonation—Propagation of a combustion zone at a velocity that is greater than the speed of sound in the unreacted medium. It is mechanical in nature, and acts through shock pressure forces. The reaction front or flame front advances into the unreacted substance at greater than the speed of sound in the unreacted substance. The progressive acceleration of reaction accounted for by the growth of the flame front area and by transition from laminar to turbulent flow gives rise to a shock wave. The increase in temperature due to compression in the shock wave results in self-ignition of the mixture and detonation sets in. The shock wave-combustion zone complex forms the detonation wave. Detonation differs from normal combustion in its ignition mechanism and in the supersonic velocity of 1.2 to 3.1 miles per second (2 to 5 km per second) for gases and 4.8 to 5.4 miles per second (8 to 9 km per second) for solid and liquid explosives. Detonation is impossible when the energy loss from the reaction zone exceeds a certain limit. *See also* **Deflagration; Explosive; Flame Speed; High Explosive.**

Detonation (Flame) Arrester—A detonation arrester is essentially a heavy-duty, thermal type flame arrester designed to handle the peak overpressure and velocities of a detonation. It is used in closed piping systems where flame speeds can increase to detonation levels. The performance requirements for detonation flame arresters are specified in US Coast Guard (USCG) regulations. *See also* **Flame Arrester.**

Detonation Limits—The range of fuel-air ratios through which detonations can propagate. *See also* **Flammable Limit** or **Flammability Limits; Upper Explosive Limit (UEL).**

Diaphragm Tank—*See* **Bladder Tank.**

Diffracted Wave—A component of a blast wave that propagates into the sheltered region behind a structure.

Diffusion Flame—A flame in which the fuel and oxygen are transported or diffused into the reaction zone of the flame from opposite sides. Most natural flaming processes produce diffusion flames. Common examples are a candle flame or a forest fire. Diffusion flames can be classified as laminar or turbulent, depending on the steadiness of the flames produced. *See also* **Laminar Flame; Premixed Flame.**

Diking—A defensive method of spill control. Provided to prevent spill runoff which may endanger other adjacent areas. May be permanent or temporary structures. *See also* **Remote Impounding.**

Dilution—The addition of an inert gas to a combustible mixture to inhibit ignition or to suppress a combustion process. The amount of dilution is dependent on the combustibility characteristics of the material involved, but its use generally achieves an atmosphere below the lower explosive limit for the particular gas.

Direct Attack—The application of firewater directly onto a burning fuel during firefighting operations. *See also* **Indirect Application** or **Indirect Attack.**

Distributor Nozzle—A firefighting spray nozzle that creates a broken stream normally used to combat basement-level fires. *See also* **Bresnan Distributor; Cellar (Pipe) Nozzle.**

Diverting—A method of spillage control, whereby non reactive materials are used to control and direct the drainage of the spill to a safe location. *See also* **Diking; Retention; Remote Impounding.**

Donut Roll—Term for the storage or handling arrangement for a length of fire hose that is not in use; it appears in the shape of a donut. Donut roll fire hose is flattened and coiled into a roll. See Figure D-5. *See also* **Accordion Horseshoe Load** or **Fold; Potato Roll.**

Double-Jacketed Hose—A fire hose constructed with an inner liner and two outer protective reinforced jackets. The inner liner is typically made from a synthetic rubber or special thermoplastic compound. The outer jackets are a synthetic fiber weave. The use of a double jacket instead of a single jacket improves the durability and strength of the hose. See Figure D-6. *See also* **Rubber-Covered Hose; Single-Jacketed Hose.**

Draft—The process of obtaining water from a static source into a pump. Drafting water is obtained by creating a pressure lower than atmospheric (vacuum) in the suction of the drafting pump, thereby allowing the atmospheric

Figure D-5 Donut roll (simplified for clarity).

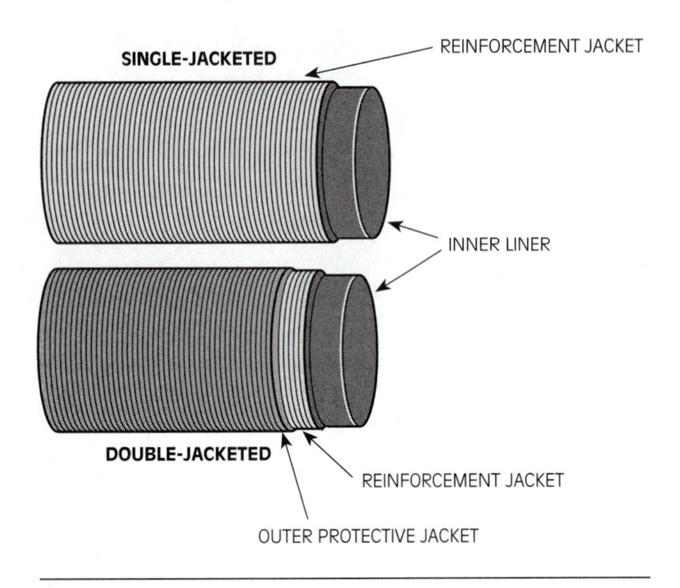

Figure D-6 Single and double jacketed hoses.

water pressure to force water into the pump. Drafting operations require the use of a hard suction hose to the pump, otherwise the hose will collapse due to the lower atmospheric pressure in the suction portion of the line. The height of drafting operations is limited to the theoretical lift of the pump. If a pump were able to produce a perfect vacuum (0 psi/kPa) then supposedly it could lift water to 33.8 ft. (10.3 m). Since ordinary liquid pumps cannot be expected to produce perfect vacuum conditions, as a rule of thumb, only two-thirds of the theoretical lift can be expected for these pumps, or a maximum of 22.5 ft. (6.8 m). *See also* **Drafting Pit; Hose, Hard Suction; Theoretical Lift.**

Draft Curtain—A partition provided to the underside of a building roof or structure to retard the spread of heat and smoke or to accelerate the actuation of a fusible head provided on a fixed fire suppression system. *See also* **Curtain Boards; Draft Wall; Smoke Stop.**

Drafting Pit—An open water tank used for fire service drafting operations or for fire pumper testing. *See also* **Cistern; Draft; Fire Pumper.**

Draft Stop—*See* **Curtain Boards; Smoke Stop.**

Draft Stopping—The provision of a barrier in a heat collection zone (such as ceiling areas) to retard the spread of hot fire gases and smoke. Draft stopping is not expected to provide a complete fire barrier to prevent fire spread. *See also* **Curtain Boards.**

Draft Wall—A partition provided to retard the spread of heat or smoke. *See also* **Curtain Boards; Draft Curtain.**

Drainage Rate, Foam—The rate at which a firefighting foam will drain from the expanded foam mass or how long it will take 25 percent ("quarter" drainage time) of the solution to drain from the foam over a given period of time. It is also called the quarter lifetime or 25 percent drain time. Foam that has a fast drain time is very fluid and mobile, spreading across the fuel surface quickly. Foams with longer drain times are normally less mobile and move across the fuel surface more slowly, but they demonstrate more heat resistance and stability. Drainage time can be affected by operating pressure, temperature, and quality of water.

Drenchers—A type of sprinkler system used to protect roofs, windows, and external openings of a building or structure from damage from the exposure to a fire in an adjacent property.

Drop and Roll—An emergency action to implement if an individual's own clothes are on fire. One immediately drops to the ground and rolls over several times back and forth—aka stop, drop and roll. This method quickly smothers the fire. This prevents the fire from continuing or allowing the effects of smoke inhalation to occur while an individual may be running for assistance or taking other protective measures.

Use of fire blankets to wrap and smother a fire should only be initially undertaken if they are immediately adjacent to the incident. *See also* **Fire Blanket.**

Drop Down—The spread of fire due to dropping or falling burning materials. Synonymous with fall down. *See also* **Flaming Debris/Flaming Droplets.**

Dry Barrel Fire Hydrant—*See* **Hydrant, Fire.**

Dry Chemical—A fire extinguishing agent principally of either sodium bicarbonate, potassium bicarbonate, or ammonium phosphate (multipurpose). Other ingredients are added to improve fluidity, noncaking ability, and water repellant effects. Dry chemical agents contain particles from 10 microns to 75 microns in size. A chemical agent extinguishes a fire by interrupting the chain reaction wherein the chemicals prevent the union of the free radical particles in the combustion process. Combustion does not continue when the flame front is completely covered with agent. Flame cooling by the surface area of the powder and the inerting effect of carbon dioxide produced by the thermal breakdown of the bicarbonate are thought to have a secondary effect on the combustion reaction. Dry chemicals are commonly used to extinguish Class B (burning liquids) or Class C (electrical) fires whereas dry powder is used to put out Class D fires (burning metals) such as magnesium and phosphorus. The two types of agents are not the same and cannot be used interchangeably. Multipurpose dry chemical may extinguish Class A fires (ordinary combustibles), however the fire may re-ignite since these fires may be deep-seated in the material and the agent may not reach these areas. Dry chemical agents are applied through fixed piping systems, through pipes to a hose station, or through portable handheld or wheeled fire extinguishers. *See also* **Fire Extinguisher, Portable; Fire Suppressant Agent; Multipurpose Dry Chemical; Potassium Bicarbonate; Sodium Bicarbonate; Twin Agent Unit** or **Twinned Agent Systems.**

Dry Chemical Fire Extinguisher—A portable fire extinguisher that contains a dry chemical powder used for fire suppression application purposes. It is usually prepressurized or pressurized by a small cartridge of pressurized gas (CO_2 or nitrogen) at the time of use. It is commonly provided with a discharge hose and nozzle with a controllable valve for application onto a fire. See Figures D-7 and D-8. *See also* **Dry Powder; Fire Extinguisher, Portable.**

Dry Chemical Fire Extinguishing System—A fixed fire suppression system that uses stored dry chemical powder that is distributed to the fire hazard through a system of pipes and application nozzles or manually directed hoses with application nozzles. It is activated by manual or automatic fire detection. The system is usually

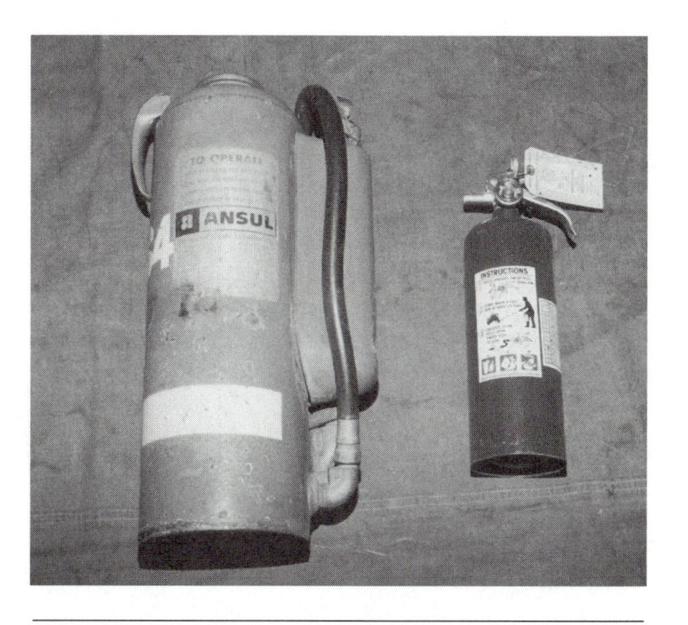

Figure D-7 Dry chemical fire extinguishers.

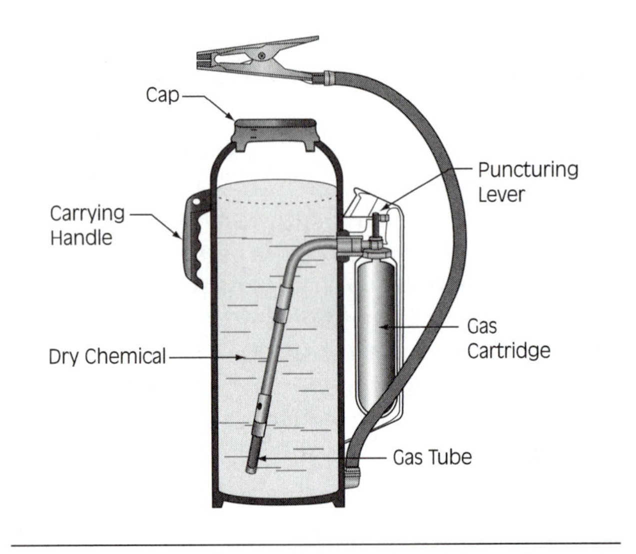

Figure D-8 Dry chemical fire extinguisher schematic.

activated by a storage bank of pressurized inert gas (such as nitrogen) high-pressure cylinders at the time of use. It is normally designated for use at locations where combustible liquids may be present, such as restaurant deep fat fryers, paint dip tanks, hydraulic systems, etc. See Figures D-9 and D-10. *See also* **Dual Agent System; Fixed Fire-Suppression System; Wet Chemical Fire Suppression System.**

Dry Pipe System—*See* **Sprinkler System.**

Dry Pipe Valve—A valve provided to control the flow of water to areas that could be exposed to freezing conditions. Water is held at the valve by air pressure in the system piping. When the air pressure is reduced, the valve operates and floods the system. Dry pipe valves are designed

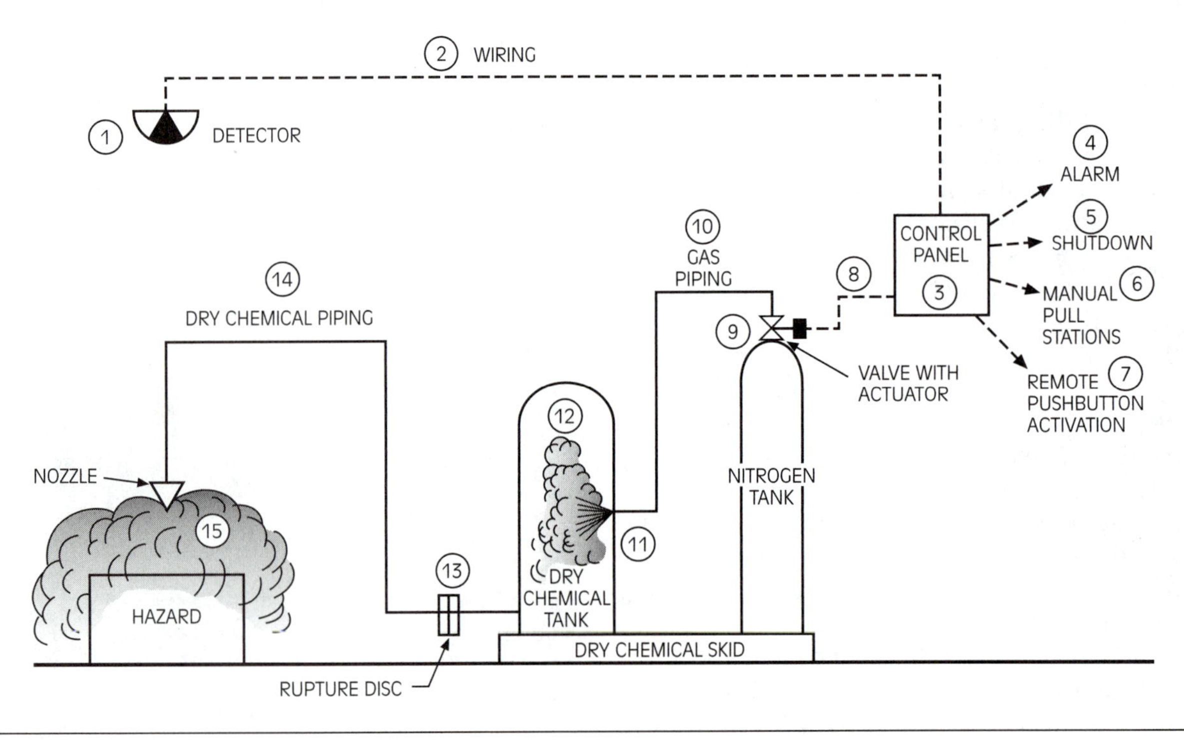

Figure D-9 Dry chemical fire extinguishing system.

Figure D-10 Dry chemical system with hose reel.

so that a moderate amount of air pressure is capable of holding a higher level of water pressure. The ratio of air pressure to water pressure that will allow the dry pipe valve to open is called its differential. It may also be called a dry valve. *See also* **Alarm Check Valve; Sprinkler System.**

Dry Powder—A fire extinguishing agent used on combustible metal fires (sodium, titanium, uranium, zirconium, lithium, magnesium, and sodium-potassium alloys). The principle of extinguishment involves forming an airtight coating over the metal material. The powder may be used to confine or localize a metal fire if it is too difficult to extinguish. A dry powder fire extinguishing agent (typically sodium chloride) is *not* the same as dry chemical agent. *See also* **Caking; Dry Chemical Fire Extinguisher; Metal Fire.**

Dry Riser—A riser pipe for water-based fire protection systems that is kept dry until needed to avoid water freezing concerns. *See also* **Riser; Sprinkler System.**

Dry Sprinkler—A sprinkler connected to a dry pipe system. *See also* **Sprinkler System.**

Dual Agent System—A fire suppression system that uses a combination of foam and dry chemical agents to control and suppress a fire. The combination affords fast knockdown, extinguishment, and prevention against reignition. AFFF foam and potassium bicarbonate dry chemicals are commonly used. When required, their storage cylinders are pressurized with high-pressure nitrogen cylinders that charge the system and expel the agents. The agents are manually directed by two nozzles and operated for sequential or simultaneous operation. The typical application is to use dry chemical to knock down the fire and then immediately apply foam to provide for vapor suppression and protection against reignition. It is usually provided for three-dimensional flammable liquid, running, and pressurized fuel fires. It is sometimes referred to as a twin agent or combined agent system. See Figure D-11. *See also* **Dry Chemical Fire Extinguishing System.**

Duct Detector—*See* **Smoke Detector Duct.**

Dust Explosion—An explosion that occurs when finely divided combustible particles are dispersed in air in a sufficient concentration and in the presence of an ignition source strong enough

Figure D-11 Dual agent system, trailer mounted.

to cause ignition. Combustible dust explosions have a slower rate of pressure rise and lower final pressure than combustible vapor explosions. Many dusts moving through air of low humidity can generate static electricity on isolated electrical conductors. Potentially, the static electricity can accumulate high enough to produce an incendiary spark. The ignition energy for dusts is significantly higher than for vapors. The explosion severity of dust depends on several factors, including dust particle size, turbulence, humidity, ignition energy, and geometry of the incident area. *See also* **Combustible Dust; Ignition Energy.**

E

Early Suppression Fast Response (ESFR) Sprinkler—An automatic fire sprinkler designed to activate quickly from fire conditions. ESFR sprinklers have a thermal element with a response time index (RTI) of 50 (meters-seconds)$^{1/2}$ or less. Standard sprinklers have a thermal element with an RTI of 80 (meters- seconds)$^{1/2}$ or more. ESFR sprinklers are used for special high hazard applications. Large drop ESFR sprinklers are specifically designed for wet pipe sprinkler systems protecting high-piled storage commodity applications. They were developed by the Factory Mutual Research Corporation (FMRC) in the late 1970s and early 1980s. *See also* **Quick-Response (QR) Sprinkler; Response Time Index (RTI); Sprinkler, Residential; Thermal Sensitivity.**

Edge Firing—A method of wildland firefighting in which fires are deliberately set along the edge of an area and allowed to burn to the center.

Eductor—A device for proportioning the specified percentage of liquid foam concentrate or wetting agent into a water stream for application onto a fire incident. The eductor is based on a venturi, where the lower pressure created in the venturi throat is used to draw the agent into the water through a pickup tube or fixed pipe connection. A sized opening at the venturi or orifice placed with the pickup tube or fixed piping meters the proper percentage of agent into the system. See Figure E-1. *See also* **Around-the-Pump Proportioner; Coupled Water-Motor Pump Proportioning; Fixed Foam Application Systems; Pickup Tube; Pressure Proportioning Tank; Proportioning.**

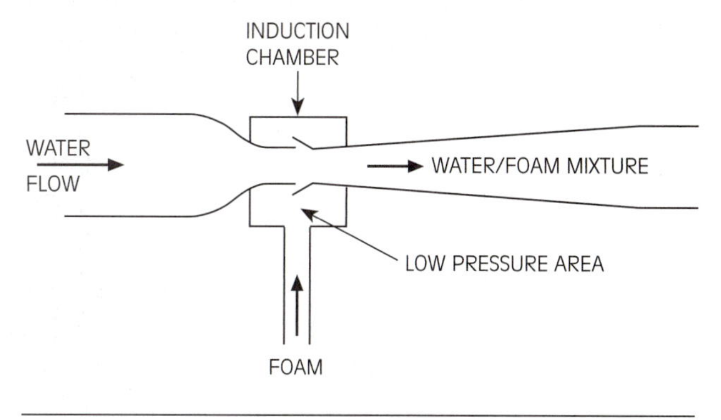

Figure E-1 Eductor indicating venturi principle drawing foam concentrate into the flow system.

Eductor, Foam Nozzle—A device for proportioning the specified amount of liquid foam concentrate or wetting agent into a foam nozzle for application onto a fire incident. Foam nozzle eductors are mainly the venturi type with a suction hose (pickup tube) to an agent storage container to draw the foam concentrate into the water stream. See Figure E-2. *See also* **Pickup Tube; Proportioning.**

Effective Width—The clear horizontal space of a stair that allows users to move up and down

E

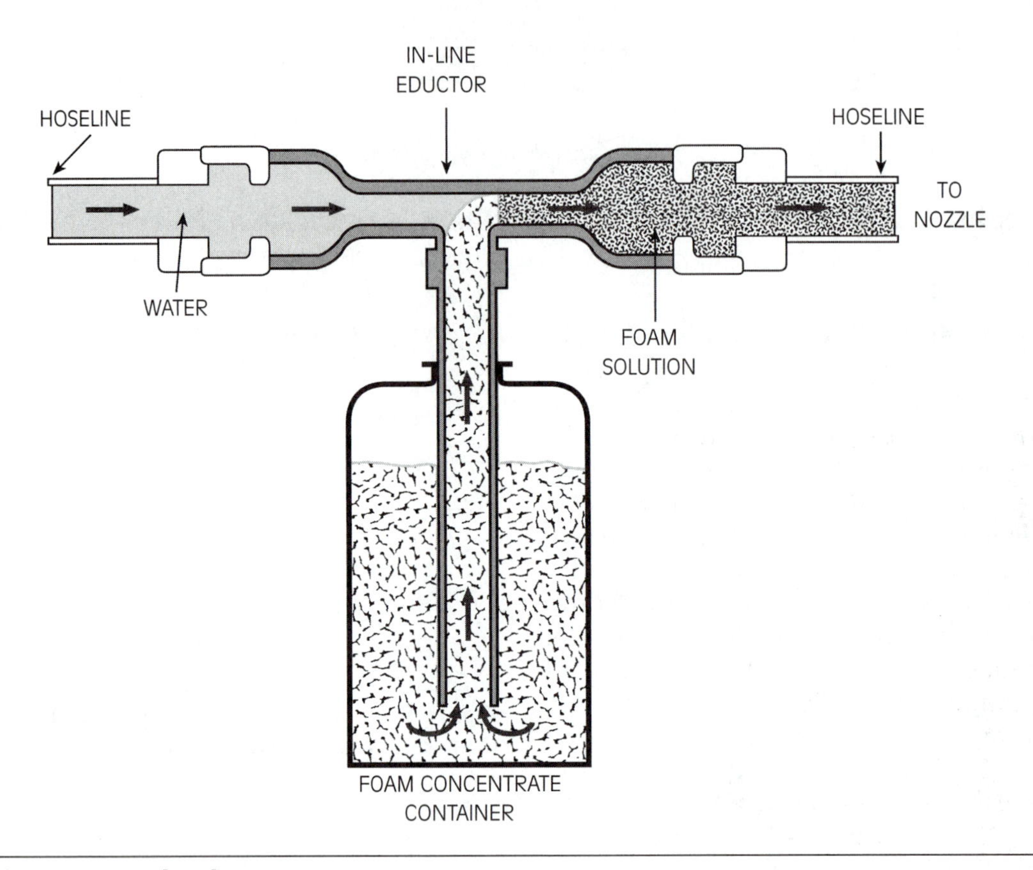

Figure E-2 Foam nozzle eductor.

unhindered especially for evacuation purposes. It is typically the measurement between stair handrails, or any wall or protective barrier where there is no handrail present.

Egress Capacity—The amount or capacity of an egress to allow orderly evacuation of a building or structure for emergency conditions. The occupancy of a structure or building depends on the egress capacity provided. NFPA 101, *Life Safety Code,* provides guidelines to determine the egress capacity and amount of egress capacity required for particular occupancies and occupant loads. *See also* **Egress Width.**

Egress Width—The unobstructed width of a door opening without projections into such width, for purposes of calculating egress capacity. Previously referred to as exit width. *See also* **Egress Capacity.**

Electrical Fire—A fire involving energized electrical equipment. They are usually propagated by electrical short circuits, faults, arcs, and sparks, and the equipment remains energized during the fire event. Due to the possibility of electrical shock, nonconductive extinguishing agents, Class C (carbon dioxide), must be used for fire control and suppression efforts. When the equipment is de-energized, Class A or B extinguishing agents may be used. *See also* **Class C Fire.**

Elevated Master Stream—A fire water nozzle rated in excess of 350 gpm (1,325 l/min) that is positioned at the end of an aerial device.

Elevating Nozzle—A water nozzle provided on a elevating or telescoping boom or platform of a fire truck to apply water at a higher level than normal grade. *See also* **Snorkel** or **Snorkel Truck.**

Elevating Platform—A work platform provided on a telescoping boom used to transport firefighters or equipment or to allow the use of a nozzle at a higher level than normal grade for firefighting and rescue activities. *See also* **Aerial Platform.**

Ember—A particle of solid material that emits radiant energy due either to its temperature or to the process of combustion on its surface. Small bits of wood fibers (brands) are commonly emitted from wood fires and can travel great distances due to a fire's thermal column causing the spread of a fire. Several major conflagrations have been started through the spread of burning embers in the air. *See also* **Ash, Fly; Firebrand; Fire Screen.**

Emergency—A condition of danger that requires immediate remedial action.

Emergency Drill—See **Fire Drill.**

Emergency Evacuation—*See* **Evacuation.**

Emergency Lighting—Lighting provided to aid in emergency evacuation in case normal lighting provisions fail. Codes for emergency evacuation normally specify the level and duration of emergency exit lighting that is required. NFPA 101, *Life Safety Code* specifies 1 ft.-candle (10 lux) at the centerline measured at the floor level of the evacuation route. Power for emergency lighting provision must not be affected by the incident requiring evacuation, therefore most emergency lighting equipment is provided with self-contained batteries. The duration specified by NFPA 101, *Life Safety Code,* is 1.5 hours.

Emergency Response Guidebook (ERG)—A reference guide issued by the U.S. Department of Transportation (DOT) containing information on the most commonly transported chemicals regulated by the DOT. It is intended for use as a guide to first responders during the initial period of a hazardous material incident.

Emergency Shutdown—A control feature to safely stop a process. An emergency shutdown generally consists of stopping equipment, closing isolation valves on the supply or discharge lines from the process, or causing the system to be depressurized. The emergency shutdown features chosen for a particular process depend on the hazards of the process materials, quantities involved, arrangement of equipment, and exposures.

Emergency Venting—Vapor pressure release capability provided to a container (tank, vessel, etc.) to prevent pressure buildup and a possible rupture of the container. Emergency venting capability is especially important for containers that contain combustible materials to avoid release of their entire contents from a rupture of the container, further escalating the incident. The capacity of the emergency relief venting device must be adequate for the expected release load. Most emergency venting for fire protection purposes provided for protection against fire exposures may input heat into the container, causing evaporation of the contents and pressure buildup. American Petroleum Institute API RP 520, *Sizing, Selection, and Installation of Pressure Relieving Devices in Refineries,* provides guidelines for the sizing of relief vents for petroleum storage tanks and pressure vessels. Depressurization is the deliberate depressurizing of a vessel due to weakening of the vessel shell from heat input, whereas emergency venting is provided to avoid internal pressure buildup.

See also **Boiling Liquid Expanding Vapor Explosion (BLEVE); Depressurization.**

Emissivity—A constant used to quantify the radiation emission characteristics of a flame. Emissivity of a perfect black body is 1.0. Emissivity values are always in the range of 0.0 to 1.0.

Encasement, Fire Protection—A layer of fire resistant material enclosing structural steelwork. It delays the heating of the steel to its critical temperature to prevent the early collapse of the structure. Concrete is the most commonly used material for structural steel fire protection encasement. *See also* **Fireproofing.**

Enclosure Integrity Test (EIT)—An examination performed on an enclosure to determine if it is adequately sealed to prevent the escape of a gaseous fire extinguishing agent when it is released into an enclosure for total flooding fire extinguishment. A common type of enclosure integrity test uses fans mounted in a frame placed in a doorway (blower door) or window to measure the amount of total holes in the protected enclosure, assuming the worst-case distribution of leaks across the enclosure surface and computing the worst-case leakage of the fire protection gas mixture through those holes. Lack of adequate pressure buildup indicates the enclosure does not have an integrity level high enough to contain a gaseous fire suppression agent. David and Arthur Saum of Infiltec were issued US patent #5,128,881, Means and Methods for Predicting Hold Time in Enclosures Equipped with a Total Flooding Fire Protection Fire Extinguishing System, for their work on this testing method. The test was later incorporated into NFPA 12, *Standard for Halon 1301,* and has also been modified for alternative gases instead of Halon in NFPA Standard 2001. It may also be called enclosure air leakage testing. *See also* **Total Flooding.**

End of Line Resistor (EOLR)—A resistor placed on fire alarm circuits to produce a supervisory signal. It reduces the voltage on the circuit by creating a resistance. This produces a supervisory voltage for the system circuit that is sensed at the fire alarm control panel. Detected voltages that are different indicate an alarm condition from a detector activation or a problem with the circuit or the circuit devices (grounding, broken wire, or connection). Since the resistor is normally placed at the end of the detection circuit, it is referred to as an end-of-line resistor. See Figures F-1 and F-2.

Endogenous Fire—A fire that originates from within an area. Such a fire may extend from its original boundaries if not controlled and suppressed.

Endothermic—A term used to describe a process reaction in which heat is absorbed, such as the manufacture of carbon disulphide from carbon and sulfur.

Endothermic Fire-Resistant Material—Material containing chemicals bonded with water in solid form to obtain the desired fire resistance. The water is released as steam during a fire exposure and cools the fire.

Engine Company—An organized fire service group that has one or more pumpers. Historically, hand pumps were called engines, which was later used for steam and internal combustion engine apparatus. The main role of the Engine Company is the provision and layout of hoses for fire attack and exposure protection.

Entry Clothing—Protective firefighting clothing designed to provide protection to an individual from conductive, convective, and radiant heat and to permit entry into flame. It may also be called a fire entry suit. *See also* **Turnout Clothing** or **Gear.**

Eutectic Solder—A metallic compound designed to soften or melt at a predetermined temperature. The fusible element of fire protection devices designed to activate at high temperatures (water sprinkler heads, fire dampers, etc.) use eutectic solders. The temperature ratings are selected based on the ambient conditions for the sprinkler installation and response time required. NFPA 13, *Standard for the Installation of Sprinkler Systems,* lists specific temperature points for sprinkler activation. *See also* **Fusible Element.**

Evacuation—The departure of occupants from a building or area due to an emergency to an area free of risk from the emergency. Failure to provide available exit facilities has been a contributing factor in the evacuation of individuals from several major fire incidents and the ensuing loss of life. Disorderly or unfamiliar evacuation can lead to panic. The term evacuation does not include the relocation of personnel to other areas within the same fire or explosion risk, that is, within the same building as part of its definition, for the purposes of fire safety. *See also* **Exit; Exit Drills in the Home (EDITH); Fire Drill; Means of Egress; Means of Escape; Panic; Safe Refuge.**

Evacuation Capability—The ability of individuals to evacuate a building or area due to an emergency to an area free of risk from the emergency. Evacuation capability is usually categorized into three levels (especially for healthcare facilities): prompt, slow, and impractical. A prompt evacuation rating indicates the ability of individuals to move reliably in a timely manner. A slow evacuation is indicative of the ability of individuals to move reliably in a timely manner but not as rapidly as the general population. The inability for individuals to move reliably in a timely manner is considered impractical evacuation.

Evacuation Plan—A prearranged set of instructions for the orderly departure of individuals from a building or area due to an emergency condition to an area that is free of risk from the emergency. The evacuation plan should include methods of alerting/communication, exit routes, accountability, alternative methods, contingency issues, and areas designated as a safe refuge.

Evolution—The predetermined operational sequence of basic firefighting tasks, for example, provision of fire equipment or hoses that are usually practiced as training drills. Fire training activities usually involve specific evolutions and fire hazards to familiarize and refresh operational firefighting techniques for firefighters. Evolutions selected for training should be commensurate with the expected fire hazards to be encountered. *See also* **Fire Drill; Hose Layout.**

Exhauster—A device provided on a dry pipe sprinkler system that causes the alarm check valve to operate more quickly. It permits the trapped system air pressure above the valve to exhaust directly to the atmosphere from an auxiliary valve, which allows the alarm check valve to trip quickly and release water to the system. It may also be called a quick opening device. *See also* **Accelerator; Quick Opening Device (QOD).**

Exit—That portion of a means of (personnel) egress separated from all other spaces of the building or structure by construction or equipment to provide a protected way of travel to the exit from fire and smoke or other impending risk to individuals. Exits include exterior exit doors, exit passageways, horizontal exits, and separated exit stairs or ramps. See Figure E-3. *See also* **Evacuation; Exit, Vertical; Fire Drill; Fire Escape; Means of Egress; Means of Escape; Secondary Exit.**

Exit Access—The portion of a means of egress that leads to an entrance or an exit. A hallway, corridor, or aisle may be considered an exit access if it leads directly to and is connected to the exit.

E

E

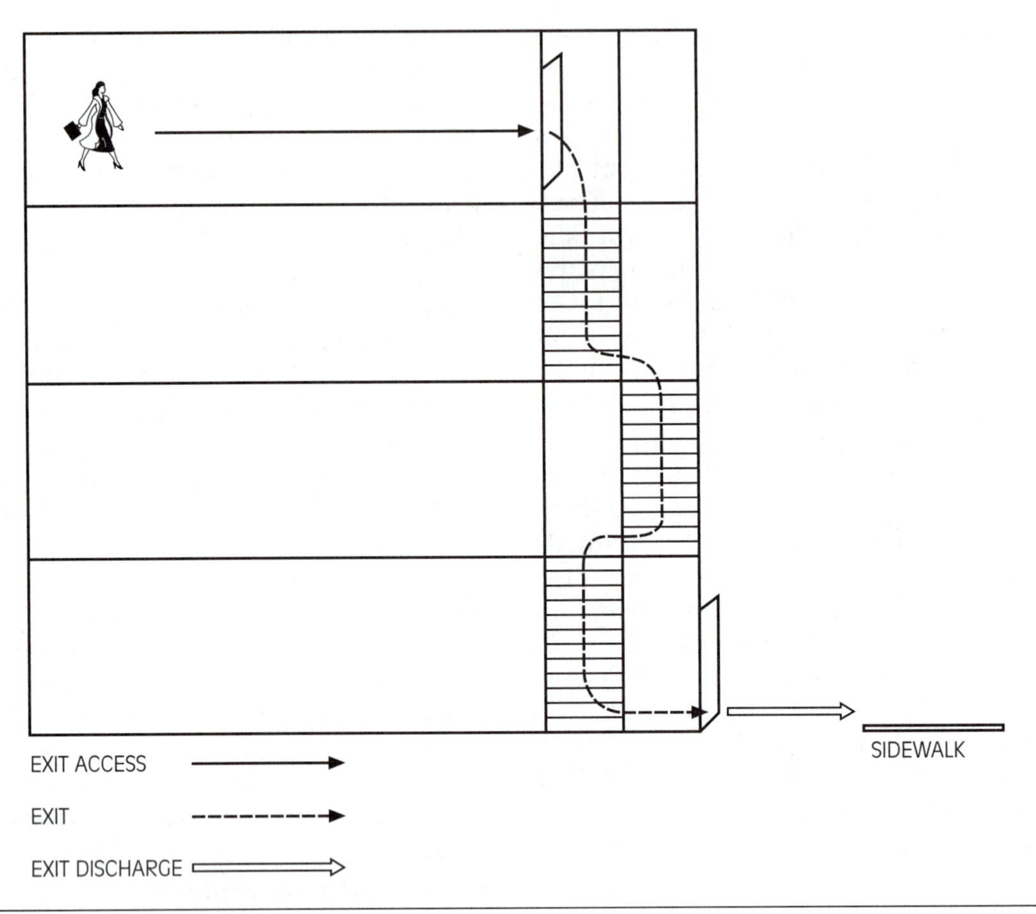

Figure E-3 The three parts of a means of egress.

Exit Capacity—*See* **Egress Capacity.**

Exit Discharge—That portion of a means of egress between the termination of an exit and a public way.

Exit Drill—Synonymous with fire drill. *See also* **Fire Drill.**

Exit Drills in the Home (EDITH)—Residential emergency evacuation drills undertaken to familiarize family members with the exits, method, and arrangements for leaving their home in case of an emergency. *See also* **Evacuation; Fire Drill.**

Exit Marking—The identification provided for an exit from a building, structure, or area to an area free from the risk exposure. *See also* **Exit Sign.**

Exit Sign—A designated identification label provided at or near an exit that is clearly recognizable and visible and that identifies the exit or the path to an exit. Some codes require red exit signs and others green. Red is usually associated

Figure E-4 Exit sign.

with fire and green is associated with safety. The *Life Safety Code* (NFPA 101) of NFPA recognizes either color as acceptable. See Figure E-4. *See also* **Exit Marking.**

Exit Sign, Phosphorescent—An exit, stairway, fire escape, or directional sign made of a luminous phosphorescent material that may be used in lieu of a powered illuminated exit sign, if allowed by local codes.

Exit, Vertical—An exit used for ascension or descension between two or more levels, including stairways, ramps, escalators, and fire escape ladders. *See also* **Exit.**

Exit Width—*See* **Egress Width.**

Exothermic—A term used to describe a process in which heat is expelled, such as the burning of carbon-based materials (paper, wood, etc.).

Exothermic Chemical Reaction—A chemical reaction in which heat is expelled when substances are combined. Exothermic heats of solution are observed when the components interact strongly with one another.

Expansion Ratio—The ratio of volume of firefighting foam bubbles to the volume of solution (water and foam concentrate) used to generate the foam. For example, an 8:1 expansion ratio means 800 gallons (3,028 liters) of foam were generated from 100 gallons (378.5 liters) of foam solution (water and foam concentrate). Expansion ratios are determined by the use of different aspirating devices with low-energy and high-energy delivery. They can also be affected by the operating pressure, temperature, and quality of the water that is used with the foam solution. Low, medium, or high expansion foams depend on the equipment and foam solutions used. The generally accepted expansion ratios are defined as:

Low Expansion: 1 : 20

Medium Expansion: 20 : 200

High Expansion: >200

The term expansion ratio is also used to refer to the ratio of burned to unburned gas volumes of a given mass of gas. *See also* **Foam; High Expansion Foam.**

Expellant Gas—Pressurized gas used to emit liquid or solid particulate fire suppression agents from storage containers, either for fixed systems or portable containers. Most gas used to expel extinguishing agents is either composed of compressed air, carbon dioxide (CO_2), or nitrogen.

Explode—To undergo a rapid chemical or nuclear reaction with the production of noise, heat, and violent expansion of gases, such as the explosion of dynamite. *See also* **Explosion.**

Explosibility Index—A dimensionless, arbitrary scale for the rating of possible dust explosions. It is the product of the explosion severity and ignition sensitivity for dust hazards. The indices are dimensionless quantities and have a numerical value of one for a dust equivalent to the standard Pittsburgh Coal. An explosibility index greater than one indicates a hazrd greater than that for coal dust. Dusts with an index level of less than 0.1 are considered weak explosions, 0.1 to 1.0 are moderate, 1.0 to 10 are strong, and greater than 10 are severe. This scale was developed by the

Bureau of Mines (BOM) of the US Department of the Interior. *See also* **Deflagration Index; Explosion Severity; Ignition Sensitivity; Index of Explosibility.**

Explosible—A term used in western Europe to describe a flammable material.

Explosion—The sudden conversion of potential energy (chemical or mechanical) into kinetic energy with the production and release of gases under pressure, or the release of gas under pressure. These high-pressure gases then do mechanical work such as moving, changing, or damaging nearby materials. *See also* **Blast; Burn; Explode; Explosion, Dust; Unconfined Vapor Cloud Explosion (UVCE).**

Explosion, Dust—Dust explosions occur when finely divided combustible particles are dispersed in air in sufficient concentration in the presence of an ignition source strong enough to cause them to ignite. Combustion dust explosions have a slower rate of pressure rise and a final lower pressure than do combustible vapor explosions. Combustible dusts have a flammability range and minimum ignition temperature. Combustible dust explosions require a higher ignition energy source than a combustible vapor explosion. Ignition energy for hydrocarbons ranges from 0.00002 to 0.001 joules, whereas for coal and chemical dusts, ignition energy ranges from approximately 0.015 to 0.1 joules. When dust particle size is above 400 micrometers, even a high-energy source cannot ignite the dust cloud. Controlled dust explosions by limited particle size is not an effective explosion control method since even 5 to 10 percent of fine dust within a material above 400 micrometers can develop into an explosive mixture. A dust explosion usually occurs with a small explosion that then causes additional dust to become airborne, resulting in a large dust explosion. *See also* **Explosion.**

Explosionproof—A common industry term characterizing an electrical apparatus designed so that an explosion of flammable gas inside the enclosure will not ignite flammable gas outside the enclosure. It is also used to refer to a device that has the capability of preventing the ignition of a specified vapor or gas by such an explosion and operating at an external temperature too low to ignite a surrounding atmosphere. Nothing is technically explosionproof, as all equipment has a relative resistance or vulnerability to the effects of explosion, depending on the force of the explosion exposure and strength of construction. *See also* **Classified Area.**

Explosion Resistant—An ability to withstand an explosion which is based on an anticipated blast wave, commonly defined in terms of peak impulse pressure and pulse duration, and the worst-case expected missile hazard (shrapnel), in terms of material, mass, shape, and velocity. *See also* **Blast Resistant.**

Explosion Severity—A factor to determine the explosibility index rating of a particular material. The explosibility index is the product of the ignition sensitivity and the explosion severity. The explosion severity is the product of maximum explosion pressure multiplied by maximum rate of pressure rise of the sample dust under consideration, divided by the product of maximum explosion pressure, and multiplied by maximum rate of pressure rise of Pittsburgh coal dust. *See also* **Ignition Sensitivity; Index of Explosibility.**

Explosion Suppression System—A protection system for enclosed vessels or other enclosures where overpressure leading to rupture is the primary concern. The explosion suppression system is designed to detect the pressure rise early in its development and to stop it before it reaches damaging pressure levels. Explosion suppression systems were developed by the

British in the late 1940s for the protection of aircraft. They were later adopted for industrial applications. They consist of a storage container for the suppressant agent, a detection system, and control devices. It operates in the early stages of an explosion to suppress the flame front combustion reaction. *See also* **Ultra High Speed Water Spray System.**

Explosion, Thermal—An explosion classified as a self-accelerating exothermic decomposition. It occurs throughout the entire mass of the material without a separate distinct reaction zone. A thermal explosion may accelerate into a detonation.

Explosion Vent—An opening used to vent explosive pressures that may occur within an enclosure. *See also* **Blast Relief Panel.**

Explosion Venting—*See* **Blast Relief Panel.**

Explosive—Chemical compounds or mixtures that undergo rapid burning or decomposition with the generation of large amounts of gas and heat and the consequent production of sudden pressure effects. The first explosive known was gunpowder, also called black powder, and was in use by the 13th century. It was the only explosive known for several hundred years. In 1846, nitrocellulose and nitroglycerin were the first modern explosives to be discovered. Since then nitrates, nitrocompounds, fulminates, and azides have been the chief explosive compounds used alone or in mixtures with fuels or other agents. Xenon trioxide, the first explosive oxide, was developed in 1962.

Explosives are grouped into two main classes; low explosives, which burn at rates of inches (centimeters) per second (deflagration), and high explosives, which undergo detonation at rates of 1,000 to 10,000 yards per second (914 to 9,140 meters per second). Explosives vary in other important characteristics that influence their use in specific applications. Among these characteristics are the ease with which they can be detonated and their stability under conditions of heat, cold, and humidity. The shattering effect, or brisance, of an explosive depends upon the velocity of detonation. Some of the newer high explosives with a detonation rate of 10,000 yards per second (9,140 meters per second) are extremely effective for military demolition and certain types of blasting. On the other hand, for quarrying and mining, when it is desirable to dislodge large pieces of rock or ore, explosives with a lower detonation velocity and lower brisance are employed.

Two types of explosives are in general use as propellants for projectiles in firearms and rockets, and both are commonly called by the generic name of smokeless powder. The term is properly applied to the low explosive, gelatinized nitrocellulose. The other type of smokeless powder, which consists of a mixture of nitrocellulose with a high explosive such as nitroglycerin, is known correctly as double-base powder or compound powder. A common double-base explosive is cordite, which contains 30 to 40 percent nitroglycerin and a small quantity of petroleum jelly as a stabilizer. The term smokeless powder applied to either type of explosive, however, is misleading, because neither is free from smoke when exploded, and neither takes the form of a true powder. The rate of burning of either type of smokeless powder is controlled by the shape of the powder grains. Because the powder grains burn from the surface inward, it is possible to produce grains that burn progressively more slowly, at an even rate, or progressively more quickly depending on the shape and dimensions of the grains. For example, spherical grains have progressively smaller surface areas as they burn, and therefore burn progressively more slowly.

High explosives are explosives that undergo detonation. Some of these, such as TNT (trinitrotoluene), have a high resistance to shock or friction and can be handled, stored, and used with comparative safety. Others, such as nitroglycerin, are so sensitive that they are almost invariably mixed with an inert desensitizer for practical use.

Detonators are used for comparatively insensitive high explosives. Compounds are used that will themselves detonate under a moderate mechanical shock or heat with sufficient force to explode the main explosive charge. Mercury fulminate, Hg $(ONC)_2$, used to be the compound chiefly employed for this purpose, either alone or mixed with other substances such as potassium chlorate. Its manufacture, however, is hazardous and it cannot be stored at high temperatures without decomposition. Consequently the fulminate has been replaced almost entirely in commercial and military detonators by lead azide, PbN_6 (diazodinitrophenol) and mannitol hexanitrate. These initiators are used in conjunction with a charge of cyclonite or PETN, which have largely replaced the tetryl (trinitrophenylmethylnitramine) used previously. These sensitive explosives have high brisance and explosive strength values. They are frequently used as booster charges between the detonator and the major charge of high explosive in large shells and bombs. A blasting cap or exploder is a small charge of a detonator designed to be embedded in dynamite and ignited either by a burning fuse or a spark.

Safety explosives are used in coal mining. Ordinary high explosives are hazardous because of the danger of igniting gases or suspended coal dust that may be present underground. For blasting under such conditions, several special types of safety explosives have been developed that minimize the danger of fires or explosions by producing flames that last for a very short time and are relatively cool. The types of safety explosives approved for work in coal mines are chiefly mixtures of ammonium nitrate with other ingredients such as sodium nitrate, nitroglycerin, nitrocellulose, nitrostarch, carbonaceous material, sodium chloride, and calcium carbonate. Another kind of blasting charge for use in mining has grown in favor because it produces no flame whatsoever. This charge is a cylinder of liquid carbon dioxide that can be converted into gas almost instantaneously by an internal chemical-heating element. The carbon dioxide charge is not a true explosive and absorbs heat rather than evolving it. It has the additional advantage that the force of the explosion can be directed at the base of the bore hole in which the charge is placed, thus lessening the shattering of the coal.[1] *See also* **Deflagration; Detonation; High Explosive; Plosophoric Material.**

Explosive Atmosphere—An atmosphere where the concentration of a combustible vapor is within the flammable range, that is, between the lower explosive limit (LEL) and the upper explosive limit (UEL). *See also* **Explosive Limits.**

Explosive Limits—The concentration range for a combustible material mixture with air that can be caused to ignite. It is expressed as the upper explosive limit (UEL) and lower explosive limit (LEL). Above the upper explosive limit (UEL), there is insufficient oxygen present, and below the lower explosive limit (LEL), there is insufficient gas or vapor. These explosive limits are normally expressed as percentages by volume in air. They may also be referred to as the upper and lower flammable limits (UFL and LFL). *See also* **Combustible Dust; Explosive Atmosphere; Flammable Limit** or **Flammability Limits; Lower Explosive Limit (LEL); Upper Explosive Limit (UEL).**

---

[1] *Blasters' Handbook,* 16th Edition, pp. 1–30, E.I. Du Pont de Nemours & Co., Inc., Wilmington, DE, 1980.

Explosive Material—Any material that can act as a fuel for an explosion.

Explosive Range—*See* **Flammable Limit** or **Flammability Limits.**

Exposed Surface—The side of a structural assembly or object that is directly exposed to the fire.

Exposure, (Fire)—The surrounding location at a fire incident that may be vulnerable to the fire itself. It includes effects from flames, radiant heat flux, convection currents, flying brands, runoff, or exposure to the harmful effects of combustion gases or smoke. The size and range of a fire exposure depends on the severity of the fire causing the exposure.

Exposure Fire—A fire that threatens to ignite nearby combustible surroundings due to radiant heat, convection currents, or flying brands or to expose individuals to the harmful effects of combustion gases.

Exposure Protection—A fire protection measure afforded to locations vulnerable to adjacent fire hazards (radiant heat, convection currents, flying brands, explosion effects, or exposure to the harmful effects of combustion gases). Exposure protection may be in the form of active (water sprays) or passive (separation distances, fireproofing) fire protection measures. *See also* **Fire Separation.**

Extended Discharge—An additional discharge period of fire suppression agent for fire hazards that have a cooling or shut-down time, such as rotating electrical equipment (electric motors, generators, converters, etc.). The extended discharge is in addition to the normal discharge required to suppress the initial fire condition. Extended discharge may also be provided to suppress surface or deep-seated fires where the initial agent release may be dissipated before fire extinguishment is completed. *See also* **Carbon Dioxide (CO_2); Fire Suppression System.**

Extinction Principle, Fire—The principle of fire extinction consists of the elimination or limitation of one of the factors in fire propagation: fuel supply, air supply, heat, or inhibition of the chain reaction of fire.

Extinguish—To cause a material to cease burning; to completely control a fire so that no abnormal heat or smoke remains. Fire extinguishment may be obtained by several methods: cooling, oxygen depletion or removal, inhibition of chemical reaction, and flame removal (blowout).

Extinguishant—*See* **Extinguishing Agent.**

Extinguisher (Fire)—*See* **Fire Extinguisher, Portable.**

Extinguishing Agent—A material used to control or terminate a fire by cooling the materials in the combustion process, obstructing the supply of oxygen to the combustion process, or chemically inhibiting the combustion process. The most common extinguishing agent is water. Other agents include inerting gases (carbon dioxide), dry chemicals or powders, foams, and specialized gases that inhibit the combustion process. Water is usually the most available, economical, and efficient agent but it is unsuitable for electrical and flammable liquid fires. An examination of a fire hazard and use of an extinguishing agent should be made before application. A fire suppression agent should not cause damage to the natural environment due to its application.

Extinguishing Agent Compatibility—The feature of separate fire extinguishing agents to be applied to the same fire (simultaneously or sequentially) without any adverse effect on the performance of either agent. *See also* **Extinguishing Agent, Complementary.**

Extinguishing Agent, Complementary—A secondary fire extinguishing agent used to support the actions of the primary fire suppression agent, which may not effect fire extinguishment if it is used alone. The secondary fire extinguishing agent must be compatible with the primary agent. *See also* **Extinguishing Agent Compatibility.**

Extinguishing Agent, Secondary—An extinguishing agent applied to support the actions of a primary agent that has been applied.

Extra (Fire) Hazard—A condition or state where the amount of combustibles present are more than would be present for what would be considered an ordinary fire hazard. *See also* **Low (Fire) Hazard.**

Extra Large Orifice (ELO) Sprinkler—A fire suppression sprinkler for automatic sprinkler systems that has an orifice size of 0.675 in. (1.59 cm). Standard sprinklers have an orifice size of 0.5 in. (1.27 cm). ELO sprinklers are used for hazards requiring a higher density of water application such as those with a high fuel loading.

Factory Insurance Association (FIA)—The Factory Insurance Association (FIA) merged with the Oil Insurance Association (OIA) to become Industrial Risk Insurers (IRI) in 1975. It consisted of a US-based organization of capital stock insurance companies engaged in the insurance of large industries, particularly those with a high degree of private fire protection and involving special manufacturing hazards. The original company was founded in 1890. *See also* **Industrial Risk Insurers (IRI).**

Factory Mutual Fire Insurance System—An outgrowth of the idea of a New England manufacturer who interested other manufacturers in a plan to share losses on their factories on a mutual basis. The originators of the plan undertook a careful study on the causes of fire, and through a program of prevention, they were able to reduce the cost of their insurance protection. A mutual company was formed and later similar companies were organized. As the size of lines increased, insurance was distributed among the different companies of the group. Companies cooperated on engineering and inspection work, adjustment of losses and other underwriting aspects.

Factory Mutual Global (FM)—A loss prevention and control service organization maintained for the policyholders of three major industrial and commercial property insurance companies, Allendale Insurance, Arkwright, and Protection Mutual Insurance. Factory Mutual provides loss control engineering, loss adjustment, insurance appraisals, building plan review, research, and education services. Fire and loss prevention equipment that has passed specific FM testing standards are considered acceptable for fire protection service and are provided with an FM listing or label. *See also* **Approved; FM Cock; Labeled; Listed.**

Fail Safe—A system design or condition such that the failure of a component, subsystem, system, or input automatically reverts to a predetermined safe static condition or state of least critical consequence. The opposite of fail safe is fail to danger. *See also* **Failure Mode; Failure Mode and Effects Analysis (FMEA).**

Fail to Danger—A system design or condition such that the failure of a component, subsystem, system, or input automatically reverts to an unsafe condition or state of highest critical consequence for the component, subsystem, or system.

Failure Mode—The action of a device or system to revert to a specified state upon failure of the utility power source that normally activates or controls the device or system. Failure modes for valves are normally specified as fail open (FO), fail close (FC), or fail steady (FS), that is, in the last operating position, which will result in a *fail safe* or *fail to danger* arrangement. Valves that are required to shut off fuel supplies are normally specified as fail close and those used to release gases to a vent or flare are specified as fail open. Electrical circuit devices either fail open or fail close. The failure mode of all the devices in a system must be known before a failure mode and effects analysis (FMEA) is undertaken. *See also* **Fail Safe.**

Failure Mode and Effects Analysis (FMEA)—A modeling tool to determine the failures frequencies and their effects on a system. It is useful in examining the failures that may occur in a fire protection system and hence its reliability. *See also* **Fail Safe; Fault Tree Analysis (FTA).**

Fall Down—*See* **Drop Down.**

False Alarm—An alarm that is not indicative of a real fire risk or event. A false alarm can be classified according to causes: environmental (temperature, lightning), animals (insects, rodents), manmade disturbances (vibration), equipment malfunction (component failure), operator error, or unknown. Alarms that are malicious in nature are normally classified as a nuisance alarm. *See also* **Accidental Alarm; Break Glass Unit.**

False Alarm Ratio—The ratio of false alarms to total alarms. It is usually expressed as a percentage or ratio.

Farm Mutual Insurance Company—A mutual company that has more than 50 percent of its risk on farm property. The oldest genuine farm mutual fire insurance company in the United States dates from 1823 and was founded by 11 Quaker farmers in Crosswick, New Jersey.

Fault Tree Analysis (FTA)—A method for representing the logical combinations of various system states that lead to a particular outcome. It is useful in examining the failures that may occur in a fire protection system. *See also* **Failure Mode and Effects Analysis (FMEA).**

Federal Emergency Management Agency (FEMA)—US government agency that maintains the United States Fire Administration (USFA), responds to national emergencies and, if requested, to emergencies outside the United States to coordinate and assist in rescue and aid efforts. *See also* **National Fire Academy (NFA); United States Fire Administration (USFA).**

Figure Eight Knot—A type of knot used in the fire service. There are several variants of the figure eight knot for various purposes, for example, follow-through figure eight, figure eight on a bight.

Fire—A combustible vapor or gas combining with an oxidizer in a combustion process manifested by the evolution of heat, light, and flame. Fire may burn either with or without flames. Fire can be categorized into four distinct categories: diffusion flames, smoldering, spontaneous combustion, and premixed flames. A flame always indicates that heat has forced a combustible gas from a burning substance. The flames come from the combination of this gas with oxygen in the air. When a coal fire flames, it does so because gas is being forced from the coal, and the carbon and hydrogen in the gas combine with oxygen. If kept from burning, such gas can be stored. Manufactured gas is forced from coal in airtight kilns, or retorts. The product left after the gas is extracted from coal is called coke. Coke will burn without flame because no gas is driven off. In order to burn, the carbon in the coke combines directly with oxygen. It is the gas given off by the heated wax in a candle that produces the bright flame. When a burning candle is blown out, for example, a thin ribbon of smoke will rise. If a lighted match is passed through this smoke one inch (2.5 cm) above the wick, a tiny flame will run down and relight the candle.

Whenever a flammable gas is mixed with air in exactly the quantities necessary for complete combination, it will burn so fast that is creates an explosion. This is what takes place in a gasoline engine. The carburetor provides the air mixture, and the electric spark sets it on fire.

Since fires are due to oxidation, they need air to burn properly, and a flame will go out after it has used up the oxygen in a closed vessel. Almost anything will combine with oxygen if enough time is allowed. Iron will rust if exposed for long to damp air; the rust is simply oxidized iron. When the chemical combination is so rapid that it is accompanied by a flame, it is called combustion.

For insurance purposes, a fire may be considered friendly or unfriendly. A fire deliberately started and contained in a location capable of withstanding the effects of the fire (stove, heater, or fireplace) is considered friendly, but a fire that causes unwanted destruction or injury is an unfriendly fire.

Ancient people commonly treated fire as a living being or spirit with good or evil intentions. The philosophies of several later societies (Greek, Indian, and Chinese) thought of fire as one of the primary elements that made up the world or even the universe. Medieval alchemists always considered fire one of their primary tools. In 1783, French chemist Antoine Lavoisier investigated the properties of oxygen and laid the foundation for modern chemistry. By doing so, Lavoisier showed that ordinary fire is due to the chemical process called oxidation, the combination of a substance with oxygen. He disproved the earlier "phlogiston" theory, which held that when an object was heated or cooled it was due to a mysterious substance (phlogiston) that flowed into or out of the object in question. *See also* **Burn; Combustion; Flame; Friendly Fire; Phlogiston Theory; Salamander; Spontaneous Combustion.**

Fire Academy—An institution that provides courses and training for the benefit of fire service personnel. Fire academies are usually sponsored by local fire departments, state governments, or through a university. *See also* **National Fire Academy (NFA).**

Fire Administration—The management of a fire organization. Most fire protection organizations are patterned after what might be considered a semimilitary framework. It is generally organized by Division, Battalions, and Companies under a classical management hierarchical pyramid. The fire chief is the leader, assisted by lower levels of intermediate commanders (captains, lieutenants). *See also* **Fire Chief.**

Fire Aerial Tankers—*See* **Fire Bombers.**

Fire Alarm—The annunciation of an unknown and undesired fire event through audio or visual means, such as a public announcement, bell, horn, siren, and flashing strobe light or beacon. A fire alarm is used to prevent or reduce the hazardous effects of smoke and fire to individuals and provide property protection. The most basic mechanism is an alarm system. It warns people to leave a building at once and can alert a fire department, initiate fire or smoke control functions, and identify the location of a fire within a structure. The latest models of fire alarm systems utilize microprocessors and can record the time and sequence of events. They may also synthesize a voice message announcing the fire location and recommended action to the personnel at risk.

Originally, onsite watchmen provided the only fire alarm system, but with the advent of electric power, boxes wired to fire departments provided a warning system from city streets and such institutional buildings as schools. While some of the latter remain in use, most modern fire alarm systems are automatic, consisting of thermostat-activated devices that at a certain temperature either sound an alarm or report to a central office, such as a municipal fire station. Some alarms are set to go off whenever the thermostat shows a rapid temperature rise. The thermostat is usually placed at or near the ceiling, where it will be most immediately affected by an increase in temperature. Another type of alarm is actuated by a photoelectric cell; when smoke darkens the room slightly, the alarm is activated. One highly sensitive device contains a small amount of radioactive material that ionizes the air in a chamber. With this device, a continuously applied voltage causes a small electrical current to flow through the ionized air. When products of combustion enter, they reduce the current flow and activate the alarm. *See also* **Automatic Fire Alarm System; Fire Drill.**

F

Fire Alarm Box—An electrical signaling device to warn a fire department of a fire event. It consists of a mechanism that transmits a specific coded signal of its location. Fire alarm boxes on city streets were commonly placed on a pedestal that was painted red, with a flashing light mounted on top of the box. Based on the fire hazards in the area of a box, a predetermined response from the fire department was arranged for each fire alarm box code. Due to the advent of telephone and radio communication, along with increased false alarms received from fire alarm boxes, most municipalities have removed their fire alarm boxes. *See also* **Firebox; Municipal Fire Alarm Box (Street Box).**

Fire Alarm Circuit—The section of wiring of a fire alarm system between the load side of the overcurrent device (or the power-limited supply) and the connected equipment of all circuits powered and controlled by the fire alarm system. Fire alarm circuits are classified as either power-limited (PLFA) or nonpower-limited (NPLFA), depending on their power supply sources. *See also* **Fire Alarm Circuit, Nonpower-Limited (NPLFA); Fire Alarm Circuit, Power-Limited (PLFA).**

Fire Alarm Circuit, Class A—A fire detection and alarm system capable of full operation over a single open or single ground fault. Class A circuits are more reliable than Class B circuits because they remain fully operational during the occurrence of a single open or a single ground fault. Class B circuits remain operational only up to the location of an open fault. A Class A circuit does not remain operational during a wire-to-wire short. See Figure F-1. *See also* **Fire Alarm Circuit, Class B.**

Fire Alarm Circuit, Class B—A fire detection and alarm circuit that remains operational only up to the location of an open fault. A Class B circuit does not remain operational during a wire-to-wire short. It is less reliable than a Class A system. See Figure F-2. *See also* **Fire Alarm Circuit, Class A.**

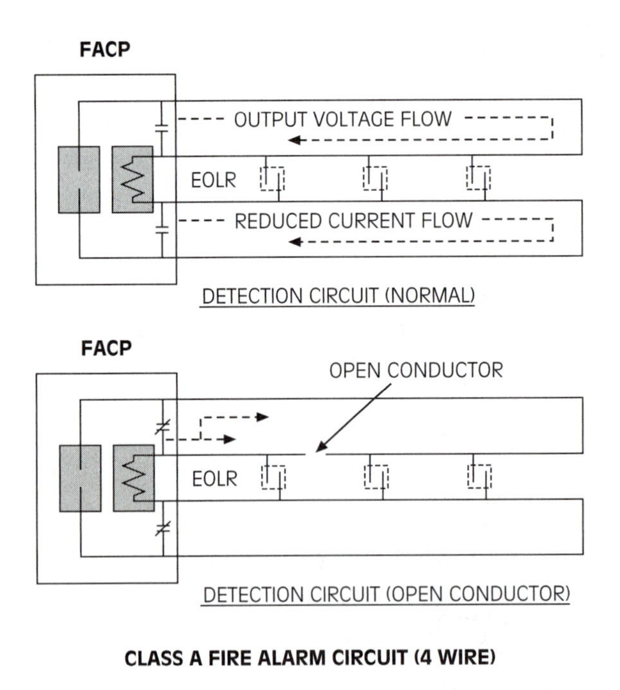

Figure F-1 Class A fire alarm circuit.

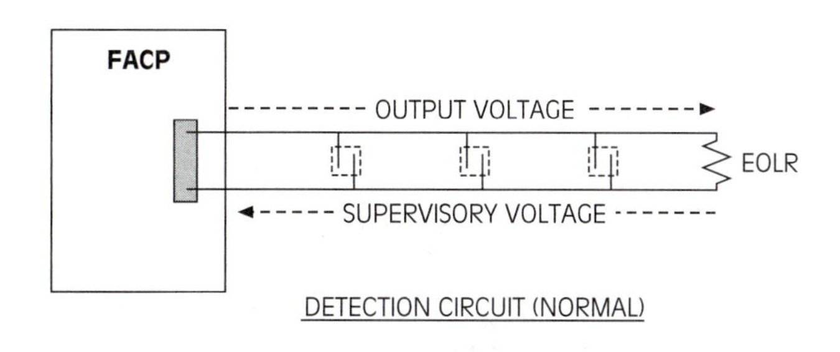

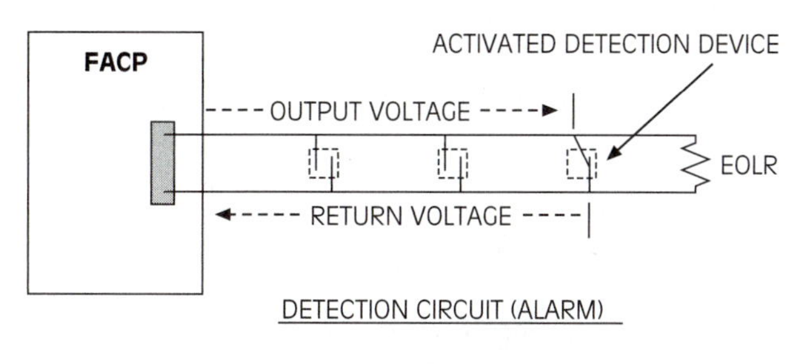

Figure F-2 Class B fire alarm circuit.

Fire Alarm Circuit, Nonpower-Limited (NPLFA)— A fire alarm circuit powered by a source that is not more than 600 volts. Nominal and overcurrent protection for electrical conductors no. 14 and larger are provided in accordance with the conductor ampacity without applying derating (i.e., wire power capacity is lowered) factors as defined by the *National Electrical Code* (NFPA 70). *See also* **Fire Alarm Circuit.**

Fire Alarm Circuit, Power-Limited (PLFA)—A fire alarm circuit powered by a listed PLFA or Class 3 transformer, a listed PLFA or Class 3 power supply, or listed equipment marked to identify the PLFA power source as defined by the *National Electrical Code* (NFPA 70). *See also* **Fire Alarm Circuit.**

Fire Alarm Control Panel (FACP)—A control system for receiving fire alarm signals and initiating actions to highlight conditions (alarms and beacons) or institute actions to automatically activate fire protective systems (fire pump startup, HVAC shutdown, etc.). The fire alarm control panel also provides an indication of the fire detection activation point through an annunciator panel or area/zone indicator lights that highlight the specific location in a facility where an alarm has been initiated. The FACP is required to meet specific performance requirements for reliability. NFPA 72, *National Fire Alarm Code* (NFAC) provides guidance in the provision and features of a fire alarm control panel. *See also* **Annunciator Panel; Fire Detection; National Fire Alarm Code (NFAC).**

Fire Alarm Riser Diagram—An engineering drawing that indicates the number and type of devices on the electrical detection and alarm circuits of a fire alarm control panel and transmission of signals to other locations. See Figure F-3.

Fire Alarm Signal—A signal initiated by a fire alarm initiating device such as a manual fire alarm box, automatic fire detector (flame, heat, smoke), waterflow switch (sprinkler activation), or other device. Its activation is indicative of the presence of a fire or fire signature (fusible link release), and sends it to an alarm monitoring station (FACP). *See also* **Heat Detector; Manual Pull Box; Smoke Detector.**

Fire Alarm System—A system or portion of a combination system consisting of components and circuits arranged to monitor and annunciate the status of fire alarm or supervisory signal-initiating devices. It is designed to initiate an appropriate response to those signals. The first electric telegraph fire alarm system, invented by Dr. William P. Channing, was installed in Boston in 1852. The idea was conceived from the invention of the telegraph by Samuel B. Morse. In 1855, Dr. Channing delivered a lecture on the system

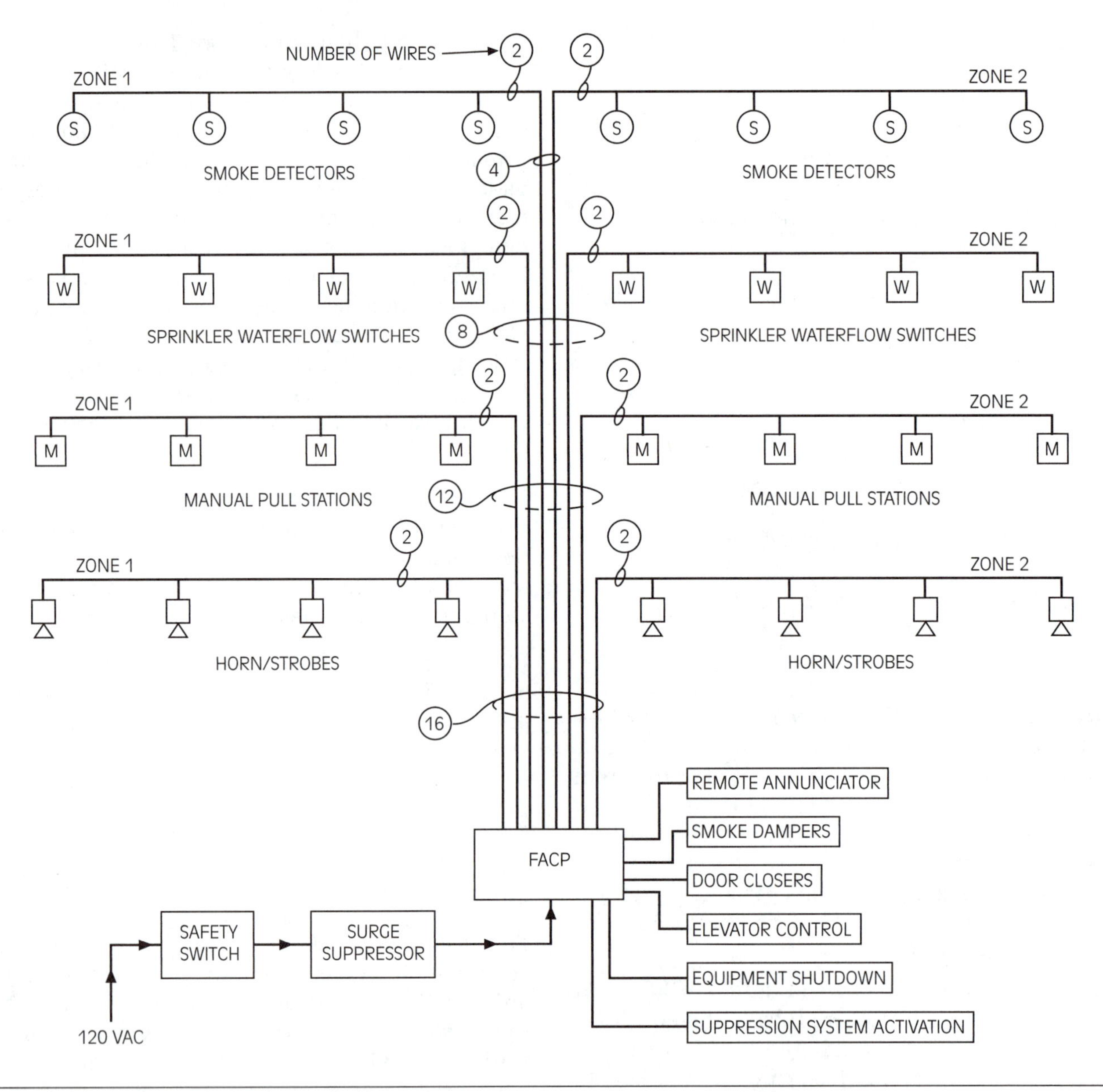

Figure F-3 Typical fire alarm riser diagram.

at the Smithsonian Institute in Washington, D.C. John N. Gamewell, a telegraph operator and postmaster, attended the lecture. He ultimately purchased the rights to the system, which became known as the Gamewell system, and virtually monopolized the municipal fire alarm industry in the United States from about 1875 until the 1950s. *See also* **Automatic Fire Detection System;**

Municipal Fire Alarm Box (Street Box); Proprietary Supervising Station Fire Alarm System; Waterflow Detector.

Fire Analysis—The process of determining the origin, cause, development, and responsibility as well as the failure analysis of a fire or explosion.

Fire and Explosion Hazard Management (FEHM)—A process industry analysis tool to review the types of credible hazardous incidents that may occur and measures that can be implemented to minimize fire and explosion effects or consequences. A fire scenario worksheet is prepared to identify and document potential fire and explosion hazards and protection methods. *See also* **Fire Hazard Analysis** or **Assessment (FHA); Fire Risk Analysis.**

Fire and Safety Committee, API—An organization of the American Petroleum Institute (API) that prepares fire and safety guidance for the petroleum industries. The committee is made up of members from individual petroleum companies who prepare consensus guidance. Its standards and practices are nonbinding and only serve to provide information and guidance, but is referred to by most petroleum companies.

Fire Apparatus—A vehicle used for fire suppression by a fire department, fire brigade, or other agency responsible for fire protection.

Fire Apparatus Charge—There may be instances in which a property is located far from established fire departments and, if called upon to provide fire protection outside their area of responsibility, may expect and demand remuneration for their efforts, termed a fire apparatus charge. These expenses are normally covered by local taxes that support a municipal fire department, but since the property is located outside its jurisdiction, it may be legally entitled to remuneration. Generally under a standard insurance policy, an insurer is not liable for expenses incurred for protection of a property from a threatening fire or from the suppression of a fire. Fire apparatus charges can be included on a fire insurance policy with an endorsement for such a risk. A specific limit must be specified for the insurance coverage to be effective.

Fire Apparatus Manufacturers Association (FAMA)—An association of companies that manufacture vehicles and equipment for the fire services and that promote safe and efficient fire protection products.

Fire Area—An area that is physically separated from another area by distance, or rated fire or explosion barriers (diking, special drainage if combustible liquids are involved), or by a combination of these based on the fire hazard present, such that the fire incident would not be expected to spread to another area.

Fire Ax—A long-handled ax with a blade on one side of the head and pick at the opposite side used to support firefighting activities (forcible entry to inaccessible areas, locked or obstructed entry, support investigations, etc.), by chopping, prying, or pulling. The handle is normally provided with a slight curve at its end to prevent slippage from wet hands or frozen gloves. See Figure F-4.

Fireback—Dating from about the 15th century, a fireback (slab of cast iron), was used to protect the back wall of a fireplace from intense heat. It was usually decorated. After the 19th century, the fireback gave way to firebrick in fireplace construction. *See also* **Fireplace.**

Fireball—An expanding cloud of flaming vapor, usually caused by a sudden release of combustible vapor under high pressure from a vessel or tank. The inner core of the cloud consists mostly of fuel that has been released, whereas the outer layers consist of a flammable fuel-air mixture. As the buoyancy force of the hot gases increases, the cloud rises, expands, and

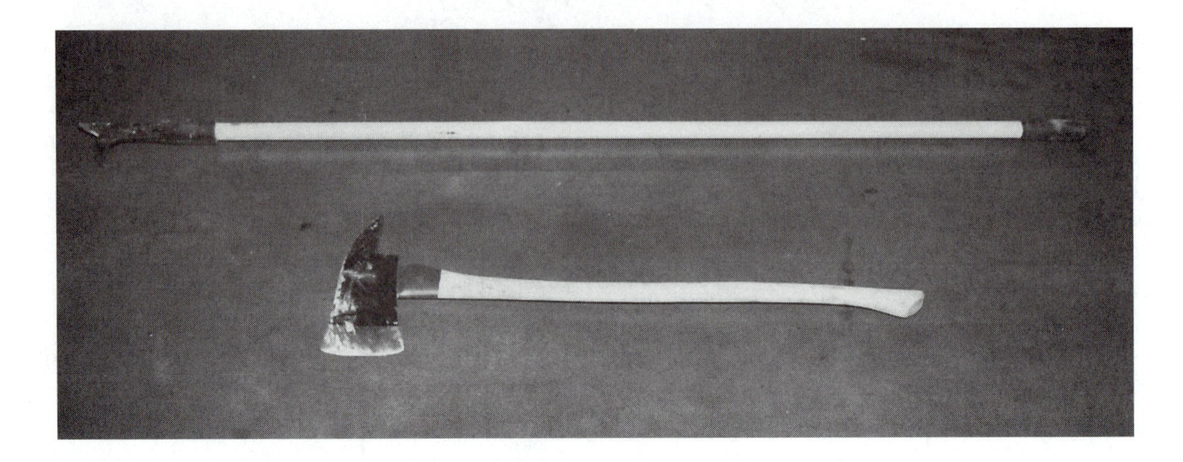

Figure F-4 Pike pole and fire ax firefighting tools.

assumes a spherical shape resulting in the term fireball.

Fire Balloon—An incendiary device used during WWII (circa 1944) by the Japanese forces against America. It consisted of a wind-blown balloon containing incendiary bombs that exploded upon contact with a solid surface and were intended to cause destructive fires. They were released into prevailing winds and were expected to be carried into a locality for wartime destruction. In reality, they had very little effectiveness due to lack of guidance and caused little if any effective destruction. The final design of the fire balloon was about 33 ft. (10 m) in diameter and carried one antipersonnel bomb and two thermite incendiary bombs.

Fire Barrier—A continuous membrane, either vertical or horizontal, such as a wall or floor assembly, that is designed and constructed with a specified fire resistance rating to limit the spread of fire and restrict the movement of smoke. Such barriers may have protected openings. In the marine and offshore petroleum industries, fire barriers or divisions are designated A, B, C (for cellulosic fire exposures), or H (for hydrocarbon fire exposures), with time specifications added to temperature limitations on the unexposed side of the fire barrier. When specifying the time period for both A and H class protection, an allied time to a required failure temperature may be specified. This would normally start at 752°F (400°C), but could be either lower or higher—as much as 1,112°F (600°C) in certain circumstances. For example, "H60 400" indicates hydrocarbon fire exposure in which the steel temperature must not reach a temperature of 752°F (400°C) in less than 60 minutes. For the general building industry, fire barriers are designated by specific time ratings: $^1/_3$, $^1/_2$, $^3/_4$, 1, $1^1/_2$, 2, and 3-hour firewalls. *See also* **Cellulosic Fire; Damper, Fire; Fire Door Assembly; Firewall; Fire Window Assembly; Hydrocarbon Fire.**

Fire Barrier Wall—A wall having a fire resistance rating but not considered a true firewall due to a lack of structural integrity. *See also* **Firewall, Fire Separation.**

Fire Behavior—The fashion in which a fire occurs or reacts—ignition, growth, spread—to the variables of fuels, weather, and topography. The study of fire behavior is primarily based on principles of combustion, fire propagation through the effects of heat conduction, convection, and radiation, influence of fuel distribution, geometry, and ambient conditions. Parameters that affect fire behavior also influence the amount of smoke produced. Normal fire behavior causes an upward extension of the fire more rapidly than it causes lateral extension. Most materials, when burned or heated, become subject to air currents, and a close examination of a burn area after a fire provides an indication of the behavior of a particular fire. Fire behavior prediction can assist in fire protection requirements for location and fire suppression activities.

Fire Bell—A bell used to signal the occurrence of a fire. Fire bells were used before modern communication devices (telegraph, telephone, fire alarm systems, etc.). Fire bells were mounted in fire towers in various locations within a city. The fire bell was rung to signal the occurrence of a fire to the local fire stations. At some locations a code was developed to indicate the location of the fire within the city, depending on the sequence and number of times the bell was rung. Church bell towers were also used. With the development of modern communication systems, the fire bell became obsolete. *See also* **Fire Tower.**

Fire Blanket—A blanket made of noncombustible materials or containing fire retardant chemicals used to smother a fire. The blanket itself can also act as a heat shield to protect the user from a fire. Industrial models are provided in a long steel cabinet, which is mounted vertically and provided with an arm loop. The blanket can be released rapidly by placing one's arm through the arm loop. This enables people to wrap themselves in the blanket in a continuous manner. Another model has two instantaneous release tapes. The tape ends act as handles after the blanket is pulled from the cabinet. Blankets should only be used if immediately available, that is, they should not be used if an individual has run to it while on fire. Individuals should be taught to drop and roll by themselves, thereby smothering a fire, rather than have the fire continue to burn when they are running toward a fire blanket, causing increased harm to the individual. Therefore, blankets are mostly recommended at laboratory and kitchen facilities where personnel are in the immediate vicinity of the blankets. *See also* **Fire-Resistant Blanket.**

Fire Blocking—Architectural trade name for a method of building design and construction in which supporting steel members are used as fire resistant barriers. Their inherent properties and arrangement help prevent or "block" the spread of fire. Blocking mostly refers to the use of solid web beams or girders in floor-ceiling or roof-ceiling construction to provide vertical fire blocking if a fire-rated ceiling or partition is in contact with the lower flange of the beam or girder. *See also* **Compartmentation.**

Fireboat—A marine vessel used in firefighting for combating fires on ships and on waterfront properties. Fireboats primarily are used as the firefighting platform, as the water supply unit, or for rescue operations. As a firefighting platform it is used to apply water, as a pumping base, as a mobile command unit, and for staging equipment and personnel. As a pumping unit it can supplement onshore facilities or provide backup capability in case the onshore systems fail. As a rescue unit it can provide an avenue of escape or access to marine facilities that may be inaccessible from shore. Modern fireboats usually take the shape of a large tugboat. They can be equipped with pumps capable of producing water streams of up to 12,000 gallons (45,000 liters) per minute. The first practical fireboats, built in the late 19th century, were steam propelled and used steam power to operate their pumps. Before that time (circa 1809, New York), they were mainly floating barges, rowed by hand, and were called “fire floats.” Modern craft are powered by internal-combustion (usually diesel) engines that also drive the pumps. A typical fireboat is about 125 ft. long with a 26-ft. beam and 7-ft. draft (38 m length by 8 m beam with a 2 m draft). It has a top speed of about 14 knots. High-speed, shallow-draft fireboats that are propelled and steered by underwater hydraulic jets are also used. Fireboats carry substantially all the same firefighting equipment found on land apparatus. These include pumps, ladders, and rescue equipment as well as special equipment necessary for marine firefighting and water rescues, including rotating and angled nozzles, portable pumps, floating booms, foam-making apparatus, and special extinguishers such as carbon dioxide systems. Other commercial marine vessels, including tug boats, barges, and offshore supply vessels, can also be used in a dire need.

The first fireboat in America was used in New York City in 1800. It was a scow-shaped flat-bottomed boat with a square stern. A hand-operated pump was provided on the boat. Two of these boats were imported from England and called floating engines (they were abandoned in 1824 as they proved too difficult to maneuver). In 1873, Boston had the first steam fireboat that was constructed with an iron hull.

Shipboard fires range from small fires in cabin cruisers to tanker fires involving thousands of metric tons of oil. Some of the special problems include complicated ship layouts, the danger of capsizing, and the difficulty of pinpointing and gaining access to the source of the fire. See Figure F-5.

Figure F-5 Typical municipal fireboat.

Fire Bomb—An incendiary device used to start a malicious fire. It is commonly used by an arsonist during civil unrest and terrorist activities. *See also* **Arson.**

Fire Bombers—Aircraft (fixed or rotary wing) used to fight fires by dropping water or fire retardant chemicals on them during flight. The first designated use of an aircraft for fighting fires is believed to be by Fire Chief Louis Algren of San Diego, California who used two hydroplanes for waterfront firefighting in the early 1910s. Aircraft were first used in fighting wildland fires in California in 1919. Airplanes are commonly used to fight forest fires where they can readily transport water to the affected area and drop it on the base of the fires. Fire bombers generally drop their materials just outside the actual fire or in a location that reinforces a natural fire barrier. The drops are made going downhill with a good escape route for the aircraft. Helicopters drop their materials from a bucket suspended below the aircraft whereas planes have built-in compartments in

Figure F-6 Air tanker making a retardant drop.

the fuselage. Most fixed-wing aircraft are amphibious to allow them to scoop up water from nearby bodies of water to shorten their refilling periods after a drop. They may also be called aerial bombers, firefighting helicopters, firefighting planes. See Figure F-6. *See also* **Aerial Attack.**

Fire Bombing—*See* **Fire Bombers.**

Fire Boss—A forest firefighting commander. He is responsible for planning, directing, and coordinating forest fire suppression activities.

Firebox—A box containing a fire alarm. It may also refer to the combustion chamber for industrial equipment (heaters or boilers). *See also* **Fire Alarm Box.**

Firebrand—A small piece or fiber of burning cellulosic material, commonly wood or paper. *See also* **Ash, Fly; Ember.**

Firebreak—A barrier of cleared or plowed land intended to prevent the spread of a forest or grass fire. Also called fireguard or fireline. Standard forest firefighting procedure includes the construction of a firebreak across the path of an advancing flame front to clear the path of all fuel and halt the fire advance. Firebreaks were also commonly employed in early firefighting efforts during the conflagration of large cities (the Great Fire of London 1666, Hamburg 1842, etc.), before the advent of modern water distribution systems and pumpers. Buildings in the path of a fire were torn down with hooks or explosives to provide an open barrier where a fire would not cross. These firebreaks were reported to be 80 percent ineffective. Buildings failed to effectively create a firebreak, they exposed more combustible materials for the fire's consumption, and they allowed a faster spread of the fire. *See also* **Back Burn; Control Line; Fire Hook; Fireline.**

Firebrick—A refractory brick capable of sustaining high temperatures. It is usually provided for the lining of furnaces and fireplaces. It is also called refractory brick. A refractory material consists of nonmetallic minerals formed in a variety of shapes for use at high temperatures, particularly in structures for metallurgical operations and glass manufacturing. The principal raw materials for firebrick include fireclays, mainly hydrated aluminum silicates; minerals of high aluminum oxide content, such as bauxite, diaspore, and kyanite; sources of silica, including sand and quartzite; magnesia minerals, magnesite, dolomite, forsterite, and olivine; chromite, a solid solution of chromic oxide with the oxides of aluminum, iron, and magnesium; carbon as graphite or coke; and vermiculite mica. Minor raw materials are zirconia, zircon, thoria, beryllia, titania, and ceria, as well as other minerals containing rare-earth elements.

Firebricks are formed by the dry-press, stiff-mud, soft-mud casting, and hot-pressing processes used in the manufacture of building bricks. Some materials, including magnesite and dolomite, require firing in rotary kilns to bring about sintering and densification before the crushed and sized material can be fabricated into refractory shapes and re-fired. Raw materials are fused in an electric furnace followed by casting of the melt in special molds. *See also* **Refractory.**

Fire Brigade—A body of firefighters. A term commonly used for industrial firefighters and, in Britain, for a fire department. *See also* **Fire Department (FD); Fire Department, Industrial.**

Fire Broom—A broom used to manually beat out (smother) a fire. It is usually employed in wildfire suppression activities. Fire brooms have been used for firefighting since ancient times. It may also be called a fire beater. *See* Figure at P-10.

Fire Bucket—A bucket designated exclusively for fire protection purposes. Usually of metal construction with a capacity of 12 quarts (11.4 liters). Its use is limited to fires that can be extinguished with limited amounts of water applied at a close distance. Water from the bucket can generally be thrown about 10 ft. (3 m). During the early evolution of firefighting, buckets constructed of leather were commonly used and required to be provided by the local inhabitants for firefighting activities. It may also be called a fire pail. *See also* **Bucket Brigade.**

Fire Buff or **Fire Enthusiast**—An individual interested in the historical or current activities of fire departments, buildings, apparatus, equipment, or personnel. The term is supposedly said to have originated from an early association of individuals who called themselves the Buffalo Corps, and who affiliated themselves with the New York Fire Departments as errand boys and helpers. Webster's Seventh New Collegiate Dictionary states the term is derived from buff-colored overcoats worn by volunteer firefighters in New York City, circa 1820. A further claim is that it is taken from buffalo-skinned coats worn by individuals watching firefighting activities. Worldwide private fire buff organizations have been formed that promote and preserve the history of fire services and allied emergency services. Famous fire buffs include George Washington, Ulysses S. Grant, Oliver Wendell Holmes, and Winston Churchill. The International Fire Buff Associates, Inc. is a group of fire enthusiasts who promote and further their ideals through meetings, publications, and web sites. Individuals may also be called fire fans or sparks.

Firebug—An individual who is an arsonist or a pyromaniac. *See also* **Arson; Pyromania.**

Fire Camp—Term for the encampment of firefighters involved in a forest fire.

Fire Canopy—A fire-resistive horizontal extension from an exterior wall. It is used to prevent flames emitted from a window from igniting the structure or floors above it.

Fire Casualty—An injury reported within one year and caused by the effects of a fire. They may be fatal or nonfatal in nature. Fire casualty statistics are collected by fire authorities and are used to research fire problems and mitigate their occurrences.

Fire Cause—The predominate and contributory reason(s) why an unwanted fire occurs. In the United States, a study conducted over a ten-year period found that the most frequent type of fire was electrical (23 percent of all fires). Other immediate causes of fire included tobacco smoking (18 percent); heat caused by friction in industrial machinery (10 percent); overheated materials (8 percent); hot surfaces in such devices as boilers, stoves, and furnaces (7 percent); burner flames (7 percent); and combustion sparks (5 percent). Underlying causes of these fires may be related to other concerns, such as lack of understanding of fire hazards.

Firecheck, Automatic—A device for stopping the advancement of a flame front in burner mixture lines (flashback protection). It automatically shuts off the fuel-air mixture. They are commonly provided as spring- or weight-loaded valves released for closure by a fusible link or by movement of bimetallic elements. They are also equipped with metallic screens for stopping the progress of a flame front.

Fire Chief—Commonly, the chief executive or highest ranking officer of a fire department who is responsible for all of its actions. The fire chief normally reports and is accountable to a municipal government. *See also* **Fire Administration.**

Fire Class—A letter designation given to a particular fire category for the purpose of generally classifying it according to the type of fuel and possible spread of the fire. The letter classification also provides a general indication of the severity and type of the fire hazard. *See also* **Class A Fire; Class B Fire; Class C Fire; Class D Fire; Class K Fire; Fire Extinguisher, Portable.**

Fire Climate—Geographical and climatic features that as a whole delineate a distinctive character of a region for the occurrence of wildfires. Primary features that influence the fire climate include annual precipitation, vegetation type and intensity, temperature variation, and winds. *See also* **Fire Weather.**

Fire Code—A regulatory document for the implementation of measures to prevent the occurrence and spread of fries and to institute suppression capability for unwanted fires. NFPA is a major source of fire codes used for adoption by legal agencies (city or state ordinances or laws). Enforcement is maintained by state fire inspectors and fire marshals. Most codes are determined by consensus agreement and therefore may not represent the highest or lowest level of protection, but rather what has been determined as appropriate by the experienced members determining the fire code specifications. Most fire codes have been prescriptive in nature (specifying exact requirements), however the more recent trend has been to provide performance-based codes that require a specific outcome and allow the detailed requirements to be determined by the premises owner as long as they meet the performance requirements. *See also* **Performance-Based Fire Protection Design; Performance-Based Regulation; Standard Fire Prevention Code.**

Fire Command Center—The principal attended or unattended location in which the status of the detection, alarm communications, or control systems are displayed and from which the system(s) can be manually controlled. See Figure F-7.

Fire Compartment—The space within a structure, usually a building, that is enclosed by fire barriers on all sides, including the top and bottom, to prevent the spread of a fire incident and

Figure F-7 Fire command center.

its effects (heat, flame, smoke, combustion gases). *See also* **Compartmentation.**

Fire Control—The stage in firefighting whereby a fire incident is controlled and not allowed to escalate in magnitude. Following fire control, suppression or extinction of the fire incident will occur. Fire control limits the growth of a fire by pre-wetting adjacent combustibles and controlling ceiling gas temperatures to prevent structural damage.

F

Firecracker—A pyrotechnic device for creating a entertaining display of burning luminous particles or cracking noise. Normally consists of a charge wrapped in heavy paper with a short fuse. Of particular interest to fire investigators as a potential ignition source for inadvertent or malicious fires.

Fire Curtain—A fire-resistive curtain or screen that can be rapidly lowered to separate a theater stage from the auditorium (audience) to resist the passage of flame and smoke, typically for 20 minutes. It is usually a legal requirement and was instituted because of several horrendous theater fires with a high loss of life that occurred in the 19th and 20th centuries.

Fire Curve—A graphical curve representation of a fire test that is intended to illustrate (but not necessarily reproduce) the temperature development of a real fire, either cellulosic or hydrocarbon (ASTM E-119, NFPA 251, or UL 1709) over the time period of the test (temperature versus time). These are typically presented in units of heat release rate or mass loss rate versus time. *See also* **Fire Performance; Hydrocarbon Fire; Time Temperature Curve.**

Fire Cut—An angle cut of floor wood joist in building construction. It is provided so that should a floor assembly collapse from a fire exposure, the floor joist will pull away from a wall instead of acting as a lever to pull the wall down. There are no recognized fire code requirements to provide installation arrangements for fire cuts.

Fire Damage—The loss caused by a fire occurrence. Losses can be both direct and indirect. Direct losses include physical destruction and business income losses. Indirect losses can be loss of aesthetics, prestige, etc.

Firedamp—A combustible gas that consists mainly of methane, and is usually associated with mining operations. The various harmful vapors produced during mining operations are frequently called damps (German Dampf, "vapor"). Firedamp is a gas that occurs naturally in coal seams. The gas is nearly always methane (CH_4), which is highly flammable and explosive when present in the air in a proportion of 5 to 14 percent. White damp, or carbon monoxide (CO), is a particularly toxic gas. As little as 0.1 percent can cause death within a few minutes. It is a product of the incomplete combustion of carbon and is formed in coal mines chiefly by the oxidation of coal, especially in those mines where spontaneous combustion occurs. Black damp is an atmosphere in which a flame lamp will not burn, usually because of an excess of carbon dioxide (CO_2) and nitrogen in the air. Stinkdamp is the name given by miners to hydrogen sulfide (H_2S), because of its characteristic smell of rotten eggs. Afterdamp is the mixture of gases found in a mine after an explosion or fire. *See also* **Afterdamp.**

Fire Damper—*See* **Damper, Fire.**

Fire Danger—*See* **Fire Risk Analysis.**

Fire Department (FD)—A professional organization, usually provided by a government but may be a private concern, for the prevention or extinguishment of fires. The Roman emperor Augustus is credited with instituting the first organized fire service, named a corps of firefighting vigilēs ("watchmen") in 24 BCE. They developed regulations for checking and preventing fires at about the same time. They had a mechanical means of throwing a continuous stream of water by using a hand pumper. Roman firefighters used axes, buckets, blankets, ladders, and poles. They also had slaves as firefighters.

In the preindustrial era, most cities had watchmen who sounded an alarm at signs of fire. In general, the principal piece of firefighting equipment in ancient Rome and into early modern times was the bucket, passed from hand to hand, or a barrel dragged on skids behind a horse team. Other important firefighting tools were axes, hooks, and ladders used to remove the fuel and prevent the spread of fire as well as to make openings that would allow heat and smoke to escape a burning building. In major conflagrations, long hooks with ropes were used to pull down buildings in the path of an approaching fire to create firebreaks. When explosives became available, they were used for the same purpose.

Following the Great Fire of London in 1666, formal fire brigades were created by insurance companies in England. The British government was not involved until 1865, when the private (mostly insurance) fire brigades became London's Metropolitan Fire Brigade. The first modern standards for the operation of a fire department were not established until 1830, in Edinburgh, Scotland. These standards explained, for the first time, what was expected of a good fire department.

After a major fire in Boston in 1631, the first fire regulations in America were established. In 1648 in New Amsterdam (now New York), fire wardens were appointed, establishing the beginnings of the first public fire department in North America. In 1659, New Amsterdam established by municipal action, the first official fire department in America under direction of Peter Stuyvesant, who was then governor of the city. He distributed 250 leather buckets and a supply of ladders and hooks imported from Holland. A tax of one guilder for every chimney was imposed for the maintenance of the equipment.

After the Great London Fire of 1666, the English developed hand-operated pumps so firefighters could spray water through a hose. Citizens began to band together in volunteer fire companies. The volunteers promised to drop everything and rush to fight a fire whenever it broke out. The city also paid for a bellman to patrol the streets at night to look out for fires and warn citizens. It also paid for firefighting equipment. Insurance companies formed their own fire brigades to fight fires in buildings they insured. They placed a fire mark on the buildings they insured, but they seldom gave any help when other fires broke out. Enthusiastic amateurs also organized manual pumps. Frederick Hodges, a Lambeth distiller, formed a brigade to protect his own distillery that had twice been destroyed by fire. He built a 120 ft. (36.5 m) tall tower as a lookout for fires so he could rush his two 40-man manual pumps in an emergency to any part of London.

In 1669, New York City (then New Amsterdam) appointed a Brent Master to be the first fire chief in America. The first paid fire department in North America was organized in 1679 in Boston after a large fire. It imported a fire engine from England and had a fire chief, Thomas Atkins, and 12 firefighters.

In 1833, London organized the firstfirefighting system, the London Fire Engine Establishment. It was assembled from the private insurance company fire brigades. The city government had not provided a fire department earlier because it claimed it would stifle free enterprise. New York City established the first paid fire patrol in 1835. There were four members who were paid about $250 a year. The following year there were 40 members, who where known as fire police. Cincinnati organized the first professional fire department in 1853. The first firehouse was built in 1855 in New York City.

In the modern sense, fire departments constitute a comparatively recent development. Their personnel are either volunteer (nonsalaried) or career (salaried). Typically, volunteer firefighters are found in smaller communities, career firefighters in cities. The modern department with salaried personnel and standardized equipment became an integral part of municipal administration only late in the 19th century.

In some cities, a fire commissioner handles the administration of the department. Other cities have a board of fire commissioners with the fire chief as executive officer and head of the uniformed force; in still other cities, a safety director may be in charge of both police and fire departments. The basic operating unit of the fire department is the company, usually commanded by a captain, assisted by a lieutenant, sergeant, or other ranks. Fire companies are usually organized by types of apparatus: engine companies, ladder companies, and squad or rescue companies.

Depending on the location and magnitude of a fire incident, specific fire stations are notified to respond. On a first alarm, more apparatus is sent to industrial sections, schools, institutions, and theaters than to neighborhoods of one-family dwellings because of the higher risk potential involved. Additional volunteer or off-duty firefighters are requested as needed. Fires that cannot be brought under control by the apparatus responding to the first alarm are called multiple-alarm fires. Each additional alarm brings more firefighters and apparatus to the fire location. Mutual aid and regional mobilization plans are agreed to in advance among adjacent fire departments for assisting each other in fighting fires that are beyond the resources of an individual department to handle. Many modern fire departments also respond to a high percentage of other emergencies besides firefighting. In the United States, approximately 70 percent of all emergency medical calls are handled by the fire service. The enormous increase in transportation of hazardous materials or dangerous goods has resulted in intensified training for firefighters, and their departments often provide them with chemical protective clothing and monitoring equipment. Fire departments also prepare and equip their members to handle emergencies that result from manmade or natural disasters (earthquakes, plane crashes, violent storms), and incidents that require the rescue of individuals trapped from fallen structures, cave-ins, and similar situations.

The organization of fire departments differs considerably in various countries. In the United States, individual state laws define the firefighting authorities. The kind and size of fire departments are determined by the individual community, fire district, fire protection district, or county. Most fire departments are organized with a commanding officer, subordinate officers, and non-officer personnel. A fire chief is usually designated as the commanding officer, though in large cities the title may be fire commissioner, director of fire, or something equivalent. Below this level, fire officers may be an assistant or deputy chief. Underneath them may be battalion chiefs in charge of several companies. Fire captains may be in charge of one or more companies. Lieutenants or sergeants usually command only one company. Non-officer personnel fire departments are usually called firefighters in paid or volunteer fire departments. These individuals may be assigned titles, such as engineer, pump operator, mechanic, dispatcher, or aide, as they are given responsibility for these duties. Where a training department is provided, a training officer and staff are maintained.

Maintenance officers, mechanics, or a larger staff is needed to keep the fire apparatus and equipment in good condition, unless these services are contracted for from outside agencies. Other fire officers and staff may be maintained for fire prevention programs. Personnel also coordinate with law enforcment in arson fire investigations. Many fire departments have changed their name to the department of fire and emergency services to better reflect the modern and multiple fire service missions and responsibilities. *See also* **Fire Brigade; Firefighter; Firefighting; Fireman; Fire Service; Volunteer Fire Department.**

Fire Department Connection (FDC)—A pipe connection for providing supplemental water flow or pressure to fixed fire suppression systems through hoses from a fire pumper. A fire department connection is the opposite of a fire hydrant. Instead of supplying water for firewater use by the fire service, it is used by the fire service to supply water to fixed fire suppression systems or standpipes. It is basically a pipe tapped into the fire main to allow for the connection of fire hoses from a mobile fire pumper. There are two basic types of connections, a single inlet and a siamese inlet (although multiple inlets have been provided in special cases). The inlets have internal clappers that allow the fire department to pump into one of the connections without the water feeding backward. This also allows the connection or disconnection of fire hose without backpressure on the connection or losing water provided to the system into which it is being pumped. There are also two styles of mounting, either wall or freestanding sidewalk post-type (standpipe). The wall mounting penetrates the exterior wall of building whereas the standpipe-type is used when the piping leaves a building and travels underground to a point near the street or fire hydrant. Thread connections for hoses must match those being used by the local fire department. The first fire department connection is credited to Sir William Congreve in 1809, in connection with his improvements to John Cavey's sprinkler system and subsequent patent on the first practical sprinkler system. See Figures F-8 and F-9. *See also* **Siamese Connection.**

Fire Department Elevators—High-rise buildings are provided with one elevator with a protected power supply available for exclusive use by the fire department, in case of emergency. It may be a normal passenger elevator fitted with a special control to allow the fire department to recall the elevator for its use, which overrides all other elevator controls. The fire department recall capability is usually located at the main entry point to the elevator area at ground level. When activated, it commands the elevator to return to ground level, where from then on it only becomes operable from the controls in the elevator itself. A "protected power supply" is defined as a source of electrical energy of sufficient capacity to permit proper operation of the elevator and its associated control and communications systems. The power supply origin, distribution, overcurrent protection, degree of isolation from other portions of the building electrical system, and amount of mechanical protection are such that it is unlikely that the supply would be interrupted except from a major fire incident or structural collapse of the facility. Section 211 of *ASME/ANSI A17.1, Safety Code for Elevators and Escalators,* provides further clarification of the requirements for elevators to be approved for firefighter service.

Fire Department, Industrial—An organization providing rescue, fire suppression, and related activities at a single facility or facilities under the same management, whether for-profit, nonprofit, or government-owned or operated, including occupancies such as industrial, commercial,

F

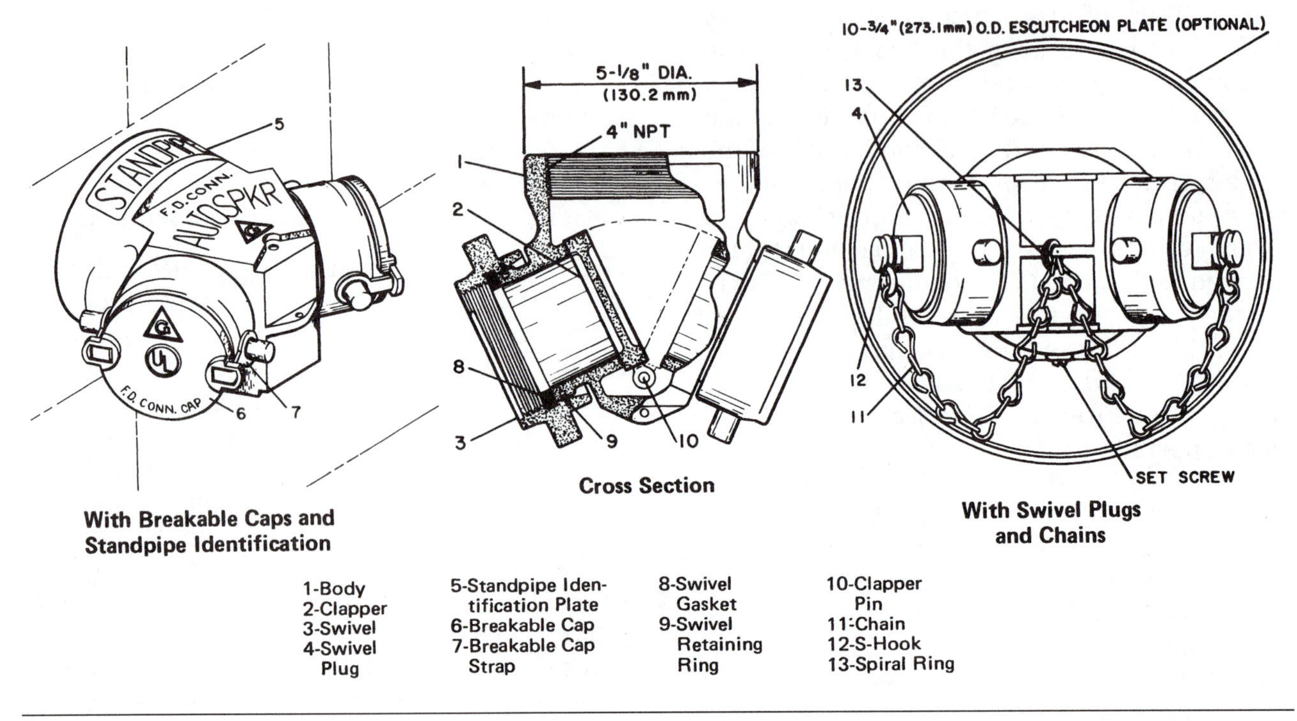

Figure F-8 Fire department connection. (*Courtesy Grinnel Fire Protection Systems Corporation, Inc.*)

Figure F-9 Advanced building fire department connection panel located in Yokohama, Japan: left to right, emergency telephone cabinets with alarm light, wall hydrants, sprinkler standpipe fire department connections and building plan placed above connections.

mercantile, warehouse, and institutional. The industrial fire department is trained and equipped for specialized operations based on site-specific hazards present at the facilities (chemical, hydrocarbon, or nuclear power plant fires). It is usually designed to carry out suppression to completion independent of, or in close cooperation with, a municipal fire department. *See also* **Fire Brigade.**

Fire Department Instructors Conference (FDIC)—A conference organized for teachers, trainers, and instructors in or servicing the fire protection profession to advance and share knowledge in the techniques and practices in fire service training.

Fire Dependent—Terminology for ecological systems that rely on the occurrence of fire to ensure their survival. For example, fire can remove decadent growth, dispose of accumulated litter, recycle nutrients, and stimulate new, vigorous growth from seeds and sprouts. *See also* **Fire Ecology.**

Fire Detection—A device for the detection and notification of a fire event. Fire alarms can be activated by people or automatic devices that can detect the presence of fire. These include heat-sensitive devices that are activated if a specific temperature is reached; a rate-of-rise heat detector is triggered either by a quick or a gradual escalation of temperature; and smoke detectors, which sense changes caused by the presence of smoke, in the intensity of light, in the refraction of light, or in the ionization of air. *See also* **Automatic Fire Detection System; Fire Alarm Control Panel (FACP); Smoke Detector.**

Fire Detection, Photoelectric—Automatic fire detectors that respond directly to visible smoke, intended for areas where it is not practical to use ionization detectors owing to normally high ambient levels of combustion gases (such as battery rooms evolving hydrogen gases) or where the material protected will produce heavy smoke.

Fire Detection System—*See* **Automatic Fire Detection System.**

Fire Detector—A device for determining the existence of a fire condition. Fire detectors sense a fire condition by detecting visible flame, IR or UV radiation, smoke, heat, or combustion gases. A fire detector may use mechanical, electrical, or electronic components to sense the presence of a fire event. A fire detector is normally arranged to cause a signal for fire alarm activation or activation of a fire protection device or system. *See also* **Automatic Fire Detection System; Combustion Gas Detector; Smoke Detector; Spot Detector.**

Fire Devil—A rapidly whirling vortex of flame commonly observed in forest fires. *See also* **Firewhirl.**

Fire District—A rural or suburban area served by a fire organization.

Fire Door—The door portion of a rated fire door assembly. *See also* **Automatic Fire Door; Fire Door Assembly.**

Fire Door Assembly—Any combination of a fire door, a frame, hardware, and other accessories that together provide a specific degree of fire protection to the opening in which the door is placed. They are given hourly fire ratings. Fire doors that have 20- or 30-minute fire-resistant ratings are primarily provided for smoke control. They are normally provided along a corridor where smoke control is required and in fire partitions of ratings of up to one hour. The degree of fire resistance of a manufactured door is determined according to the fire test requirements specified in ANSI/NFPA 80, *Standard for Fire Doors and Fire Windows.* See Figure F-10. *See also* **Fire Barrier; Automatic Fire Door.**

Fire Door, Automatic—*See* **Automatic Fire Door.**

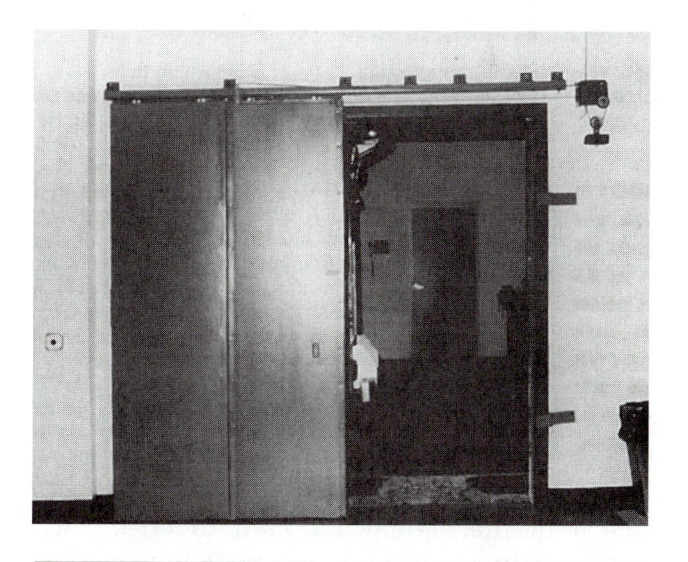

Figure F-10 Rolling fire door assembly.

Fire Door, Power Operated—Fire doors that are normally kept in an open position and are closed by power. They are equipped with a releasing device that automatically disconnects the power operator (electromagnetic catch) at the time of fire (through a fire detection and alarm system). This allows a self-closing or an automatic closing device to close the door regardless of power failure or manual operation. *See also* **Automatic Fire Door.**

Fire Door, Self-Closing—A fire door that is fitted with a closer that immediately forces a door closed when it is released, after it is opened. Most fire doors are fitted with a self-closure device required by fire safety codes. This is required to prevent fire spread through a opening (open fire door). *See also* **Automatic Fire Door.**

Fire Drill—The periodic practice and familiarization of fire alarm signals, evacuation procedures and exits in case of a real fire emergency are commonly referred to as fire drills. They are a prime life safety-training requirement in buildings, large, unusual, or congested structures and complex facilities. The performance of a fire drill is often required by local fire ordinances or company loss prevention policies. Individuals are instructed to leave a building by designated emergency fire exits in an orderly fashion and assemble at a prearranged point for accountability. Drills that are performed in a residence are also an important safety feature. NFPA promotes a fire drill for the home called EDITH (**E**xit **D**rills **i**n **t**he **H**ome). A fire drill can also be an exercise in the use of firefighting equipment (evolutions) done by the fire service. By performing drills, individuals will be less frightened or confused in a fire situation and will have more confidence in the actions required to effectively combat a fire or when to evacuate from it. *See also* **Evacuation; Evolution; Exit; Exit Drill; Exit Drills in the Home (EDITH); Fire Alarm; Fire Training.**

Fire Dynamics—Deals with the fundamentals of combustion of materials and energy balances, chemical thermodynamics, premixed and diffused burning, theory of combustion, flame propagation and efficiency of combustion, as well as the physical and chemical properties of combustion of a material. Typically involves the use of heat and mass balance.

Fire Ecology—The study of a fire's relationship to the physical and biological environment of plants and animals. Initially a fire may kill plants and animals, but over time it can give birth to more types of plants and animals in an environment. Scientists believe wildfires may be a natural and necessary occurrence for the ecology of some environments—pine cones may open from the heat of a fire, freeing the seeds inside and allowing them to germinate from the remains of a forest fire. Ecosystems are dynamic in that the populations constituting them do not remain the same. This is reflected in the gradual changes of the vegetational community over time, known as succession. Fire can be a contributing factor to successive ecosystems. Complete fire exclusion may bring about changes in vegetation patterns and may also allow dangerous accumulations of fuel, with increased potential for feeding catastrophic fires. In some parks and wilderness areas, where the goal of management is to maintain completely natural conditions, lightning-caused fires may be allowed to burn under close surveillance. *See also* **Fire Dependent; Fire Effects; National Interagency Fire Center (NIFC).**

Fire Effects—The consequences of a fire incident. Fire effects may be detrimental, inconsequential, or in some cases, beneficial. *See also* **Fire Ecology.**

Fire Emergency—An incident that contains the presence of or possibility of a fire, explosion, or development of smoke or fumes and requires immediate measures to prevent or control and suppress such an occurrence.

Fire Endurance—The length of time that a structural element can resist fire either up to the point of collapse, or alternatively, to the point when the deflection reaches a limiting value.

Fire Engine—A mobile piece of automotive equipment used in firefighting, also called fire truck or pumper. It is used to supply water at adequate pressures and volumes for the needs of fire control and extinguishment. The first fire engines, which appeared in the 17th century, were simply tubs carried on runners, long poles, or wheels; water was still supplied to the fire site by a bucket brigade. The tub of the engine functioned as a reservoir. It sometimes housed a hand-operated pump that forced water through a pipe or nozzle. They were moved to the scene of a fire by human or animal power. In large fires, the bucket brigade method of water supply was inefficient, and the short range of the stream of water necessitated positioning the apparatus dangerously close to the fire. The introduction of more powerful pumps and flexible hose solved this problem. In 1654, Joseph Jencks was commissioned to build the first hand-operated fire engine in America.

In 1721 and 1725, Richard Newsham, a British engineer, took out patents that resulted in much more powerful wheeled fire engines. They produced a continuous stream of water and allowed men to pump while simultaneously working the handles or a treadle. It was a reciprocating pump with two pistons driven alternately up and down by hand. It could produce a jet of water approximately 160 ft. (50 m) high. See Figure F-11.

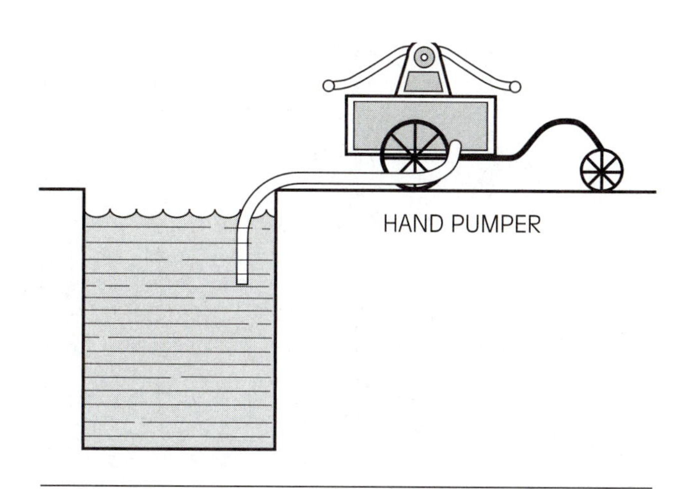

Figure F-11 Typical early American fire pumper arrangements.

A great advance in fire engine technology was made with the introduction of steam power. The world's first steam-driven fire engine was introduced in London in 1829 by Captain John Ericsson (1803–1889) and John Braithwaite (1797–1870). By the 1850s, it was in use in many large cities. Jon Braithwaite's 10 hp, two-cylinder steam pump of 1829 delivered 150 gallons a minute, was coal-fired, and was drawn by a single horse. America's first fire engine arrived from London in 1679 after a major fire in Boston. The first steam fire engine was built in the United States in 1840 by Paul R. Hodge, an Englishman living in New York. His fire engine also moved by steam power. Most steam pumpers were equipped with reciprocating piston pumps, although a few rotary pumps were used. Some were self-propelled, but most used horses for propulsion, conserving steam pressure for the pump. With the development of the internal-combustion engine early in the 20th century, pumpers became motorized. At first these engines were only used to drive the vehicle to the scene of a fire where the stream-driven pump was then employed. See Figure F-12.

F

Figure F-12 American mobile steam-driven fire pump.

Also, because of problems in adapting geared rotary gasoline engines to pumps, the first gasoline-powered fire engines had two motors, one to drive the pump and the other to propel the vehicle. The first pumper using a single engine for pumping and propulsion was manufactured in the United States in 1907. By 1925, the steam pumper had been completely replaced by motorized pumpers. The pumps were originally of the piston or reciprocating type, but these were gradually replaced by rotary pumps and finally by centrifugal pumps, used by most modern pumpers.

At the same time, the pumper acquired its main characteristics: a powerful pump that can supply water in a large range of volumes and pressures; several thousand feet (meters) of fire hose, with short lengths of large-diameter hose for attachment to hydrants; and a water tank for the initial attack on a fire while firefighters connect the pump to hydrants or for areas where no water supply is available. In rural areas, pumpers carry suction hose to draw water from rivers and ponds. Current standards for pumper fire apparatus in the United States require that a fire pump have a minimum capacity of 750 gallons (2,840 liters) per minute at a pump pressure of 150 psi (10.35 bar). They also call for a water tank capacity of at least 500 gallons (1,893 liters). *See also* **Fire Steamer.**

Fire Engineering Technician—Usually individual with an Associate of Science (AS) degree in fire and safety engineering technology or similar discipline or a recognized equivalent or an individual highly experienced in most of these disciplines. They typically have studied fire dynamics and fire science; fire and arson investigation; fire suppression technology, tactics, and management; fire protection; fire protection structures and systems design; fire prevention; hazardous materials; mathematics and computer science; fire-related human behavior; safety and loss management.

Fire Engineering Technologist—An individual with a Bachelor of Science (BS) degree in fire engineering technology, fire and safety engineering technology, or a similar discipline, or a recognized equivalent. Such an individual would have typically studied fire dynamics; fire science; fire and arson investigation, fire suppression technology, fire extinguishment tactics, and fire department management; fire protection of structures and systems design; fire prevention; hazardous materials; applied upper-level mathematics and computer science; fire-related human behavior; safety and loss management.

Fire Equipment Manufacturers and Services Association (FEMSA)—An association of companies that manufacture fire protection products. It promotes the use of fire protection devices and development of fire codes and standards.

Fire Enthusiast—*See* **Fire Buff.**

Fire Escape—A means of rapid egress from a building, primarily intended for use in case of fire. Several types have been used: a knotted rope or rope ladder secured to an inside wall; an open iron stairway on the building's exterior, an iron balcony; a chute; and an enclosed fire and smoke-proof stairway. The iron stairway is the most common because it can be added to the outside of nearly any building of modest height, although it has certain drawbacks—unless built against a blank wall, it may be rendered useless by smoke from windows, and a means must be provided for keeping it in readiness while denying its use to thieves and prowlers. The iron balcony extends around the exterior of a building to provide a corridor along which persons can flee from fire-imperiled rooms to safety behind a firewall or in an adjacent building. The chute, or slide escape, is either a curved or a straight incline and may be open or enclosed; it is well suited to such buildings as hospitals, from which patients can be evacuated on their mattresses. The best fire escape, however, is a fully enclosed fireproof stairway in the building or in an adjoining tower. Elevators are not considered safe because fire damage may cause them to fail and heat-sensitive call buttons may stop the car where the fire is hottest. Over 1,000 patents have been issued by the US Patent Office for fire escapes. A fire escape that consisted of a wicker basket on pulleys and chains was patented in 1766 by the London watchmaker David Marie. *See also* **Exit; Fire Escape Ladder.**

Fire Escape Ladder—A ladder specifically designed and used to escape from the hazard of an unwanted fire. *See also* **Fire Escape.**

Fire Exit—*See* **Exit.**

Fire Exit Hardware—Labeled devices for swinging fire doors installed to facilitate safe egress of persons. It generally consists of a cross bar and various types of latch mechanisms that cannot hold the latch in a retracted locked position.

Fire Exposure—*See* **Exposure, (Fire).**

Fire Extinguisher, Portable—Portable or movable apparatus used to put out a small fire by directing onto it a substance that cools the burning material, deprives the flame of oxygen, or interferes with the chemical reactions occurring in the flame. They are intended as the first line of defense against fires of limited size. Portable extinguishers can be water-based, gaseous, or dry chemical types. Most portable fire extinguishers are small tanks provided with an expelling gas that has been compressed (compressed air or carbon dioxide) to propel the extinguishing agent through a nozzle and onto the fire. This method supersedes the previous method used in the soda-acid fire extinguisher whereby carbon dioxide (CO_2) was generated by mixing sulfuric acid with a solution of sodium bicarbonate.

The type of portable fire extinguisher depends primarily on the nature of the burning materials. Secondary considerations include cost, stability, toxicity, ease of cleanup, and the presence of electrical hazard. Small fires are classified according to the nature of the burning material.

Class A fires involve wood, paper, and similar cellulosic materials. Class B fires involve flammable liquids, such as cooking fats and paint thinners. Class C fires are those in electrical equipment and Class D fires involve highly reactive metals, such as sodium and magnesium. Water is suitable for putting out fires of only one of these classes (A), though water is the most commonly used because it can cool and protect exposures as well. Water converts to steam when it absorbs heat and the steam displaces the air from the vicinity of the flame. The water may contain a wetting agent to make it more effective against fires in upholstery, an additive to produce stable foam that acts as a barrier against oxygen, or antifreeze to prevent freezing in cold ambient temperatures. Fires of classes A, B, and C can be controlled by carbon dioxide (CO_2), halogenated hydrocarbons such as environmentally friendly Halon substitutes, or dry chemicals such as sodium bicarbonate or ammonium dihydrogen phosphate. Class D fires are ordinarily fought with dry chemicals.

The CO_2 extinguisher is a steel cylinder filled with liquid carbon dioxide, which, when released, expands suddenly and causes so great a lowering of temperature that it solidifies into a powdery "snow." This snow volatilizes (vaporizes) on contact with the burning substance, producing a blanket of gas that cools and smothers the flame.

Early fire extinguishers in the 1730s were just glass balls of water or saline solution that were thrown on fires. They were invented by a German physician, M. Fuches, in 1734. Although they were widely advertised and sold, in general they were not really used (primarily because they were too small to be effective). In 1816, George Manby, an English army captain, invented the first practical extinguisher similar to modern models. It used compressed air to force water (pressurized water) out of a cylinder through a control valve. It delivered 3 gallons (11.4 liters) of water from a cylinder that was pressurized with compressed air and was three-quarters full of water. A more efficient portable extinguisher was invented by Francois Carier, a French doctor, in about 1866. He mixed sodium bicarbonate with water and fixed a glass bottle of sulfuric acid inside the extinguisher near the neck. The bottle was broken by striking a pin and the chemical mixed. This produced carbon dioxide (CO_2) gas, which forced out the water. In 1909, Edward Davidson of New York patented the use of carbon tetrachloride (CCl_4). It was ejected out of the extinguisher by pressurized carbon dioxide. It vaporized immediately to form a heavy, noncombustible gas that smothered the fire. Four years before this, foam extinguishers had been invented in St. Petersburg, Russia, by Professor Alexander Laurent. He mixed a solution of aluminum sulfate and sodium bicarbonate with a stabilizing agent. The foam bubbles that were formed contained carbon dioxide gas. They were able to float on burning oil, paint, or petrol and smother a fire. NFPA's standard on portable fire extinguishers was developed in 1921.

In addition to fire class, fire extinguishers are grouped by the means of expelling the agent. Five methods are commonly employed. These include self-expelling, gas cartridge or cylinder, stored pressure, mechanically pumped, and hand propelled or applied. *See also* **Carbon Dioxide; Carbon Tetrachloride (CCl4); Dry Chemical; Dry Chemical Fire Extinguisher; Fire Class; Fire Extinguish Rating; Fire Suppression System; Gas Cartridge or Cylinder Extinguisher; Mechanically Pumped Extinguisher; Self-Expelling Extinguisher; Stored Pressure Extinguisher; Water; Wheeled Fire Extinguisher.**

Fire Extinguisher Rating—A rating of relative extinguishing effectiveness of a portable fire extinguisher along with the class of fire it is rated for, as specified by NFPA 10, *Standard for Portable Fire Extinguishers*. Fire extinguisher ratings consist of a relative number followed by a fire class letter (A, B, C or D). Color coding is also a part of the identification scheme. A green triangle is used for Class A, a red square for Class B, a blue circle for Class C, and a yellow five pointed star for Class D. Fire extinguishers classified for use on Class C or D hazards are not required to have a number preceding those specific classification letters. A Class C rating is not provided unless a Class A or B rating has been established. Fire extinguisher ratings began around 1948. They were developed in order to classify the appropriate extinguishing agent to the fire hazard and quantify the relative effectiveness

Fire Extinguishers, Portable

Extinguishing Material	Water and Antifreeze	Wetting Agent	Loaded Stream	AFFF and FFFP	Multi-Dry Chemical	CO_2	Dry Chemical	Halons	Dry Powder
Self-expelling						✔		✔	
Cartridge or GN_2 cylinder			✔	✔	✔		✔		✔
Stored pressure	✔	✔	✔	✔	✔		✔	✔	✔
Pump	✔								
Hand applied	✔								✔

of portable fire extinguishers. See Figure F-13. *See also* **Fire Extinguisher, Portable; Fire (Safety) Symbols.**

Fire Extinguishment—The complete suppression of a fire until there is no burning of combustible materials. *See also* **Fire Suppression.**

Fire Factors—Features that contribute to ignition, fire development, or its spread.

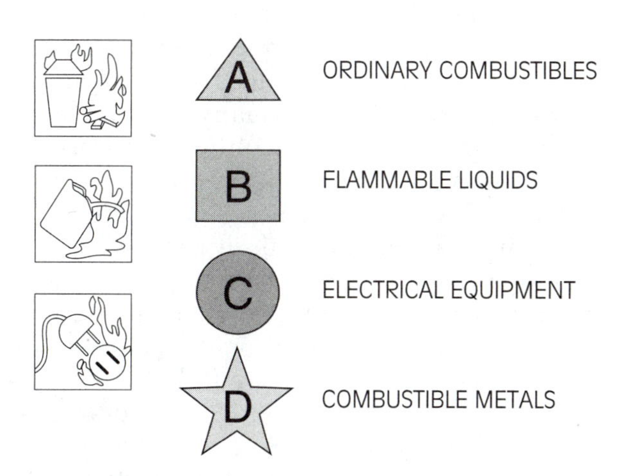

Figure F-13 Portable fire extinguisher symbols.

Firefighter—An individual qualified by training and examination to perform activities for the control and suppression of unwanted fires and related events. A firefighter is normally a member of an organization dedicated to fire protection activities, such as a fire department. *See also* **Fire Department (FD); Fireman; Volunteer Firefighter; Volunteer Fire Department.**

Firefighter Assist and Search Team (FAST)—A fire service company assigned to search for and rescue trapped or lost fire fighters. May also be called a Rapid Intervention Team (RIT).

Firefighters' Smoke-Control Station (FSCS)—A control station for use by the fire service to monitor or override a building's smoke control system and equipment. Operational firefighting requirements specific to an individual fire incident may dictate the smoke control system be inhibited to facilitate fire and rescue activities.

Firefighting—The activity directed at limiting the spread of fire and extinguishing it, particularly as performed by members of organizations (fire services or fire departments) trained for the purpose. When possible, firefighters rescue persons endangered by the fire, if necessary, before turning their full attention to putting it out.

Firefighters, skilled in the use of specific equipment, proceed as rapidly as possible to the site of the fire. In most urban areas, fire stations housing a company of firefighters and their equipment occur frequently enough that an alarm receives a response within two or three minutes. Most fire services in towns inhabited by 5,000 persons or more will dispatch an engine company (pumper), a truck company (ladder truck), and a rescue vehicle to the scene. If the fire involves a structure occupied by many persons, two or more companies may respond to the first alarm. The first firefighters to arrive assess the fire to determine the techniques to be used in putting it out, taking into account the construction of the burning building and any fire protection systems within it.

Firefighting is a battle against time. The initial priority is rescuing any occupants that may be in a burning building. Precedence is then given to any location from which the fire may spread to a neighboring structure. A typical method of firefighting is the over-and-under system. Working from inside the building, if possible, the bulk of the firefighting takes place from below, while further attack is carried out from above in an effort to prevent the fire from spreading upward.

Systematic firefighting involves the following steps: (1) protection of currently uninvolved buildings and areas, (2) confinement of the fire, (3) ventilation of the building, (4) extinguishment of the fire, (5) overhaul and (6), salvage. Additionally, rescue operations are simultaneously undertaken for victims of the fire incident as a primary objective. Pathways by which the fire could spread are closed off, and flames are controlled by the application of water or other cooling agents. For enclosed structures, openings are made to permit the escape of toxic combustion products and hot air. This operation (ventilation) must be conducted with keen judgment to permit firefighters access to the fire without causing its intensification or risking a smoke explosion (the result of admitting fresh air to a space in which a high concentration of unburned fuel particles is present in a hot, oxygen-depleted atmosphere). Extinguishment is accomplished with the use of water streams mixed with appropriate extinguishing agents where applicable to quench the remaining flames. When this is accomplished, the firefighters initiate overhaul of the fire, checking for hot spots or undetected flames. Salvage of a structure is undertaken by removing smoke and water from the interior and protecting undamaged materials.

The basic tactics of fighting a fire can be divided into the following categories: rescue operations, protection of buildings exposed to the fire, confinement of the fire, extinguishing the fire, and salvage operations. The officer in charge, usually designated as the incident commander, surveys the area and evaluates the relative importance of these categories. The commander also estimates what additional assistance or apparatus may be needed. Rescue operations are always given priority. Firefighter safety has also assumed increasing importance.

Once the incident commander has appraised the situation, firefighters and equipment are deployed. Pumper, ladder, and other truck companies, as well as rescue squads, are assigned to different areas of the fire, usually in accordance with the number and types of hose streams the incident commander considers necessary to control the fire and prevent its spread. In accordance with standard procedure for first alarms, fire companies go immediately to their assigned locations without waiting for specific orders. Special plans cover contingencies such as a fire covering a large area, a large building, or a particularly hazardous location. Usually on a first alarm, one of the pumpers attacks the fire as quickly as possible using preconnected hose lines supplied by the water tank in the truck, while larger hose lines are being attached to the hydrants. Members of the ladder and rescue companies force their way into the building, search for victims, ventilate the structure—break windows or cut holes in the roof to allow smoke and heat to escape—and perform salvage operations. Ventilating the structure helps advance the hose lines with greater safety and ease. It also serves to safeguard persons who may still be trapped in the building by providing a path for hot gases and smoke to escape.

Brightly burning fires principally generate heat, but smoldering fires also produce combustible gases that need only additional oxygen to burn with explosive force. The hazards to which firefighters and occupants of a burning building are exposed include breathing superheated air, toxic smoke, gases, and oxygen-deficient air, as well as burns, injuries from jumping or falling, broken glass, falling objects, collapsing structures, and overexertion. Handling a hose is difficult even before the line is charged with water under pressure. Nozzle reaction forces can amount to several hundred pounds, requiring the efforts of several people to direct a stream of water. *See also* **Fire Department (FD).**

Firefighting, Approach—Limited, specialized exterior firefighting operations at fire incidents where very high levels of conductive, convective, and radiant heat, such as bulk flammable gas and bulk flammable liquid fires occur. Specialized thermal protection from exposure to high levels of radiant heat is necessary for the firefighters involved in these operations due to the limited scope of these operations and the greater distance from the fire that these operations are conducted. Not considered entry, proximity, or structural firefighting.

Firefighting, Entry—Extraordinarily specialized firefighting operations that can include the activities of rescue, fire suppression, and property conservation at incidents involving fires producing very high levels of conductive, convective, and radiant heat, such as aircraft fires, bulk flammable gas fires, and bulk flammable liquid fires. Highly specialized thermal protection from exposure to extreme levels of conductive, convective, and radiant heat is necessary for persons involved in such extraordinarily specialized operations due to the scope of these operations and that direct entry into flames is made. Usually these operations are exterior operations, not structural firefighting. *See also* **Firefighting, Proximity.**

Firefighting, Proximity—Specialized firefighting operations that can include the activities of rescue, fire suppression, and property conservation at incidents involving fires producing very high levels of conductive, convective, and radiant heat, such as aircraft fires, bulk flammable gas fires, and bulk flammable liquid fires. Specialized thermal protection from exposure to high levels of radiant heat, as well as thermal protection from conductive, and convective heat, is necessary for persons involved in such operations due to the scope of these operations and the close distance to the fire that these operations are conducted, although direct entry into flame is not made. Usually these operations are exterior operations but might be combined with interior operations. *See also* **Firefighting, Entry; Firefighting, Structural; Proximity Suit.**

Firefighting, Structural—The activities of rescue, fire suppression, and property conservation in buildings, enclosed structures, aircraft, vehicles, vessels, or like properties that are involved in a fire or emergency situation. *See also* **Firefighting, Proximity.**

Fireflood—An in situ combustion process used to recover or extract additional petroleum in underground reservoirs. Compressed air or oxygen is injected into the petroleum reservoir. A flame front is ignited at the reservoir that heats the oil, allowing it to facilitate or improve oil flow to a production well cavity for recovery.

Fire Flow, Available—The actual amount of water available from a municipal supply system for firefighting as determined by water flow testing from a fire hydrant. The available fire flow should be equal to or more than the fire flow required.

F

Fire Flow (Required)—A common term in the fire protection profession and insurance industry for the required firewater delivery rate in a municipal supply system for a particular area in addition to normal water usage. It is usually expressed as a volume flow rate for specific duration. A fire flow is commonly estimated at a residual pressure of 20 psi (137.9 kPa), which is sufficient to supply pumpers, unless direct use for hose streams from hydrants is required.

Fire Fog—*See* **Water Fog.**

Fire-Gas Detector—A device that detects gases produced by a fire. Primarily applied to underground coal mining operations.

Fire Gases—The gaseous products released from a combustion process that may include a wide variety of gases depending on the material undergoing combustion and the ambient environment. Of primary concern and attention are toxic fire gases, which can be classified into three main components: asphysicants, irritants, and toxicants that produce other effects. Asphysicants produce unconsciousness and can lead to death. Carbon monoxide and hydrogen cyanide are two primary asphysicants found in fire gases. Irritants affect sensory organs (nose and upper respiratory tract) and may also affect pulmonary functions (lungs). Examples include hydrogen chloride (HCl) and nitrogen oxides (NO_2). *See also* **Smoke.**

Fireground—The specific area involved in a fire incident and used by all emergency services conducting or supporting operations for the incident. Although most fireground areas are not well defined, some may be marked by barrier tape to restrict public access or prevent operation of nonemergency vehicles in the area. It may also be called a fire scene. *See also* **Fire Perimeter.**

Fireground Commander—The individual in charge and responsible for all the actions of individuals and equipment in the fireground. The term fireground commander has generally been replaced by incident commander to encompass all emergency incidents, not solely fire incidents. *See also* **Command, Incident Commander.**

Fire Growth Curve—A graphical representation of the growth of a fire. *See also* **Fire Growth Rate; Fire Signature.**

Fire Growth Rate—The periodic increase in a fire, dependent on the ignition process, flame spread, and mass burning rate over the area involved. Fire growth rates are used for the design of smoke detection systems and smoke control systems (Ref. NFPA 72, *National Fire Alarm Code,* and NFPA 92 B, *Guide for Smoke Management Systems in Malls, Atria, and Large Areas*). NFPA classifies fire growth rates to achieve a 1 MW release (Q) as slow (t_1 = 600 seconds), medium (t_1 = 300 seconds), fast (t_1 = 150 seconds), ultrafast (t_1 = 75 seconds). See Figure F-14. *See also* **Fire Growth Curve.**

Fireguard—A metal screen placed in front of an open fireplace to catch embers. It may also be called a fire screen. *See also* **Fireline; Fire Watch.**

Fire Hazard—Any situation, process, material, or condition that, on the basis of applicable data, can cause a fire or explosion or provide a ready fuel supply to augment the spread or intensity of a fire or explosion and that poses a threat to life, property, continued business operation, or the

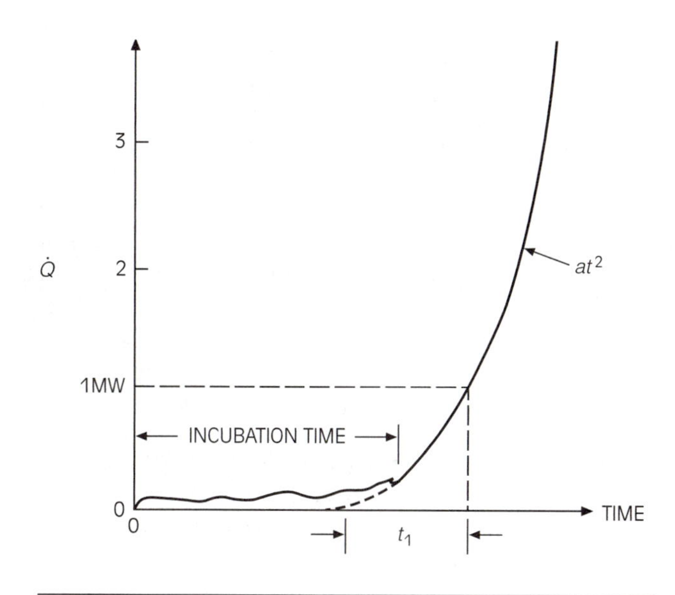

Figure F-14 Fire growth rate example.

environment. The relative degree of hazard can be evaluated and appropriate safeguards provided.

Fire Hazard Analysis or **Assessment (FHA)**—A systematic review to assess the fire hazards present and identify fire protection measures and adequacy of these devices to prevent injuries, damage, loss of production, or environmental impacts.

1. Identify the facility and its fire prevention and fire protection systems. List fire hazards that can exist and the loss-limiting criteria being used.
2. Identify the applicable codes and standards used.
3. Define and describe the characteristics associated with potential fires (maximum fire loading, hazards of flame spread, smoke generation, toxic contaminants, and contributing fuels).
4. List the fire protection system criteria and design basis for the water supply, water distribution systems, and fire pump.
5. Describe the performance criteria for the detection systems, alarm systems, automatic suppression systems, manual systems, chemical systems, and gas systems for fire detection, confinement, control, and extinguishment.
6. Develop the design considerations for suppression systems; smoke, heat, and flame control; combustible and explosive gas control; and toxic and contaminant control. Select the operating functions of the ventilating and exhaust systems to be used during the period of fire extinguishment and control. List the performance criteria for the fire and trouble annunciator warning systems and the auditing and reporting systems.
7. Consider the qualifications and abilities necessary for personnel performing inspection checks and the frequency of testing needed to maintain reliable alarm, detection, and suppression systems.
8. Use the features of building and facility arrangements and the structural design features to generally define the methods for fire prevention, fire extinguishing, fire control, and control of hazards created by fire. Define and plan the fire barriers, egress, fire walls, separation distances and the isolation and containment features that should be provided for flame, heat, hot gases, smoke, and other contaminants. Outline the drawings and list of equipment and devices needed to define the principal and auxiliary fire protection systems.
9. Identify the dangerous and hazardous combustibles and the maximum quantities estimated to be present in the facility. Consider where these materials can be appropriately

located in the facility (or a reduced quantity or nonhazardous materials substituted).

10. Review the types of potential fires, based on the expected quantities of combustible materials, their estimated severity, intensity, duration, and the hazards created. For each of the potential fires, indicate the total time involved and the time for each step from the first alert of the fire hazard until safe control and extinguishment is accomplished. Describe the facility systems, functions, and controls that will be provided and maintained during the fire emergency.
11. Define the essential electric circuit integrity needed during a fire incident. Evaluate the electrical and cable fire protection, the fire confinement control, and the extinguishing systems that will be needed to maintain their integrity.
12. Carefully review and describe the control and operating room areas and the protection and extinguishing systems provided for these areas. Do not overlook the additional facilities provided for maintenance and operating personnel, such as kitchens, maintenance, storage, and supply cabinets.
13. Analyze the available forms of backup or public fire protection that can be considered for the installation. Review the backup fire department, equipment, number of personnel, special skills, and training needed.
14. List and describe the installation, testing, and inspection necessary during construction of the fire protection systems that demonstrate the integrity of the systems as installed. Evaluate the operational checks, inspection, and servicing needed to maintain such integrity.
15. Evaluate the program for training, updating, and maintaining competence of the facility firefighting and operating crew. Provisions should be required to maintain and upgrade the firefighting equipment and apparatus during facility operation.
16. Review the qualifications for the fire protection engineer or consultant who will assist in the design and selection of equipment. This individual will also inspect and test the physical features of the completed system and develop the total fire protection program for the operating facility.
17. Evaluate life safety, protection of critical process and safety equipment, provisions to limit contamination, and restoration of the facility after a fire.

A fire hazard assessment is primarily used in the nuclear power industry.[2] *See also* **Fire and Explosion Hazard Management (FEHM); Fire Risk Analysis or Fire Risk Assessment (FRA).**

Fire Hazard Classification—A relative rating of fire risk for a particular occupancy based on the materials fire hazards and their quantity at a particular location. It is used to determine the level of fire protection features that should be provided to protect the occupancy and fire spread. Fire hazard classification levels are usually unique for a particular fire code and may not be similar for universal application.

Fire Hazard Data—Physical properties that contribute to a material's potential degree of combustibility (flash point, auto-ignition temperature, etc.). NFPA 325, *Fire Hazard Properties of Flammable Liquids, Gases, and Volatile Solids,* lists fire hazard data for most common materials.

---

[2] *NFPA 803, Standard for Fire Protection for Light Water Nuclear Power Plants,* pp. 2–7, National Fire Protection Association, Quincy, MA, 1997.

Fire Hazard Identification—A system of labeling established by the National Fire Protection Association (NFPA) to provided a readily identifiable means to ascertain material hazards. The system identifies fire hazards in three main areas: health, flammability, and reaction or instability. The relative ranking in severity of each hazard category is indicated with a numerical value from zero to four (no risk to severe risk). The system is primarily provided for emergency situations and is not intended to apply to normal hazard evaluations. The specific features of the system are outlined in NFPA 704, *Standard System for the Identification of the Fire Hazards of Materials.* Generally, it consists of a diamond-shaped placard, divided into four smaller diamonds or quadrants. Each quadrant is color coded and specifically arranged for the three main hazard areas. The health rating is provided on the left at the 9 o'clock position and is blue. The flammability rating is provided at the top or 12 o'clock position and is red. The reactivity hazard is provided on the right at the 3 o'clock position and is yellow. The relative rankings for each hazard are indicated in each quadrant. Special hazard identifiers are provided in the bottom quadrant at the 6 o'clock position, which is usually white. Special hazard qualifiers generally include radioactivity, explosives, corrosive, water reactive, oxidizer, etc. The NFPA fire hazard identification scheme is somewhat limited as it only identifies relative potential hazards with the individual material. It does not identify the material itself or all of its potential reactions with other materials. The ICC building code has adopted the use of the NFPA 704 system. See Figure F-15. *See also* **Hazardous Materials Identification System (HMIS).**

Fire Hazardous Equipment—Equipment that produces a significant fire risk such that the adjacent area requires an increased level of fire protection measures (a fire hazardous zone is created). This is commonly encountered in the process industries. For example, normally high capacity pumps and compressors handling combustible materials are classified as fire hazardous equipment due to the leakage potential and ability to spread combustible materials from the source of leak. *See also* **Fire Hazardous Zone (FHZ).**

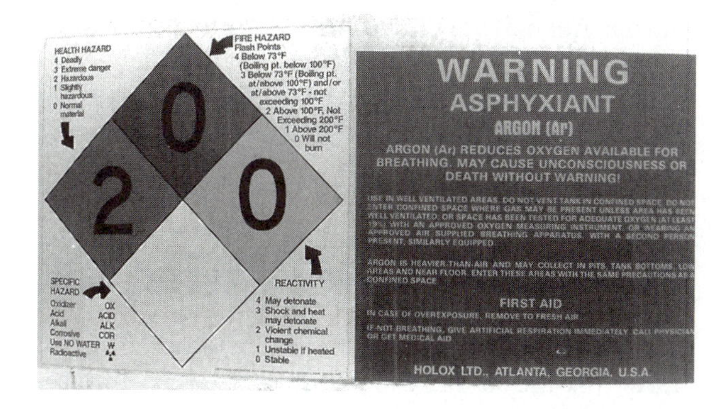

Figure F-15 Hazardous materials label (From NFPA 704).

Fire Hazardous Zone (FHZ)—As defined in the petroleum process industries, a location in which a high fire hazard is likely to occur and the need for additional fire protection measures are necessary (primarily fireproofing nearby primary structural elements or emergency process system isolation control systems). They are typically defined around pumps, compressors, or heaters handling flammable materials or combustible liquids above their flash points. *See also* **Fire Hazardous Equipment; Fireproofing; Hazardous Fire Area.**

Fire Helmet—Head protection personal protective equipment, worn by firefighters to protect against head injury, fire effects, and hot water during firefighting activities. Originally, helmets were made of leather. Jacobus Turck, a caretaker of New York's first fire engines, is credited with creating the first leather fire cap in America sometime before 1740. Fire helmets evolved from leather, steel, plastic, fiberglass, and lately composite materials that are strong, lightweight, heat resistive, and nonconductive. Identification (introduced circa 1830s) is normally provided on the front of the helmet to indicate the organization that the individual represents (Engine Company No. 1) and fire service rank (Captain). Modern helmets are also provided with a protective face shield. NFPA 1972, *Standard for Helmets for Structural Fire Fighting,* provides guidance in the design and manufacture of firefighting helmets. See Figure F-16.

Figure F-16 Fire helmets, evolving from leather, metal, and fiberglass to modern plastic.

Fire Hook—A metal hook used in the early development of firefighting to pull down buildings or other structures to create a firebreak. It is used to stop the advance of fire through a city by creating a firebreak or for overhauling ruins. Fire hooks were used until about the 19th century, when technological advances in water supply distribution systems and pumping devices allowed sufficient quantities of water to be available for firefighting so that major city fires could be contained and suppressed with water application instead of the creation of firebreaks at the time of the fire incident. See Figure F-17. *See also* **Firebreak.**

Fire Horn—A horn-shaped speaking trumpet or megaphone used on firegrounds by fire commanders to annunciate orders to other members of the fire department before the advent of wireless communication devices. The fire horns were used to direct and raise the speaking voice above the ambient noise levels at a fire scene. Generally two types of fire horns were produced,

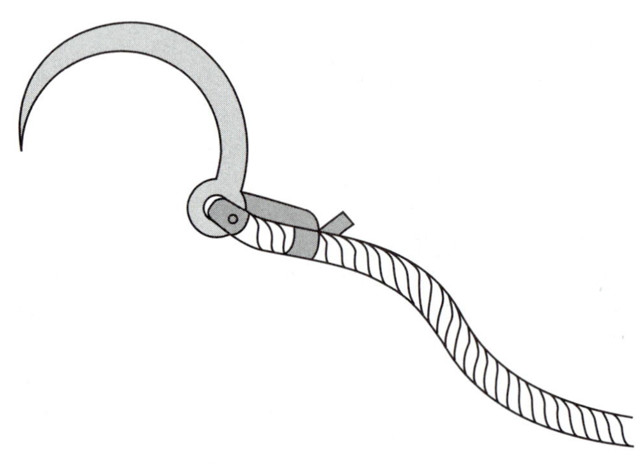

Figure F-17 Fire hook used for pulling down buildings to create a firebreak.

one for actual use at a fire scene and another as an ornamental type for presentations, gifts, or rewards. Fire horns are now obsolete but are valuable as collectibles or historical items by fire buffs or museums.

Fire Hose—*See* **Hose, Fire.**

Firehouse—*See* **Fire Station.**

Firehouse Dog—A dog originally kept at a firehouse with horse-drawn apparatus to keep other dogs from pestering the horses during a response to a fire. They were also used to direct the fire service horse-drawn wagons and guard the firehouse. They were traditionally kept by firehouses during the 19th century. Firehouse dogs eventually developed into a mascot for the firefighters, and Dalmatian dogs were used most often. A firehouse dog was also known as a coach dog or carriage dog. *See also* **Sparky the Fire Dog.**™

Firehouse Pole or **Fire Pole**—A device to allow individuals to quickly slide from one floor to another by gripping a pole provided in a firehouse, in order to rapidly respond to fire event. It was invented in 1878 by Captain David B. Kenyon of the Chicago Fire Department. He devised a wooden pole to slide from the sleeping quarters to the fire wagons to speed response times to fires. A brass pole was first installed in a firehouse located in Worcester, Massachusetts in 1880.

Fire Hydrant—*See* **Hydrant.**

Fire Hydraulics—Term for the science or study of water in motion (fluid mechanics) as applied to fire protection applications (firefighting, fire suppression, fixed water-based suppression systems, etc.) In the late 1800s, John R. Freeman (1855–1932), a prominent American civil engineer and expert in hydraulics, investigated the flow of water through nozzles, hoses, and features of pumps applicable to fire protection aspects. His findings were published in 1889 and 1891 and provided the basics of fire hydraulics. The science of water in motion was originally called hydrodynamics. The foundations of hydrodynamics were laid in the 18th century by mathematicians Leonhard Euler (1707–1783) and Daniel Bernoulli (1700–1782). They began to explore a continuous medium, like water, based on the dynamic principles that Sir Isaac Newton had enunciated for systems composed of discrete particles. By the end of the 19th century, explanations had been found for the flow of water through tubes and orifices, the waves that ships moving through water leave behind them, raindrops on windowpanes, and the like. However, there was still no proper understanding of problems as fundamental as that of water flowing past a fixed obstacle and exerting a drag force upon it. The theory of potential flow, which worked so well in other contexts, yielded results that at relatively high flow rates were grossly at variance with experiment. This problem was not properly understood until 1904, when the German physicist Ludwig Prandtl introduced the concept of the boundary layer. Since that time, the flow of air has been of as much interest to physicists and engineers as the flow of water, and hydrodynamics has become fluid dynamics. The term fluid mechanics embraces both fluid dynamics and the subject still generally referred to as hydrostatics.

Fire Incident Report—A report, usually prepared by the fire services, summarizing the details of a fire event, causes, probably ignition, location and materials affected, fire service response, and actions.

Fire Injury—Physical or psychological injuries caused by a fire event and suffered by those affected by the fire or by firefighters. Fire injury statistics are compiled in the United States by the National Fire Incident Reporting System

(NFIRS) and NFPA. *See also* **Burn Injury; Fire Trauma; Heat Exhaustion; Heat Stroke; National Fire Incident Reporting System (NFIRS); Smoke Inhalation.**

Fire Inspection—Commonly an inspection for the purpose of ensuring that fire prevention regulations or ordinances are being enforced. It may also be used to familiarize the fire service with an occupancy as part of pre-fire planning.

F

Fire Inspector—An individual who is qualified as a fire prevention specialist. *See also* **Fire Marshal.**

Fire Insurance—A legal financial reimbursement agreement generally against losses caused by fire, lightning, and the removal of property from premises endangered by fire between two parties. The insurer agrees, for a fee, to reimburse the insured in case of such an occurrence. The standard policy limits coverage to the replacement cost of the property destroyed, less a depreciation allowance. Indirect loss, such as that resulting from the interruption of business, is excluded but may be covered under a separate contract. Insurance rates are influenced by the quality of fire protection available where the building is located, the type of building construction, the kind of activity conducted within the building, and the degree to which the building is exposed to losses originating outside it. Certain kinds of property, such as accounting records, currency, deeds, and securities, are frequently excluded from fire insurance coverage or are declared uninsurable. Loss from such causes as war, invasion, insurrection, revolution, theft, and neglect by the insured are also customarily excluded. Coverage is suspended if the insured does anything that increases the hazard or if the property is vacant beyond a specified period. The policy may be cancelled by either party for any reason, but the insurer must give the insured prior notice of cancellation. The policy may specify in addition that the insurer may replace or rebuild the damaged property rather than make a cash settlement.

Although the first insurance was devised for ships' cargo, merchants began to band together to share other kinds of risk, including that of fire. The first real insurance company was founded in 1667, the year after the Great Fire of London destroyed some 13,000 homes and left 100,000 people homeless. Nicholas Bardon, an entrepreneur engaged in real estate development, began a speculative business operation involving residential construction and sales. Each sale included an agreement to repair or rebuild any house damaged or destroyed by fire. So successful was this scheme that Bardon opened a private commercial fire insurance business in 1667 called "The Phoenix Fire Office." It was also the origin of the fire mark and private fire brigades. It maintained a group of men trained and equipped to fight fires. The Fire Office placed a distinctive metal insignia or mark on each insured dwelling.

In America, the first insurance company was founded by an association of storeowners in 1735 to share the risk of fires destroying their wooden buildings. It lasted only five years. Benjamin Franklin founded the Philadelphia Contributionship for the Insurance of Houses from Loss by Fire in 1752. This company exists as the Insurance Company of North America and is still based in Philadelphia. Similar groups formed and split into various companies. Many companies, such as the Hartford Fire Insurance Company, Aetna Life and Casualty Company, and the Travelers Corporation insurance companies came to be based in Hartford, Connecticut, giving it the informal title of the insurance capital of the United States.

Insurance is obtained by owners of homes and commercial properties to provide reimbursement in case of losses resulting from fire.

Such insurance is supplied in exchange for the payment of a premium. The five types of insurers who write policies are stock companies, mutual companies, reciprocal exchanges, Lloyd's organizations, and advance premium cooperatives. Most of the fire insurance in the United States is underwritten by stock companies. Some business firms, however, are self-insurers; that is, they set aside funds to be used exclusively for indemnifying losses resulting from fire.

The first scientific system of obtaining funds to compensate for fire loss was developed after the Great Fire of London in 1666, which devastated some 13,000 buildings. In the system inaugurated the following year by the London merchant Nicholas Bardon, small sums of money were collected from many individuals, and a fund was established as compensation for the losses sustained by the few whose property was subsequently destroyed by fire.

The concept of fire insurance in America came about in the 18th century in the form of firefighting organizations. Groups of neighbors banded together to fight local fires. Other groups formed in hope of a reward by the owner of the building whose property was saved. The Union Fire Company, of which Benjamin Franklin was a founding member, was the first of the community-supported type to be established, circa 1730. The first effective fire insurance company established in the United States was the Philadelphia Contributorship for the Insurance of Houses from Loss by Fire, which is still in operation. It was operated like a local fire brigade. Each member provided a sum of money to reimburse any member who suffered a fire loss. It was organized in 1752 by Benjamin Franklin. The use of fire insurance became widespread during the 19th and 20th centuries. The standard fire insurance policy in the United States was adopted in New York State in 1943. It became the prototype of such policies in most other states, either through statute or through regulation by state insurance departments. The resulting standardization has helped reduce litigation on disputed claims by making the insurance coverage more understandable to the policyholder and by simplifying adjustment of losses.

The basic fire insurance policy covers losses resulting directly from damage or destruction by fire or lightning. In the early 1900s, insurance companies in the United States first offered, for an additional premium, to extend the coverage of fire insurance policies to other perils by using an endorsement on the policy. By the late 1920s, these additional perils were incorporated into a so-called extended-coverage endorsement. Extended coverage at present includes the perils of damage by windstorm, hail, explosion, riot, strikes, civil commotion, aircraft, vehicles, and smoke. The endorsement may also be extended further.

In fire insurance, the premium rates are of two kinds: class rates and schedule rates. Dwellings are largely class rated; that is, they are grouped into homogeneous categories according to the type of occupancy, type of construction, and type of community fire protection. A uniform rate is applied to all risks in the same category. Commercial and industrial properties, which vary greatly with respect to degree of hazard, are usually schedule rated. In schedule rating, the individual physical characteristics of each risk are appraised according to a schedule of charges and credits. The elements considered in the rating include occupancy, construction, internal protection, community fire protection, and exposure from neighboring buildings. The frequently used homeowners package policy combines fire, extended coverage and other perils, theft, and comprehensive personal liability into one policy. This results in savings to the policyholder and eliminates overlapping coverages. Similar package policies are available for commercial use.

In recent years, many owners of residential and business properties in congested inner-city areas have been unable to obtain adequate insurance coverage for fire, vandalism, burglary, and theft. The Federal Housing and Urban Development Act of 1968 encouraged insurance companies and state regulatory authorities, through a federal program of reinsurance, to set up Fair Access to Insurance Requirements plans, through which residents of low-income areas may obtain the insurance they need. If damage is caused by riot or civil commotion, the federal government will reimburse the insurance company for its losses. *See also* **Fireline; Rating Bureau.**

Fire Intensity—A measurement of Btus produced by a fire. Sometimes measured in flame length in the wildland fire environment.

Fire Investigation—The process of determining the origin, cause, and development of a fire or explosion. Fire investigations are undertaken by the state fire marshal in conjunction with local law enforcement agencies for the purpose of determining if arson has been involved. *See also* **Arson.**

Fire Investigator—An individual knowledgeable and trained in fire origins and growth aspects who determines fire causes and the possibility of malicious intent to ensure that the due processes of law are enforced. A fire investigator is usually employed when an arson fire is suspected. *See also* **Fire Marshal.**

Fire Laddie—An archaic term for someone associated with the fire services.

Fire Lane—A clear space usually provided in front of a building or special road usage that is suitable for firefighting operations by motorized fire apparatus and commonly restricted for use by other activities (parking) by city traffic ordinances. Fire lanes have generally not less than 20 ft. (6.1 m) of unobstructed width and have a minimum of 13.5 ft. (4.1 m) of vertical clearance. They must be constructed to withstand the live loads of the heaviest piece of fire apparatus likely to be driven upon it and be properly marked for public observance.

Fire Liability—Liability of a firm or person for fire damage caused by negligence of and damage to property of others.

Fireline—(1) A strip of land cleared of vegetation to prevent the spread or advance of a forest fire. It may also be referred to as a firebreak or fireguard. *See also* **Back Burn; Control Line; Firebreak; Fireguard; Wildland Fire.** (2) The maximum value that is written on an insured or a class of insured by an underwriter. *See also* **Fire Insurance; Underwriting.**

Fire Load or **Fire Loading**—The amount of combustibles present in a given area of a room or building enclosure. It may also refer to the measure of the heat content of the combustible materials present in an area expressed in Btu/ft^2 (kiloJoules/m^2). The heat content per unit area is called the fire load density. It is used to determine fire severity and therefore to calculate the amount of fire resistance needed to withstand a fire or the application rate for fire control or extinguishment by manual or fixed systems for the subject area. The fuel load density is a function of the quantity, arrangement, combustibility, and rate of heat release of the material. It may also be called fuel loading. *See also* **Fuel Loading; Heavy Content Fire Loading.**

Fire Load Density—*See* **Fire Load.**

Fire Lookout—A fire tower or individual responsible for surveying a designated area for the occurrence of a fire. It may also be called a lookout tower or fire tower. *See also* **Fire Tower.**

Fire Main—The primary piping conduit through which firewater is supplied and distributed. It may be horizontal or a vertical riser. Ground level fire mains are normally buried to protect them from freezing, to prevent accidental impacts, to avoid exposure to fire conditions, to provide unobstructed access, and for aesthetic reasons. The layout, sizing, and materials of construction for fire mains must be thoroughly examined to ensure they meet firewater demands and reliability requirements. They usually supply water to fire hydrants, fixed water-based fire suppression systems (sprinkler and standpipe systems), and hose reels. Fire mains were originally constructed of hollowed-out logs with tapered ends fitted into each other. *See also* **Riser; Water Distribution System.**

Fireman—An individual qualified to control and suppress unwanted fires or perform other firefighting support activities. The term fireman has been commonly replaced by firefighter to avoid a gender classification. *See also* **Fire Department (FD); Firefighter.**

Fire Mark—A plaque or symbol on a building indicating that it was insured for fire damage by a certain underwriting company. It was used in the early stages of fire insurance arrangements in the underwriting industry throughout the world. Fire marks were in use from 1700 to the 1950s. They are sometimes referred to as the heraldry signs of insurance companies. Buildings provided with a fire mark would be assisted in firefighting efforts by the respective insurance company firefighters during an incident.

The first fire marks displayed a lead plate with a phoenix rising from the flames. Fire marks were generally not placed upon a building until it was actually insured. They were promptly removed if an insurance policy was cancelled or not renewed. They were made of lead, brass, cast iron, copper, tin, or zinc. Often they were painted in vivid colors accompanied by intricate designs. Some included a policy number. The first American fire mark was four hands clasped and crossed issued by the Philadelphia Contributorship for Insurance of Houses from Fire in 1752. This design was used by both American and English companies (Hand-in-Hand Fire and Life Insurance Society and the Friendly Insurance Society). Other popular designs included the green tree on the shield of the Green Tree Mutual Insurance Company and the eagle emblem of the Insurance Company of North America. The Green Tree Mutual Insurance Company became known because it was willing to insure houses near trees "provided the trees were not allowed to grow higher than the eaves of the adjacent houses." Ben Franklin, one of the founders of the first insurance company in America, felt that trees attracted lightning and posed a fire risk near buildings. Fire marks are now obsolete but are considered a collector's item with historical value. See Figure F-18.

Fire Marshal—The head of a governmental department or an office that is charged with the prevention and investigation of fires. Each state usually has its own state fire marshal (SFM). Most state fire marshals were established at the turn of the last century. The first state fire marshals in the United States were appointed in Maryland and Massachusetts in 1894. Canada appointed its first provincial fire marshal in Manitoba in 1876. Some were established as the result of major tragedies in the state. A fire marshal may also hold a rank in a local fire department responsible for inspections and public education programs. The earliest establishment of an office similar to a fire marshal, was that of Brent Masters in New York City (then New Amsterdam), in 1689.

Figure F-18 Fire marks.

In a few states (Arkansas, Michigan, Oregon, and Pennsylvania), the state fire marshal's office is an arm of the state police. A fire marshal may also refer to a person in charge of firefighting personnel and equipment at an industrial plant. *See also* **Fire Inspector; Fire Investigator.**

Fire Marshal's Association of North America (FMANA)—An association that organizes activities and coordinates exchange of information for fire marshals responsible for fire prevention and arson investigation. The association was founded in 1906.

Fire Modeling—Mathematical modeling that is used to predict one or more effects of a predefined fire or fuel configuration. Fire event equations and quantitative risk analysis methods can be used to formulate or predict an outcome based on a defined set of input variables and assumptions. ASTM E1472, *Standard Guide for Documenting Computer Software for Fire Models,* provides guidance in the preparation of fire modeling programs. *See also* **Quantitative Risk Analysis (QRA).**

Fire Monitor—*See* **Monitor; Monitor Nozzle.**

Fire Museum—An institution devoted to the procurement, care, study, and display or exhibit of historical firefighting organizations, apparatus, equipment, artifacts, and accounts of disastrous fire events of lasting interest or value. Major fire museums are located in every state and in major cities in the United States. The American Museum of Firefighting is located in Hudson, New York. It contains 66 pieces of hand-pulled, horse-drawn, steam, and motor-drawn apparatus, and the oldest mobile fire apparatus in America, a Newsham engine built in London that was imported to America in 1731.

Fire-Packed—Normally refers to a fire packed within a cotton bale as a result of a handling process. Cotton ginning is a frequent cause of fire packing.

Fire Pail—*See* **Fire Bucket.**

Fire Performance—The results of a material, building component, or design to withstand exposure to a fire in either actual fire conditions or under test environments. *See also* **Fire Curve; Fire Resistance Rating; Flame Spread Index (FSI); Performance-Based Fire Protection Design.**

Fire Perimeter—The edge of a fire event. *See also* **Fireground; Unburned Island.**

Fire Pillow—*See* **Intumescent.**

Fireplace—A hearth, fire chamber, or similarly prepared area and a chimney for safely disposing of flue gases. A fireplace is used for heating, cooking, or decorative purposes. The first fireplaces developed when medieval houses and castles were equipped with chimneys to carry away smoke. Experience soon showed that the rectangular form was superior and that a certain depth was most favorable. A grate provided better draft, splayed sides, and increased reflection of heat. Early fireplaces were made of stone; later, brick became more widely used. A medieval discovery revived in modern times is that a thick masonry wall opposite the fireplace is capable of absorbing and re-radiating heat. From early times, fireplace accessories and furnishings have been objects of decoration. Since at least the 15th century, a fireback (a slab of cast iron) protected the back wall of the fireplace from the intense heat; these were usually decorated. After the 19th century, the fireback gave way to firebrick in fireplace construction. *See also* **Chimney; Fireback.**

Fireplug—Terminology for the first fire hydrants, it is based on early arrangements for a firewater supply from wooden water distribution mains. Early fire hydrants were commonly referred to as fireplugs because wooden water mains needed a plug in the opening after use. The plug was removed and water formed a pool from which buckets were used to supply the tank of a hand-pumped fire engine. Primitive versions of modern fire hydrants on public water mains in America began to be installed in the 1830s and 1840s. Prior to this, the wooden fireplugs were used. The first fireplug in New York City was installed in 1807. They were later recommended for installation in all of the city streets. Modern fire hydrants may be unconventionally (slang) referred to as a fireplug due to this early association of a hydrant as a means to plug the water distribution main. See

Figure F-19 Fire plug in hollow wooden log water main.

Figure F-19. *See also* **Hydrant, Fire; Water Distribution System.**

Fire Plume—The visible profile of a diffusion flame. *See also* **Plume; Thermal Convection.**

Fire Point—The lowest temperature at which a liquid fuel in an open container will give off sufficient vapors to support combustion once ignited. It generally is slightly above the flash point. It may also be called burning point. *See also* **Flash Point (FP).**

Fire Pole—*See* **Firehouse Pole** or **Fire Pole.**

Fire Police—Individuals assigned to support positions in emergencies to facilitate traffic control, crowd control, and securing the scene. They are typically members of the fire service and are used when regular police officers are not available.

Fire Prevention—Measures directed toward avoiding the inception of fire. Fire prevention and control are the prevention, detection, and extinguishment of fires, including such secondary activities as research into the causes of fire, education of the public about fire hazards, and the maintenance and improvement of firefighting equipment.

Up until the 1920s, little official attention was given to fire prevention, because most fire departments were concerned only with extinguishing fires. Since then most urban areas have established some form of a fire prevention unit. The staff of this unit concentrates on such measures as heightening public awareness, incorporating fire prevention measures in building design and in the design of machinery and the execution of industrial activity, reducing the potential sources of fire, and outfitting structures with such equipment as extinguishers and sprinkler systems to minimize the effects of fire.

The importance of increasing public understanding of the causes of fire and of learning effective reactions in case of fire is essential to a successful fire prevention program. To reduce the impact and possibility of fire, the building codes of most cities include fire safety regulations. Buildings are designed to separate and enclose areas so that a fire will not spread. They incorporate fire prevention devices, alarms, and exit signs to isolate equipment and materials that could cause a fire or explode if exposed to fire, and to install fire extinguishing equipment at regular intervals throughout a structure. Fire retardant building materials such as paints and chemicals have also been developed. They are used to coat and impregnate combustible materials such as wood and fabric.

Perhaps more important than firefighting itself in many modern industrial countries is fire prevention. In Russia and Japan, for example, fire prevention is treated as a responsibility of citizenship. Firefighters in the United States are trained in basic fire prevention methods, and fire companies are assigned to inspection districts in which they attempt to prevent or correct unsafe conditions. Fire departments are charged with enforcement of the local fire prevention code and of state fire laws and regulations.

A fire prevention bureau in the fire department usually directs fire prevention activities. It handles the more technical fire prevention problems, maintains appropriate records, grants licenses and permits, investigates the causes of fires, and conducts public education programs. All commercial or multiple-dwelling buildings are inspected regularly, and orders are issued for the correction of violations of fire laws. If necessary, court action is taken to compel compliance.

In some communities protected by volunteer or part-time paid fire departments, fire prevention is the responsibility of a state or county fire marshal or of a professional fire staff in an otherwise voluntary organization. In addition, fire departments usually inspect commercial buildings for what is called pre-fire planning. Private dwellings may also be inspected as part of a fire department's educational program to impress the importance of fire safety on the inhabitants and to check for any unsafe conditions.

Many public buildings are equipped with automatic sprinkler systems that release a spray of water on an affected area if a fire is detected. The effectiveness of these systems has been proved in data accumulated from throughout the world. In buildings protected by sprinkler systems that had fires, the system extinguished fires in 65 percent of the cases and contained fires until other firefighting measures could be taken in 32 percent of the cases. A major problem with sprinkler systems is the potential for water damage, but in most cases this threat is minimal compared with the damage that a fire could cause.

There exists a considerable variety of firefighting equipment, ranging in sophistication from buckets and extinguishers to the elaborate yet portable apparatus used by fire departments. The most common of these is the fire engine, equipped with hoses, ladders, water tanks, and tools. Ladder and rescue trucks work in conjunction with trucks equipped with platforms that can be elevated by hydraulic lifts to carry out rescue efforts.

Fire extinguishing agents other than water are used to fight various types of fire. Foaming agents are employed to handle oil fires. "Wet" water, formed by the addition of a chemical that reduces surface tension, can be used in clinging foam to protect the exterior of a structure near the source of a fire. Ablative water, made by mixing water with additives, forms a dense, heat-absorbing blanket. Steam is used to control fire in confined areas, whereas inert gas is employed to extinguish gas, dust, and vapor fires.

In rural areas, water-tank trucks are usually needed, thus the time factor becomes even more critical. Bush, grass, and forest fires are frequently fought using the same equipment used on structural fires. Aircraft are sometimes employed to dump fire-retardant slurries or water mixtures on these blazes.

It has also become necessary to combat fires in pressurized chambers, including spacecraft. The combustion rate in these environments is much higher than under normal atmospheric pressure. Strict construction guidelines are followed to keep fire hazards to a minimum, and highly pressurized sprinklers are installed that act immediately upon any combustion. *See also* **Fire Protection; Loss Control.**

Fire Prevention Week—A designated week in the United States that highlights the awareness of the role of fire prevention to avoid the occurrence of unwanted fires. Usually the observance of Fire Prevention Week occurs in first week of October. The idea for Fire Prevention Day was first suggested in 1911, to occur on October 9 to commemorate the Great Chicago Fire of 1871, which had taken place 40 years earlier. The suggestion of a Fire Prevention Day was not effective. The real impetus for a designated day for fire prevention observance was not undertaken until President Harding issued the first Fire Prevention Week proclamation in 1922 as a result of efforts by the National Fire Protection Association.

Fire Products—*See* **Combustion Products.**

Fire Progress Map—A real-time map that depicts a fire event to indicate the property destroyed by fire, direction of fire growth (or possible growth), locations affected by fire exposures, areas of fire suppression activities (equipment and manpower), and progress made in fire control and suppression. A fire progress map is usually prepared for large forest fire incidents.

Fireproof—Common trade name for materials used to provide resistance to a fire exposure. Essentially nothing is fireproof, but some materials are resistant to the effects of a fire (heat, flame, etc.) for limited periods. Independent testing agencies such as UL and NIST test submitted materials for a standard fire test exposure for fireproof ratings. NFPA recommends the term fire resistive in place of fireproof. *See also* **Fireproof Construction; Fireproofing.**

Fireproof Construction—Term coined early in the building construction industry for buildings that supposedly were able to resist fire damage and prevent the spread of fire. Fireproof buildings technically do not exist, because most non-combustible materials employed in construction suffer some damage under the action of heat and flame. This was amply demonstrated after several disastrous fires in which fireproof buildings were destroyed. Therefore "fire resistive" or a "level of fire endurance" is commonly used and "fireproof construction" is no longer recognized as a descriptive label. A so-called fireproof building containing only non-burning components, such as steel, terra cotta, plaster, and concrete, may be destroyed by an intense fire in adjacent buildings. It may also be gutted by an interior fire that feeds on fixtures and trim only but spreads through the building if the building is improperly designed.

The two primary considerations in fire-resistant construction are design and materials.

F

A building should be subdivided by fire-resisting walls, floors, and partitions to limit the spread of fire. Elevator and stair shafts, walls, light wells, and other vertical structures must be isolated for the same reason and because vertical openings act as chimneys, increasing the intensity of a fire. Stairwells or shafts that must be continuous are isolated by heavy, fire-resistant walls that are either solid or, if hollow, are provided with a number of horizontal partitions, or fire stops. All doorways or other wall openings should be provided with doors or covers that are self-closing or that close automatically in case of fire. Materials used for the primary structure typically should provide one to three hours of fire resistance. Materials used for interior finish should be able to resist the spread of fire. Local building codes usually specify minimum flame-resistance requirements. Consideration must also be given to the poisonous products formed by materials in a fire; some are more hazardous than the fire itself. *See also* **Fireproof; Fire Resistive.**

Fireproofing—A common industry term used to denote materials or methods of construction that provide fire resistance for a defined fire exposure and specified time. Essentially nothing is fireproof if it is exposed to high temperatures for an extended time period. The process of fireproofing consists of treating or coating a material so that its tendency to burn is reduced. The term is therefore actually misleading, since no process can completely stop a material from burning under all circumstances. More appropriately, the term fire retardant is applied to substances used in treating materials chemically to reduce the likelihood of their catching fire. The degree of retardation is measured by standard flammability tests. Successful fire retardants contain compounds that include phosphorus, nitrogen, antimony, chlorine, or boron, depending on the application and the material to be treated. These agents work in different ways. Some reduce the amount of burnable fuel vapors produced by the burning substance. Others release gases that do not burn, such as water vapor, to dilute the fuel vapors. Still others slow the rate of oxidation, thus reducing the rate at which energy is released and the amount of heat that is generated. The improper use of chemical retardants on fabrics can result in a loss of strength or a breakdown of the fibers over time. Some chemicals can cause skin rashes and cannot be used in clothing, especially in infants' clothing. Some retardants wash out. Proper fire retarding of fabrics should be done by the manufacturer, not at home. Even materials generally considered safe will burn under the right conditions. Metals, for example, can burn in the presence of high concentrations of oxygen. Large piles of iron metal filings and small steel scrap have been known to catch fire. Even the steel structure of a tall building can crumble at very high temperatures from a loss of strength unless the steel is protected. This is achieved by adding concrete around load-bearing columns.

Proper surface preparation and application is essential for the optimum performance of a fireproofing material. Fireproofing materials may blister or crack if not properly bonded to the surface of the material to be protected. In the case of steel structures, this may allow surface corrosion effects to occur. The corrosion may be undetected for a long time and lead to premature structural failure caused by the application of the fireproofing that does not allow the corrosion to be seen.

Commonly applied fireproofing materials include cement, intumescent coatings, refractory fibers, composite materials, purpose-made penetration seals, and specialized coatings. These materials are described as follows:

Cementatious Materials—These include a hydraulically setting cement, such as Portland cement, as a binder with a filler of good insulation properties (vermiculite, perlite, mica, etc.). They are sprayed or spread directly onto the surface to be protected or can be precast sections or panels. A wire grid is required for in situ applications to ensure adherence to the surfaces, and proper surface preparation is essential. The surface of the material should also be sealed to prevent the absorption of water, oil, and protection against weather effects. Cement is commonly applied to steel supports where weight is not a concern. Performance relies on two properties: insulation and dehydration (causing cooling). Oxychloride and oxysulphate cements can cause corrosion of the steelwork to which they are attached. Portland cements normally do not directly cause corrosion.

Intumescent Coatings—These materials have an organic (epoxy) base that expands when exposed to a fire, producing a stable char with good thermal insulating properties. Their main advantage is the use of a thinner coating than cement. They are also lightweight and less prone to absorb water or oil. They provide good corrosion protection to the substrate. Application is by spray and repair is easy. Intumescents give off toxic smoke and fumes as they char. Erosion of the char may occur by jet fire impingement or by the use of a firewater hose stream during the fire.

Refractory Fibers—These are fibrous materials with a high melting point that are made into fire-resistant board and mats. The fibers are derived from glass, mineral, or ceramics. They may be woven into cloths or used to form felts or blankets that can wrap around an item. Blankets are held in place by straps or wires. Semi-rigid board can be made by binding the fibers with a resin. They are widely used for the protection of piping, isolation valves, sleeves for wall penetrations, etc. Their low physical strength and absorbency requires that they be protected externally where they may be subject to impact. They are sometimes used as a component in a sandwich-type construction panel.

Composite Materials—These designate a material made of a combination of materials, usually a resin and a fiber. Most composite materials are made of fiberglass and ceramic materials of proprietary design that are lightweight and may be flexible or in panel form. Due to the chemicals involved, composite materials may give off toxic vapors in fire conditions.

Seals and Sealants—These are purpose-made penetration seals that form a barrier at penetrations in a fire barrier. Sealants include intumescent mastics. Silicone elastomers are used to seal joints where movement may be expected.

See also **Encasement, Fire Protection; Fire Hazardous Zone (FHZ); Fireproof; Heat Resistant; Intumescence; Intumescent; Passive Fire Protection (PFP); Refractory.**

Fire Propagation Index—A rating of the heat output from a wall lining material when subjected to a fire test designated in British Standard 476, Part 6 (fire in a furnance test).

Fire Protection—In general terminology, fire protection refers to the prevention, detection, and extinguishment of fire and reduction or avoidance of loss in lives, assets, business activities, environmental impact, and prestige. In a specific application, it is providing fire control or extinguishment. It may also be used to signify the degree to which protection from fire is applied. *See also* **Fire Prevention.**

Fire Protection Engineer (FPE)—An individual who by education, training, and experience is familiar with the nature and characteristics of fire and explosions and the associated products of combustion. He or she understands how fire originates and spreads whether within or outside of buildings or structures; is knowledgeable about how fire can be detected, controlled, and/or extinguished; and is able to anticipate the behavior or materials, structures, machines, apparatus, and processes as related to the protection of life, property, business interest, and environmental impacts from fire or its effects. The fire protection engineer should also know about fire detection and suppression systems (smoke detectors, automatic sprinklers, or similar systems). Additionally, he or she should be familiar with building and fire codes, fire testing methods, fire performance of materials, computer simulation/modeling of fires, and reliability or failure analysis. *See also* **Professional Engineer (PE), Fire Protection.**

Fire Protection Engineering—The discipline of engineering that applies scientific and technical principles to safeguard life, property, loss of income, and threat to the environment from the effects of fires, explosions, and related hazards. It is associated with the design and layout of buildings, industrial properties, structures, equipment, processes, and supporting systems. It is concerned with fire prevention, control, suppression, and extinguishment and provides for consideration of functional, operational, economic, aesthetic, and regulatory requirements. *See also* **Fire School.**

Fire Protection Engineering, Department of, University of Maryland—A four-year college-level course of study in the aspects of fire protection engineering. Degrees of BS and MS in Fire Protection Engineering are conferred upon individuals successfully completing the prescribed academic requirements. The curriculum and research activities address the causes, consequences, and mitigation of unwanted fires in the constructed environment. It is the only ABET-accredited program in Fire Protection Engineering in the United States.

The idea of a fire protection curriculum at the University of Maryland originated in 1950 from the Maryland State Firemen's Association in response to the university's suggestion of closer ties with training aspects and the university. They suggested a program similar to the one then in existence at Oklahoma A & M College (now Oklahoma State University). In 1956, the University of Maryland started the four-year curriculum leading to a Bachelor of Science degree in the College of Engineering. In 1976, the curriculum achieved accreditation from the ABET and the department was renamed in 1977 from Fire Protection Curriculum to Department of Fire Protection Engineering. Dr. John L. Bryan was instrumental in establishing the department and ensuring its success. He continued with the department for 37 years as its professor and chair (1956–1993). A graduate program in Fire Protection Engineering was started in 1990. The undergraduate enrollment averages 100 students and graduate enrollment approximately 35 students.

Fire Protection Handbook (FPH)—A reference handbook providing guidance and practices in the science and technology of fire prevention and protection. The Fire Protection Handbook originated from the *Handbook of the Underwriters' Bureau of New England,* which was published in

1896. It was primarily authored and edited by Everett U. Crosby, manager of the Underwriters' Bureau of New England (UBNE). It incorporated rules and regulations of contemporary insurance organizations for fire barriers, automatic sprinkler installations, fire pumps, water supply, and building light, heat, and power systems. It was intended as a reference book for insurance inspectors and other interested parties. Since 1935, the publication of the handbook has been undertaken by the National Fire Protection Association (NFPA) and its staff. The handbook is now considered the most widely read fire protection publication in the world, providing life safety and property protection guidance, fire protection application and design principles, and fire hazard identification in a wide variety of subjects and detail. The handbook is updated periodically to encompass the latest technologies and retire outdated practices.

Fire Protection Rating (FPR)—A designation of fire resistance duration for a material or assembly when exposed to standard test conditions, having met all acceptance criteria.

Fire Protection System—An integrated system that affords protection against fire and its effects. It may be made up of either active or passive fire protection measures. The fire protection system should be commensurate with the level of hazard it is protecting.

Fire Pump—A pump specifically designated, designed, and installed to provide adequate and sufficient water supplies for controlling and suppressing unwanted fires. Firewater pumps are required to be constructed and installed to recognized standards, such as NFPA 20, *Installation of Centrifugal Firewater Pumps,* for reliability in emergency situations and levels of performance. A firewater pump may be driven by an electric motor, a diesel engine, or a steam turbine. Gasoline engines are not recommended due to their reduced reliability and their inherent fire hazard. Fire pumps may be mobile or stationary. Mobile pumps used by a fire department and mounted on trucks are called fire pumpers. Heron of Alexandria, the technical writer of antiquity (circa 100 CE), describes in his journals a two-cylinder pumping mechanism with a dirigible for fighting fires. Devices akin to this were used in the 18th and 19th centuries to provide firefighting water in Europe and the Americas. See Figure F-20.

Modern fire pumps are generally of two types, centrifugal or rotary. Centrifugal pumps consist of a rotating impeller encased in a housing. The force of rotation imparts a velocity pressure to the fluid, causing it to flow. Rotary pumps consist of a pair of encased intermeshing gears that force fluid by mechanical action. Fluid is directed through the gears from which positive displacement action provides pressure to the fluid. See Figure F-21. *See also* **Centrifugal Pump; Firewater Pump; Fire Pumper; Firewater Pump, Booster; Foam Pump; Pump, End Suction and Vertical In-Line; Pump,**

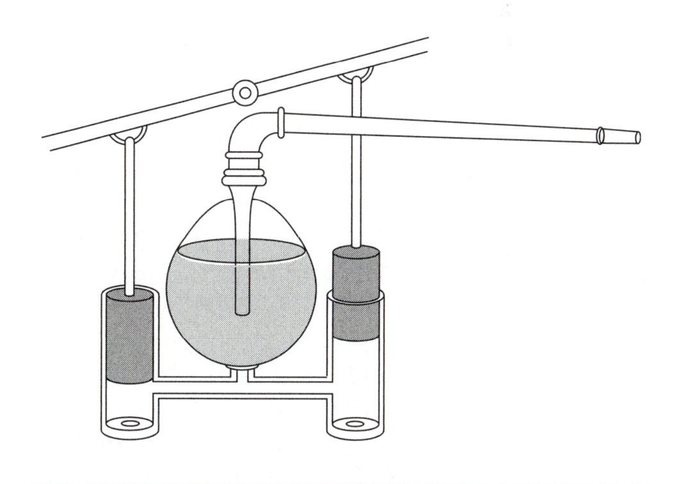

Figure F-20 Ancient fire pumping device using double-acting piston pump.

Figure F-21 Modern diesel-driven centrifugal fire pump.

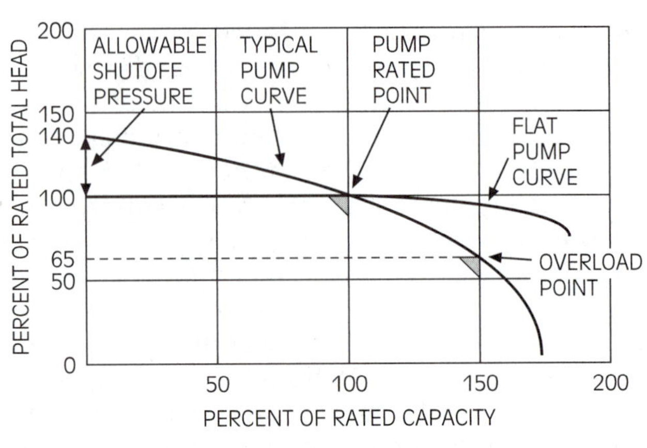

Figure F-22 Fire pump curve.

Horizontal Split Case; Pump, Vertical Shaft, Turbine Type; Rotary Pump.

Fire Pump Controller—A control cabinet, motor starter, circuit breaker, disconnect switch, or other device that starts or stops a firewater pump driven by a fixed electric motor or an internal combustion engine. Fire pump controllers are considered a critical support item for the system and are required to be listed or approved for such service by a recognized testing agency capable of assessing the performance and reliability of the device.

Fire Pump Curve—The characteristic flow performance curve for firewater pumps. Individual fire pumps are rated for a specific flow and corresponding pressure, such as 1,000 gpm (3,785 l/min) at 150 psi (1,033 kPa). NFPA 20, *Standard for Installation of Centrifugal Fire Pumps,* specifies that firewater pumps should furnish not less than 150 percent capacity at not less than 65 percent of rated flow. Additionally the total shutoff head should not be more than 140 percent or less than 100 percent of the rated pressure. This produces a flat pump performance curve, which achieves a relatively standard flowrate over a wide range of pressure output. See Figure F-22. *See also* **Overload Point; Shutoff Pressure.**

Fire Pumper—A fire pump on an automotive truck used to deliver water at sufficient pressures and quantities for fire service operations. A fire pumper usually receives water from public water supply source (water distribution mains, lake, etc.) through a 4-inch (10.2 cm) suction hose, increases the flowrate and pressure using its onboard pump, and discharges the water through delivery hoses for direct firefighting operations or further distribution. Some pumpers may carry a limited supply of onboard water. NFPA 1901, *Standard for Automotive Fire Apparatus,* provides guidance in the features required for a fire pumper.

In 1912, the International Association of Fire Engineers (IAFE), with the assistance of engineers from the National Board of Fire Underwriters (NBFU), conducted tests on seven pumping engines discharging under net pumping pressures of 120 psi (827 kPa), 200 psi (1,378 kPa), and 250 psi (1,723 kPa). From these tests, the IAFE developed a fire pumper test procedure that consisted of six hours pumping at a 100 percent capacity at 120 psi (827 kPa), three hours pumping at 50 percent capacity at 200 psi (1,378 kPa), and three hours pumping at 33 percent capacity at 250 psi (1,723 kPa). This test procedure was used beginning in 1913. In the 1939 and 1941 editions of the *Suggested Specifications for Motor Fire Apparatus* issued by the National Board of Fire Underwriters, they specified the IAFE pumper tests, which were later termed Class B requirements. An optional specification, termed Class A requirements, consisting of a delivery of 100 percent capacity at 150 psi (1,033 kPa), 70 percent capacity at 200 psi (1,378 kPa), and 50 percent of capacity at 250 psi (1,723 kPa), were also provided. The 1947 edition still defined the two classes, but the 1956 edition eliminated the Class B type. As a result, most fire service pumpers built prior to 1939 were specified as Class B, those built from 1939 through 1956 were specified as Class A or B, and those built since 1957 are specified as Class A. The fifth edition of *Fire Department Pumper Tests and Fire Stream Tables* (1950) listed the requirements for the test of Class A pumpers, which specified the delivery of 100 percent of rated capacity at 150 psi (1,033 kPa), 70 percent of capacity at 200 psi (1,378 kPa), and 50 percent of capacity at 250 psi (1,723 kPa). This is the pumper test currently in use. Class B pumpers are no longer recognized for original purchase. See Figure F-23. *See also* **Drafting Pit; Fire Pump; Pumper.**

Figure F-23 Modern fire pumper.

Fire Quenching Pit—Gravel-filled pit for the collection of a combustible liquid. By collecting the combustible liquid in the voids of the semi-filled pit, it prevents the formation of a spill fire. The entrapment of the combustible liquid below the gravel material surface decreases the available free air space above the liquid surface. Fire quenching pits are provided where accidental spills of combustible liquids may occur. They are recommended by the IEEE for the collection of liquid spills from oil-filled electrical transformers using combustible fluids for insulation, Ref. IEEE 979, *Guide for Substation Fire Protection. See also* **Quench.**

Fire Rated—*See* **Fire Resistance Rating.**

Fire-Rated Cable—Electrical cable with a fire resistance rating based on maintaining functionality when exposed to fire tests in NFPA 251, *Standard Methods of Tests of Fire Endurance of Building Construction and Materials. See also* **Fire-Resistant Cable.**

Fire-Rated Internal Conduit Seal—A conduit seal that has been tested and approved as a rated fire seal per ASTM E 814, *Fire Tests of Through-Penetration of Fire Stops.* A fire-rated internal conduit seal is normally provided where the conduit penetrates a rated fire barrier. *See also* **Firestop.**

Fire Rating—*See* **Fire Resistance Rating.**

Fire Report—An account of a fire event that details the nature of the fire, source of ignition, fire spread, materials involved, type of facility or location, amount, type of damage, injuries or fatalities, method of fire control and suppression, fire equipment involved, etc.

Fire Reserve—Usually refers to the water kept in combined-use water storage tanks for firefighting activities and not available for any other use. Take-off points in the water tank for other services are usually located above the reserve amount of water for firefighting purposes to prevent its unauthorized use.

Fire Resistance Rating—A rating in minutes or hours that material or assemblies have withstood a fire exposure. Normal fire resistance is tested in accordance with procedures of NFPA 251, *Standard Methods of Tests of Fire Endurance of Building Construction and Materials,* ANSI/UL 263, ASTM E-119, ISO 834, SOLAS Reg. 3 (b) of Chap. II-2, etc. A fire exposure is obtained from cellulosic materials such as wooden furniture, buildings, or paper in a domestic or industrial fire. The test method requires that samples be representative of the floor, roof/ceiling assembly, and wall assembly being investigated. Horizontal samples are a minimum of 250 sq. ft. (23.2 m^2) in size and vertical samples are a minimum of 100 sq. ft. (9.3 sq. m^2). The test methods require the samples to support a maximum load conditions allowed by limiting conditions of design under nationally recognized structural design criteria. The sample must support this load during the entire fire endurance rating period. Walls must support this maximum load during the application of a water hose stream as part of the test. The furnace environment simulates a building with a fully involved fire. The furnace temperature reaches 1,000°F (537°C) after five minutes, 1,700°F (927°C) at one hour, and 2,000°F (1,093°C) at four hours.

The acceptance criteria for horizontal and vertical samples require the test sample to:

1. Prohibit the passage of flames through the test sample.
2. Limit the average temperature rise on the surface of the sample away from the fire exposure to 250°F (121°C) above ambient.
3. Limit the maximum temperature rise on the surface away from the fire exposure to 325°F (163°C) above ambient.

The acceptance criteria measure the effectiveness of the test sample to restrict the flow of heat. Wall assemblies are also required to withstand the impact and erosion of a water hose stream. For one- to three-hour assemblies, the hose stream consists of a stream at 30 psi (206.7 kPa) that is applied for one minute for one-hour ratings and for 2.5 minutes for two- and three-hour rated assemblies. The hose stream application simulates the ability of an assembly to withstand the impact of loads. The ability to resist the impact of a load is important because during a fire, the contents of a room may fall onto the wall and potentially create an opening, permitting the passage of a flame. Horizontal samples are not subject to the same hose stream test because they are uniformly live loaded during the test. It may also be called a fire-resistive rating. *See also* **Cellulosic Fire; Fire Performance; Fire Resistant; Firestopping; Fire Test, Building Assemblies; Hydrocarbon Fire Test.**

Fire Resistant—A generic term indicating the ability of a material to withstand the effects of a fire for a limited period. Specific fire-resistant capabilities are dependent on the properties of the material involved and fire exposure. Various standardized tests have been developed to classify the fire resistance of materials. *See also* **Blast Resistant; Fire Resistance Rating; Flame Resistant; Time Temperature Curve, Cellulosic Fire.**

Fire-Resistant Belting Materials—Materials used for the construction of conveyor belts that satisfy the Mine Safety and Health Administration (MSHA) Standard 2G flame test. Conveyors are commonly used to transport mining products such as coal, and can easily initiate or spread a fire within the mining tunnels.

Fire-Resistant Blanket—Flexible matting (asbestos fire-resistive blankets) that has fire-resistive properties used to protect hot work operations and aircraft fire/rescue operations. *See also* **Fire Blanket.**

Fire-Resistant Cable—Electrical cable that has been tested and found resistant to the spread of flame. Various test methods are available to classify electrical cables as fire resistant based on their use and configuration. Electrically insulated cable is classified as fire resistant if it has a flame travel distance of not more than 5 ft. (1.52 m) when tested in accordance with NFPA 262, *Standard Method of Test for Fire and Smoke Characteristics of Wires and Cables.* Cables capable of preventing carrying fire from floor to floor satisfy UL 1666, *Test for Flame Propagation Height of Electrical and Optical-Fiber Cable Installed Vertically in Shafts.* Cables that do not spread fire to the top of the tray in the vertical tray flame test in UL 1581, *Reference Standard for Electrical Wires, Cables, and Flexible Cords* or the damage (char length) does not exceed 4.9 ft. (1.5 m) when tested to vertical flame test cables in cable trays in CSA C22.2 No. 0.3-M-1985, *Test Methods for Electrical Wires and Cables. See also* **Fire-Rated Cable; Low Smoke-Producing Cable.**

Fire-Resistant Clothing—Clothing able to provide limited resistance to the effects of exposure to fire or flames (burns to the wearer's skin). These garments are normally chemically treated and made of natural fibers. Performance requirements and test methods for fire-resistant clothing is mentioned in NFPA 1975, *Standard on Station/Work Uniforms for Fire Fighters.* The flame resistance for clothing is determined by Federal Test Method Standard 191A, *Textile Test Methods, Method 5903.1, Flame Resistance of Cloth: Vertical. See also* **Fire-Retardant Coveralls (FRC); Flame Resistant.**

Fire-Resistant Fluid—A combustible fluid that is difficult to ignite due to its high fire point and autoignition temperature. It does not sustain combustion due to its low heat of combustion.

Fire-Resistant Partition—A partition having a fire-resistance rating of 20 minutes or more when tested in accordance with NFPA 251, *Standard Methods of Fire Tests of Building Construction and Materials.* Usually specified for use in a cooling tower environment.

Fire-Resistant Safes and Cabinets—A container provided to protect records and valuables that is rated for various fire durations by segregating them from the surrounding fire exposure. UL 72, *Standard for Tests for Fire Resistance of Record Protection Equipment,* contains requirements for the fire tests of these containers. Records protection equipment is classified in terms of an internal temperature limit and an exposure time (rated in hours). Two internal temperature and humidity performance ratings or conditions are specified: 150°F (65.6°C) with 85 percent relative humidity, which are the limiting conditions for preservation of photographic, magnetic, or similar nonpaper records; and 350°F (196°C) with 100 percent relative humidity, which are the limiting conditions for preservation of paper records. The fire exposure limits are four hours, three hours, two hours, and one hour. Rating approval means that the specified interior temperature and humidity limits are not exceeded when the record protection equipment is exposed to a standard fire test for the length of time specified. Ratings are assigned as follows:

F

Insulated Record Containers:	*Class 150—4 hours*
	Class 150—3 hours
	Class 150—2 hours
	Class 150—1 hour
	Class 350—4 hours
	Class 350—2 hours
	Class 350—1 hour
Fire-Resistant Safes:	*Class 350—4 hours*
	Class 350—2 hours
	Class 350—1 hour
Insulated Filing Devices:	*Class 350—1 hour*
Insulated File Drawer:	*Class 350—1 hour*

Insulated records containers and fire-resistant safes are effective in withstanding exposure to a standard test fire before and after an impact from a fall of 30 ft. (9.1 m). Insulated filing devices and file drawers are not subjected to an impact test and are not required to have the strength to endure such an impact. Insulated records containers, fire-resistant safes, and insulated filing devices can withstand a sudden exposure to 2,000°F (1,093°C) temperature without exploding. Noncombustible cabinets with cellular or solid insulation of less than 1 in. (2.5 cm) thickness have less than a 20-minute fire resistance rating under standard test conditions for insulated filing devices. The exact fire resistance rating depends on the thickness and character of the insulation and other factors. Noncombustible, uninsulated steel files and cabinets have about a five-minute rating under standard test conditions for insulated filing devices.

Fire-Resistant Tank—A tank constructed so that it prevents the release of its storage liquid, failure of the tank, failure of its supporting structure, or impairment of its venting for a period of not less than two hours when tested using a fire exposure that simulates a high-intensity pool fire, as described in UL 2085, *Standard for Insulated Aboveground Tanks for Flammable and Combustible Liquids.*

Fire-Resistive—Properties of materials or designs that are capable of resisting the effects of any fire to which the material or structure may be subjected. Commonly refers to materials when tested in accordance with NFPA 251, *Standard Methods of Fire Tests of Building Construction and Materials. See also* **Fireproof Construction; Flame Resistant.**

Fire-Resistive Construction—*See* **Building Construction Types.**

Fire-Resistive Rating—*See* **Fire Resistance Rating.**

Fire Retardant—The property of a material constructed or treated so it will not support but rather inhibit or resist combustion and the spread of fire. In general, a term that denotes a substantially lesser degree of fire resistance than "fire resistive." It is frequently used to refer to materials or structures that are combustible but have been subjected to treatment or surface coatings to prevent or retard ignition or the spread of flame. A fire retardant can be a liquid, solid, or gas and can be mixed with, applied on, or combined with (impregnate) a material to achieve the desired fire retardant properties. *See also* **Fire-Retardant Coating; Fire-Retardant Coveralls; Fire-Retardant Impregnated Wood; Fire-Retardant Paint.**

Fire-Retardant Coating—A coating material that reduces the flame spread of Douglas fir and all other tested combustible surfaces to which it is applied, by at least 50 percent or to a flame spread index of 75 or less, whichever is the lesser value, and has a smoke developed rating not exceeding 200. Fire-retardant coatings can be classified as Class A or Class B. A Class A fire-retardant coating reduces the flame spread index to 25 or less, and has a smoke developed rating not exceeding 200. A Class B fire-retardant coating reduces the flame spread index to between 25 and 75, and has a smoke developed rating not exceeding 200. *See also* **Fire Retardant.**

Fire-Retardant Clothing or **Coveralls (FRC)**—Clothing or one-piece outer garment that is chemically treated to provide limited protection against the effects of heat and flame (burns to the wearer's skin). They are used as personal attire at work locations or specialized activities that have a high potential for fire exposure, such as by firefighters, aircraft servicing and test pilots, race car drivers, oil and chemical plant workers, etc. Flame resistance for textiles used as clothing is determined by *Federal Test Method Standard 191A, Textile Test Methods, Method 5903.1, Flame Resistance of Cloth: Vertical. See also* **Fire-Resistant Clothing; Fire Retardant; Nomex.™**

Fire-Retardant Impregnated Wood—Wood and plywood products treated with fire-retardant materials to increase their level of fire resistance. Wood is commonly made fire retardant by a pressure-reated method (pressure-impregnated, fire-retardant treated lumber). The wood is placed in a vacuum chamber and all the moisture and sap is drawn out. Mineral salts (chromium-copper arsenate, mono-ammonium phosphate, diammonium phosphate, and zinc chloride) are then forced into the wood. The wood may be sometimes pricked to improve penetration of the mineral salts. The absorbed mineral salts essentially lower the flame spread, smoke developed, and fuel contributed ratings for the wood being treated. Pressure-treated wood is given a fire hazard classification (Interior Type A, Interior Type B, or Exterior Type) depending on moisture limitations. It has a flame spread index of 25 or less and no evidence of significant progressive combustion when tested for 30 minutes. The flame front must also not progress to more than 10.5 ft. (3.2 m) beyond the centerline of the burner. Wood impregnation and fire classifications are provided in American Wood Preservers' Association Standard C1, *All Timber Product-Preservative Treatment by Pressure Process* and NFPA 703, *Standard for Fire Retardant Impregnated Wood and Fire Retardant Coatings for Building Materials.* It may also be called pressure-treated fire rated wood. *See also* **Fire Retardant; Fire-Retardant Paint; Fire Test.**

Fire-Retardant Paint—An intumescent type of paint used to improve the fire resistance of a combustible surface. Protection by such paint that has been properly applied can prevent the actual ignition of cardboard. Intumescent paint does not intumesce effectively under approximately 400°F (204°C). *See also* **Fire Retardant; Fire-Retardant Impregnated Wood; Flame-Retardant Paint.**

Fire Riser—*See* **Riser.**

Fire Risk Analysis or **Fire Risk Assessment (FRA)**—A systematic study to identify potential fire hazards, the start and spread of fire, the generation of smoke, gases, or toxic fumes; the possibility of explosion or other hazardous occurrences; and to assess possible consequences arising from these hazards, the probability of those hazards occurring, and the need and type of protective measures. Quantitative probability techniques are usually employed. A fire risk analysis may be used to evaluate the cost/benefit of proposed fire protection measures where prescriptive fire protection requirements are not in effect. *See also* **Fire and Explosion Hazard Management (FEHM); Fire Hazard Analysis or Assessment (FHA).**

Firesafe—Trade terminology for something that has inherent resistance to fire exposure.

Firesafe Shutdown (FSSD)—The actions, components, capabilities, and design features necessary to achieve and maintain safe shutdown of a nuclear power generation reactor after a fire is detected in a specific fire area.

Fire Safety—Free from the risk of an unwanted fire event.

Fire Safety Concepts Tree—A comprehensive qualitative guide to fire safety. It uses a systems approach to fire safety. Instead of considering each feature of fire safety separately, the fire safety concepts tree examines all of them and demonstrates how they influence the achievement of fire safety goals and objectives. The tree components are concepts rather than events. It was developed as a branch of system safety applications by the defense and aerospace industries in the 1970s on the pattern of a logic tree, and indicates successful achievements rather than failures.

Fire (Safety) Symbols—Graphical symbolic representations provided on architectural/engineering drawings, investigation maps, public fire safety diagrams, or insurance surveys that represent fire protection devices, equipment, or systems. NFPA 170, *Standard for Fire Safety Symbols,* provides guidance in the use and application of fire safety symbols. *See also* **Fire Extinguisher Rating; Fire Tetrahedron; Fire Triangle; Foam Tetrahedron.**

Firesafe Valve—An isolation valve tested against specific fire conditions that maintains minimal leakage and operation capability. Several test standards are used within the industry to specify test requirements for a firesafe valve. The most widely known is API Standard 607, *Fire Test for Soft-Seated Quarter Turn Valves.* Firesafe valves are normally specified in the process industries for emergency isolation valves to ensure their operation if exposed to a fire event.

Fire Scenario—A description of a specific fire from ignition or established burning; to the maximum extent of growth, fire, and smoke effects; until its extinguishment. It includes ambient environmental conditions.

Fire Scene Reconstruction—The process of recreating the physical scene during fire scene analysis and investigation. It is accomplished by the removal of debris and the replacement of contents or structural building elements in their locations before the fire event.

Fire School—A class of instruction for fire prevention and fire protection. Most fire schools provide hands-on fire suppression training and related fire service activities. *See also* **Fire Protection Engineering; Fire Science.**

Fire Science—The knowledge and study of fire prevention and fire protection. The foundation of fire science is drawn from chemistry, physics, engineering, mathematics, computer science, historical knowledge, and management techniques. *See also* **Fire School.**

Fire Screen—A protective (and sometimes ornamental) screen in front of a fireplace to contain flying embers within the fireplace. The containment of embers to the inside of the fireplace prevents the spread of fire to combustible materials outside the fireplace. Firewood sometimes contains water, which vaporizes and expands within the fibers of the wood. The wood often breaks apart because of the water vaporization and expansion, causing small burning bits of wood or embers to fly out. These embers can easily spread a fireplace fire to the interior contents of a house or building if not contained. The fire screen was developed early in the 19th century and has been ornamented and shaped to serve decorative as well as functional purposes. *See also* **Ember; Spark Arrester.**

Fire Seal—*See* **Fire Stop.**

Fire Season—The portion of the year that predisposes a particular area to a higher risk from the outbreak and spread of wildfires due to climatic and vegetative conditions that readily support combustion. Primary features that influence the duration of a fire season include precipitation, vegetation moisture content, temperature, and winds. Forest fires seldom occur in tropical rain forests or in the deciduous broadleaved forests of the temperate zones. But all coniferous forests and the evergreen broadleaf trees of hot, dry zones frequently develop conditions ideally suited to the spread of fire through standing trees. For this to happen, both the air and the fuel must be dry, and the fuel must form an open matrix through which air, smoke, and the gases arising from combustion can quickly pass. Hot, sunny days with low air humidity and steady or strong breezes favor rapid fire spread. Some areas have a fire season only in dry summer months whereas others (southern California) can have a fire season nearly all year long. *See also* **Fire Tower.**

Fire Separation—A fire-resistive barrier to restrict the spread of fire, provided in a horizontal or vertical orientation. *See also* **Exposure Protection; Fire Barrier Wall; Firewall.**

Fire Service—Career or volunteer personnel, usually provided as part of a governmental administration, that are organized and trained for the prevention and control of loss of life and property from fire and other disasters. It may also refer to equipment or systems used for fire protection. *See also* **Fire Department (FD).**

F

Fire Service Response Time—The amount of time required for a local fire department to arrive at a fire incident after notification. Factors that influence response time include dispatch handling time, distance to the incident, apparatus condition, weather conditions, traffic conditions, and training.

Fire Severity—The maximum effects that can be caused by a fire event, usually described in terms of temperature and duration, and may be used to describe the potential for fire destruction for a particular location. The rate of heat release has also been accepted as a guide to fire severity.

Fire Shelter—A protective tent shelter used by wildland firefighters for protection against heat radiation in an emergency. A tent pack is carried on the belt, which is made of aluminized fabric that can be quickly unfolded. See Figure F-24. *See also* **Shelter-In-Place (SIP).**

Fire Shutter—A fire barrier used in an opening in a fire-rated partition that drops into place to protect it from the effects of fire at the time of exposure. Fire shutters can be of three general forms: a swinging door, a horizontally or vertically sliding door, or a rolling steel door of metal slats. They can be arranged to operate automatically through fire detection means such as signals from automatic fire detection systems, local fusible links mounted on the shutter assembly, or manually from pull cords.

Figure F-24 Fire shelter.

The fire rating of the fire shutter should meet the requirements for the fire rating of the partition for which it has been provided. *See also* **Flame Baffle.**

Fire Signature—A graph that indicates the amount of heat release per unit of time for a particular combustible material. The amount of heat released per unit of time varies in accordance with available combustible materials, arrangements of available combustibles, the availability of oxygen, and other factors. A fire signature determination is useful in arson investigations to determine if an accelerant has been used. See Figure F-25. *See also* **Fire Growth Curve.**

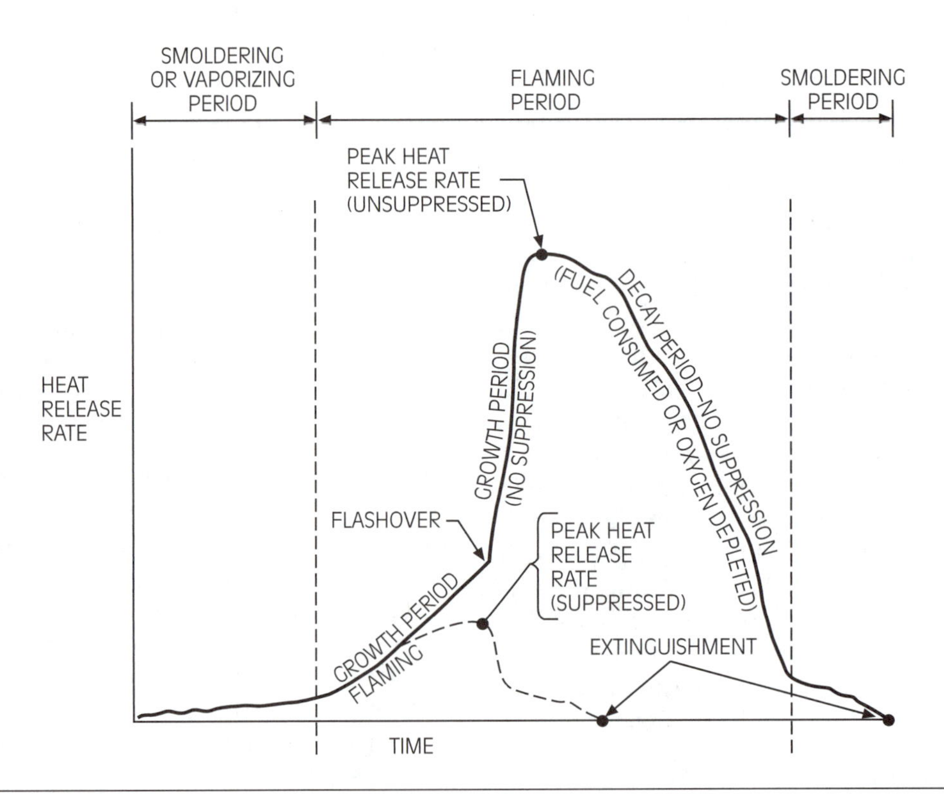

Figure F-25 Fire signature.

Fire Simulation—*See* **Burn Building; Fire Modeling.**

Fire Spread—The process of an advancing flame front though smoldering or flaming. The primary factors in fire spread are direct contact with flames or hot combustion gases, exposure to radiated heat, hot gases, or hot surfaces. Secondary causes include inadequate shutoff or separation of fuel supplies.

Fire Stages—Ordinary combustible fires (cellulosic) generally start and progress through four stages: incipient (precombustion), visible smoke (smoldering), flaming fire, and intense heat. Hydrocarbon-based fires normally do not have several stages to fire growth, but almost instantaneously achieve high flame temperatures and heat radiation after ignition. *See also* **Incipient Stage (Fire); Smoldering.**

Fire Stair—Any enclosed stairway that is part of a fire-resistant exitway and provides a path of escape from a fire exposure. *See also* **Exit; Exit, Vertical.**

Fire Station—A building designed for fire apparatus, equipment, and firefighters. It may have the capability to receive fire alarms from the areas it is intended to protect and also facilities for maintaining or repairing equipment, storing supplies, training firefighters, and performing related activities. The quantity of personnel, amount of equipment, and location of a fire station is primarily dependent on the fire risk that is present in the community it serves. Supplemental support services, such as training facilities, may also be provided. It may also be called a firehouse or firehall (Canada).

Fire Station Alerting System—An electronic communications system used to transmit emergency response information to the fire station personnel by voice or digital transmissions.

Fire Steamer—Name of a fire engine that used a steam-driven pump and was usually pulled by horses (some early models were pulled by hand) and is now obsolete. They were used from 1829 to the early 1900s, before internal combustion engines were applied to vehicular transport and as a prime mover for pumps. They consisted of three main components: a boiler (a large vertical stack), a steam chest (usually a rectangular box), and a water pump (commonly a square box with hose fittings). The boiler had two parts: a fire chamber in the lower portion and an upper section of boiler tubes or water pipes to heat water into steam. While in the fire station, the water in the pipes was kept warm from the station's heating stove. The fire chamber was provided with quick-starting coal for burning, which was lit as the steamer left the station for a fire. The engineer in charge of the steamer kept the boiler's draft wide open at the start of the fire and then regulated it once the boiler temperature was stabilized. While en route to the fire or while pumping, the engineer kept the boiler fire chamber stocked with coal. Flame and smoke heated the tubes in the top of the boiler stack and converted the water into low-pressure steam. The steam was collected in a manifold and then routed to the steam chest. The steam chest contained two large pistons that were driven downward by the incoming steam pressure. These pistons in turn drove other pistons in the water pump located below the steam chest via a system of connecting rods and a crankshaft. Water for firefighting was not carried by the steamer; instead, it was positioned in locations where water was available, such as in cisterns. Firefighting hoses were laid from the steamer to the fire location. Water was sucked into the pump by large drafting hoses that were carried on the steamer. Water was pressurized by the pump and forced from the ends of the firefighting hoses. A second-class steamer was capable of providing 600 gallons per minute (37.8 liters per second), a first-class steamer provided 900 gallons per minute (56.8 liters per second), and an extra-first-class steamer 1,100 gallons per minute (69.4 liters per second) at 100 psi (689 kPa).

The first steam-driven fire pump was introduced in London in 1829 by John Ericsson (1803–1889) and John Braithwaite (1797–1870). They were used in many large cities by the 1850s. Jon Braithwaite's 10 hp (7.5 kW), two-cylinder steam pump of 1829 delivered 150 gallons per minute (9.5 liters per second), was coal fired, and was drawn by a single horse. America's first fire engine arrived from London in 1679 after a major fire in Boston. The first steam engine was built in 1840 in the United States by Paul R. Hodge, an Englishman living in New York. His fire engine could also move by steam power. Most steam pumpers were equipped with reciprocating piston pumps, although a few rotary pumps were used. Some were self-propelled, but most were driven by horses, conserving steam pressure for the pump. Steam fire engines were used to fight the Chicago Fire of 1871. By 1925, the steam pumper had been almost completely replaced by motorized pumpers (internal combustion engines). A steam engine was kept in service with the New York Fire Department until 1932. Although thousands of steamers were made, only several hundred now survive and have been restored. Preserved fire steamers can be seen at most fire museums. *See also* **Fire Engine.**

Firestone—A fire-resistant stone, such as certain sandstones.

Firestop—Blocks of solid construction at regular intervals in concealed spaces to restrict passage of fire and smoke. Interior fires in buildings spread by breaching walls or floors. A fire can bypass these physical barriers through openings created for people, utilities, or through concealed spaces in walls, ceilings, attics, or under floors. Additionally, concealed penetrations or voids can act as flues through which a fire or smoke can spread undetected. Normal construction framing methods leave voids between members and membranes on each side of a wall. If these voids run the height of the building or connect with similar horizontal voids between the floor and ceiling, hot gases and flames can spread to areas far from the area of origin. Firestop systems consist of the fire-resistive assembly being penetrated, the penetrating item, and the firestop material. Firestop performance ratings are measured by the standard ANSI/UL 1479, *Fire Tests of Through Fire Penetration Firestops.* It may also be called a penetration seal or fire-rated penetration seal. *See also* **Fire-Rated Internal Conduit Seal; Firestopping; Penetration Seal; Unprotected Opening.**

Firestopping—The provision of a material to prevent the spread of fire through an opening in a fire-rated barrier. The material should afford the same level of fire resistance to the opening as provided by the fire barrier. Firestopping is commonly required around cable and pipe penetrations through fire barriers. Electrical bus bar duct penetrations may also be required to be firestopped internally due to the large cross-sectional area they represent. The performance of firestopping systems are measured by ANSI/UL 1479, *Fire Tests of Through Fire Penetration Firestops,* also known as ASTM E-814. The fire exposure used in these tests is the same as that used for the hourly fire ratings for floors and walls (ANSI/UL 263 or ASTM E-119). Ratings developed by this test evaluation are stated in terms of hours and identified as "F" and "T." The hour duration of T is always equal to or greater than an F rating. Both ratings prohibit the passage of flame. The acceptance criteria for T ratings also place limits on the temperature rise permitted on the penetrating items. Firestopping may also prevent the passage of smoke, toxic vapors, or water. Pressure gradients are created by a fire in a building due to heat generated from a fire. This pressure causes smoke and toxic vapors to be spread through gaps in building firewalls and floors that are left unsealed. *See also* **Firestop; Fire Resistance Rating; Fire Test; Penetration Seal; Smoke Management System (SMS).**

Firestorm—A violent convection caused by a continuous area of intense fire and characterized by destructively violent surface indrafts. Sometimes it is accompanied by tornado-like whirls (firewhirls) that develop as hot air from the burning fuel rises. Conflagration conditions can cause numerous buildings on fire to burn together as one fire, creating firestorm conditions. Such a fire is generally beyond human intervention and subsides only upon the consumption of everything combustible in the locality. Firestorms were purposely created during World War II by the aerial dropping of incendiary bombs for the mass destruction of cities. *See also* **Blowup Fire; Firewhirl.**

Fire Stream—*See* **Hose Stream.**

Fire Suppressant Agent—A material used to control or extinguish an unwanted fire. Water is the most common fire suppression agent, but may be unsuitable, unsafe, or ineffective for some types of fires. Other fire suppression agents commonly used include foamwater, carbon dioxide (CO_2), Halon, dry powders, and dry or wet chemicals. Their relative effectiveness is determined by the type of fire hazard and method of application. *See also* **Carbon Dioxide (CO_2) Fire Suppression Agent; Dry Chemical; Halon; Water; Wet Chemical.**

Fire Suppression—Firefighting activity concerned with controlling and reducing a fire prior to its actual extinguishment. Fire suppression is generally taken as the sharp reduction of the rate of heat release of a fire and the prevention of its regrowth. Fire extinguishment activities encompass the actual direct fire extinction process. *See also* **Fire Extinguishment.**

Fire Suppression Systems Association (FSSA)—A nonprofit trade association organized to share ideas and strategies for the benefit of the fire suppression industry. It is dedicated to the highest levels of safety, reliability, and effectiveness of special hazards fire suppression. Its objective is to be the recognized authority on the proper application, design, installation, and maintenance of fire systems employing existing and new technologies.

Fire Test—A simulation of specific fire conditions to investigate fire development, spread, extinguishment, material fire resistance, material flame spread, material smoke development, performance of fire protection devices, and similar studies concerning fire protection. Fire test conditions are varied to simulate various fuel loadings, different fuel geometry, heat inputs, and ambient conditions as required for the specific fire exposure test requirements. *See also* **Crib; Fire-Retardant Impregnated Wood; Firestopping; Hydrocarbon Fire Test; Standard Fire Test.**

Fire Test, Building Assemblies—A test designed to quantify the resistance to fire of elements of construction and/or materials used in construction in both time and type of fire exposure. Conventionally, such tests are performed in specifically built furnaces operating to a defined fire curve (heat input). A specific acceptance criterion is specified for the individual test methods. ASTM, UL, and NFPA all have similar test standards to classify fire resistance (UL 263, ASTM E-119, and NFPA 251). *See also* **Fire Resistance Rating; Firewall.**

Fire Test, Roof Coverings—A fire test intended to measure the relative fire characteristics of roof coverings under a simulated fire originating outside the building. NFPA 256, *Standard Methods of Fire Tests of Roof Coverings,* provides test methods for roof coverings applied in the United States. It includes (1) an intermittent flame exposure test, (2) a spread of flame test, (3) a burning brand test, (4) a flying brand test, (5) a rain test, and (6) a weathering test. Three classes of fire test exposure are prescribed: Class A, Class B, and Class C. A Class A test is applicable to roof coverings that are effective against severe test exposure, afford a high degree of fire protection to the roof deck, do not slip from position, and do not present a flying brand hazard. Class B tests are applicable to roof coverings that are effective against moderate test exposure, afford a moderate degree of fire protection to the roof deck, do not slip from position, and do not present a flying brand hazard. Class C tests are applicable to roof coverings that are effective against light test exposure, afford a light degree of fire protection to the roof deck, do not slip from position, and do not present a flying brand hazard. The test procedure was originally developed prior to 1920 by Underwriter's Laboratories, Inc., and was provided in a standard form by the American Society for Testing and Materials (ASTM) E5 Committee on fire standards. It was adopted by ASTM in 1955 and by NFPA in 1958. The contents of NFPA 256 is also published as ASTM E108, *Standard Test Methods for Fire Tests of Roof Coverings* and UL Standard for Safety No. 790, *Test for Fire Resistance of Roof Covering Materials.*

Fire Tetrahedron—A graphical, symbolic representation of the four factors needed for the propagation of combustion or fire. Each side of the tetrahedron is representative of a factor. The four factors include fuel, oxidizer, ignition source, and chain reaction. Removal or blockage of one of the elements prevents the combustion process from occurring or continuing. See Figure F-26. *See also* **Combustion; Fire (Safety) Symbols; Fire Triangle.**

Fire Tower—Commonly a term for an observation tower, provided mainly in forested areas, for an individual to watch for the occurrence or spread of undesired fires. Fire surveillance is essential during seasons of high risk. The tower affords a greater view of the surrounding area and possible smoke plumes. Information on the nature of the fire is related to fire control and suppression organizations. Fire towers are set on hilltops where observers equipped with binoculars, maps, and a direction scale determine the compass direction of smoke and notify the fire control base via telephone or radio. If a fire can be seen from two or more towers, its precise position is quickly determined by mapping the intersection of cross bearings. Tower lookouts are the mainstay of nearly all forest fire detection systems, although the use of aircraft and satellites has supplemented this system in countries with an advanced fire control program. Developments in remote-control television, high-resolution photography, heat-sensing devices, film, and radar make fire detection by aircraft and satellite more efficient and location more accurate. Satellites provide a rapid means of collecting and communicating highly precise information in fire detection, location, and appraisal. Fire towers were also established in large cities prior to

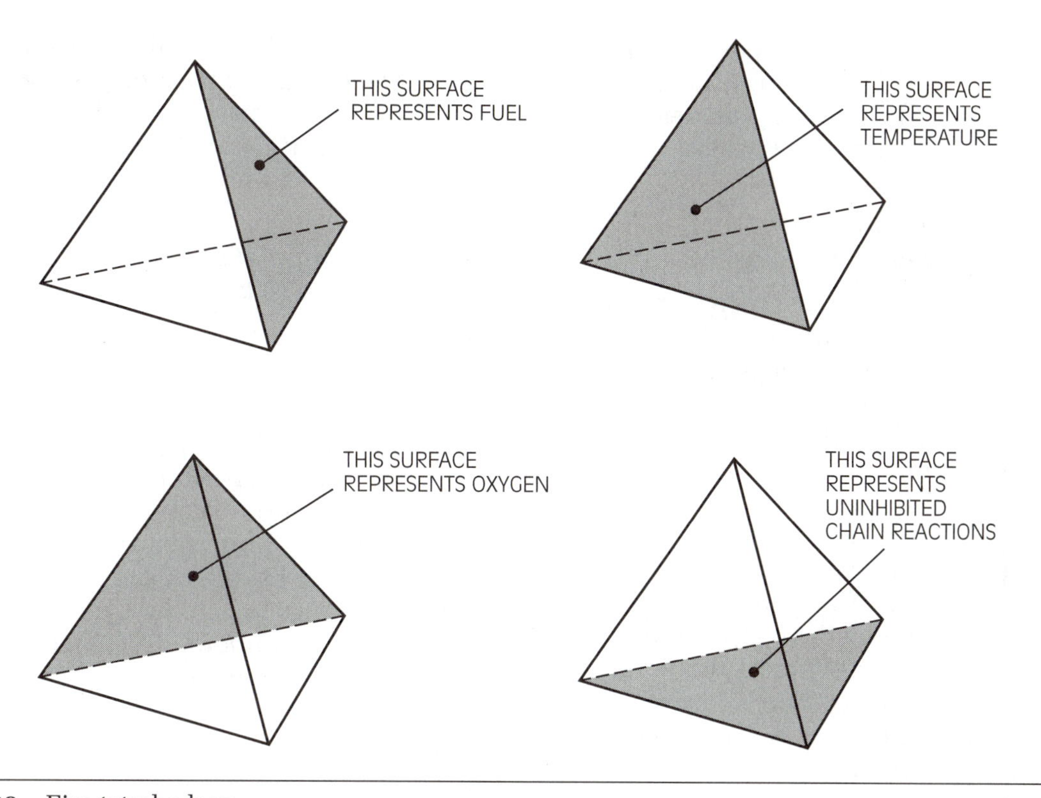

Figure F-26 Fire tetrahedron.

modern fire alarm and communication systems. Fire tower may also refer to a tower used by the fire service for training. *See also* **Fire Bell; Fire Lookout; Fire Season.**

Fire Training—In all industrial countries, firefighters undergo training beginning with an introductory firefighters school and continuing throughout a firefighter's career. In the United States, each municipal or county fire district maintains a fire training section and additional fire training is available from specialized schools, colleges, and institutions. Private fire brigades and personnel also receive fire training commensurate with their duties and responsibilities. *See also* **Fire Drill.**

Fire Trap—Common term (slang) for the condition of a place (including outdoor locations) or a building that due to its fire hazards can catch fire easily, is highly susceptible to fire, or would be difficult to escape from in case of a fire due to inadequate protective equipment, access to exits, or the lack of provision of exits. Therefore, such locations are considered a "trap" for an individual or the occupants during a fire condition.

Fire Trauma—A critical incident emotional stress reaction that may be experienced by firefighters or fire victims following a fire event that is considered devastating to the individual. Symptoms may appear immediately after the incident, hours later, or sometimes even days or weeks later, and may last for a few days, weeks, or months. *See also* **Critical Incident Stress Debriefing (CISD); Fire Injury; Post Incident Trauma.**

Fire Trench—A ditch dug into the earth to serve as a firebreak for ground fuel fires.

Fire Triangle—Geometric symbolic representation of the combustion process whereby each side of the triangle is one independent element of the process, namely, fuel (usually in vapor form), oxidizer, and ignition source (of sufficient energy and high temperature to initiate a combustion process). Removal of one element of the triangle stops the combustion process. Fuel must generally be in a vapor form for combustion. Liquid mist and finely divided particles that are readily converted to vapor have combustion characteristics much the same as vapors. Carbon and some metals and dusts are an exception that fuel must be in a vapor form for a combustion process. See Figure F-27. *See also* **Combustion; Fire (Safety) Symobls; Fire Tetrahedron.**

Fire Tub—Term used to describe the early cart hand pumps or "engines" used in the late 1600s and 1700s. They were called "tubs" because a water tank shaped like a tub was provided at the base of the cart that provided the water using a hand piston suction pump. The tub was later replaced on hand pumpers by a suction hose that was lowered into cistern, lake, or river.

Fire Valve—A valve designed to automatically close off the supply of fuel to an oil-fired boiler in case of fire. *See also* **Fusible Valve.**

Fire Venting—*See* **Venting, Fire.**

Firewall—A designated partition, which by nature of its construction and certification status, is warranted to resist a cellulosic or hydrocarbon fire test for a particular time period and is able to withstand structural collapse on either side and remain standing. Firewalls are used prevent the spread of fire or its effects and are provided fire-resistance ratings to standardized fire exposures (ASTM E119, *Standard Test Methods for Fire Tests of Building Construction and Materials;* UL 263, *Fire Tests of Building Construction and Materials;* NFPA 251, *Standard Methods of Fire Tests of Building Construction and Materials;* and UL 1709, *Rapid Rise Fire Tests of Protection Materials for Structural Steel*) and standard test periods (20, 30, 45, 60 90, 120, and 180 minutes). *See also* **Fire Barrier; Fire Barrier Wall; Fire Separation; Fire Test, Building Assemblies.**

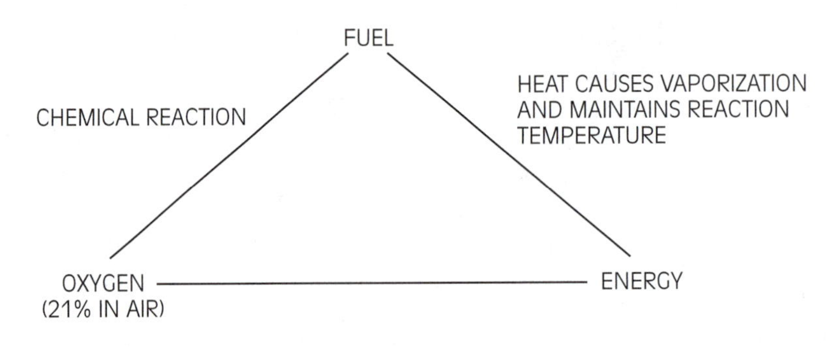

Figure F-27 Fire triangle.

Fire Warden—A building staff member or tenant trained to perform assigned emergency duties in case of a fire incident, or an individual assigned the responsibility of prevention and suppression of wildland fires. Fire wardens were first created early in the 1700s to patrol city streets, looking for unwanted fires. They sometimes were provided with staffs that had an ornamental flame attached to the end.

Fire Warp—Wire rope or other cable made of fireproof materials and of sufficient strength to tow a marine vessel in the event of a fire. A fire warp is normally hung from the forward and aft ends of the vessel at a position for convenient retrieval by a towing vessel.

Fire Watch—The appointment of an individual to remain in an area unprotected by normal protective systems or that has temporarily increased its fire hazards (welding operations). A fire watch individual observes an area for fires, notifies the fire department of an emergency, prevents a fire from occurring, extinguishes small fires, or protects the public from fire or life safety dangers. Specific requirements for fire watch personnel, training, and any equipment must be defined prior to the assignment of any individual. *See also* **Fireguard.**

Fire Watcher—*See* **Fire Watch.**

Firewater—Water used or reserved and stored solely for the purpose of fire protection activities (cooling, control, suppression). Firewater supplies can be from public or private, fresh or saltwater, potable or nonpotable sources. It should not contain particulates or marine invertebrates that would plug or cause deterioration of the firewater distribution system. The characteristics of firewater sources determines the specification of materials used for fixed firewater systems. *See also* **Water; Wetting Agent.**

Firewater Distribution System—A piping network used and specified exclusively for the provision of firewater to transport the water from a supply source to application devices or systems. In order to be useful, the firewater distribution system must have adequate capacity and high reliability. The first municipal firewater distribution systems were made of hollow wooden logs. The first water distribution network in America was installed in Bethlehem, Pennsylvania in 1755. Water was pumped into a water tower through a series of hollow hemlock logs fashioned into pipes. Wooden logs, however, tend to rot, cannot hold much pressure, and held back water distribution technology. Screwed cast iron pipes eventually replaced wood in the late 1700s and early 1800s. Steel pipes were later used, and in the 1920s, steel pipes with welded joints made it possible to construct a distribution system that was leaktight, large in diameter, and could withstand high pressures. Today, reinforced fiberglass piping is commonly used due to its corrosion-resistant properties. *See also* **Water Distribution System.**

Firewater Pump—A designated pumping system designed to supply adequate water volume and pressure to fire incidents. Firewater pumps are normally are required to be designed and built to nationally recognized standards to meet reliability and functionality requirements. In America, NFPA 20, *Installation of Centrifugal Fire Pumps,* is commonly referred to. *See also* **Fire Pump.**

Firewater Pump, Booster—A firewater pump is provided where additional pressure is required on the firewater distribution system. They are usually provided where firewater supply system pressures are less than required (although the quantity may be adequate) or where there are extreme elevations in the facility being protected that are beyond the capability of the regular firewater pumps (high-rise buildings). The booster pumps provide the additional pressure required for fire protection activities at such a location. Booster firewater pumps meet the same criteria as regular firewater pumps. Firewater pumps performing suction activities from public water supplies for storage tanks may be technically considered booster firewater pumps where the quantity of water has been considered adequate. Mobile firetrucks or even fireboats, where suitable arrangements have been provided, may serve to boost the pressure in a firewater distribution network, in an emergency. Firetrucks may sometimes have a small (less than 500 gallons per minute/2.2 liters per second) power take-off pump, used with a small diameter fire hose line (0.75 or 1.0 in./1.9 or 2.54 cm) that is referred to as a booster pump. *See also* **Booster Hose or Line; Fire Pump.**

F

Firewater Pump Unit—An assembled unit consisting of a fire pump, driver, controller, and accessories. Firewater pumps are installed per the requirements of locally adopted fire codes and normally, NFPA 20, *Installation of Centrifugal Fire Pumps,* is referenced. *See also* **Fire Pump.**

Firewater Reliability Analysis (FRA)—A mathematical analysis of a firewater supply system to determine its capacity to supply adequate water on demand to meet fire protection requirements without a component failure, that is, the mean time between failure (MTBF) analysis.

Firewater System (FWS)—A water delivery system consisting of a water supply and distribution network to provide firefighting water for the control and suppression of fire incidents and for exposure cooling requirements.

Fire Weather—Ambient conditions that reduce moisture content in surface vegetation and make the fire potential for the region high (summer months). Major changes in vegetation moisture are generally seasonal in nature. Vegetative moisture content is influenced by precipitation, air moisture, air and surface temperatures, wind, and cloudiness as well as fuel factors, which include surface to volume ratios, compactness, and arrangement. *See also* **Fire Climate.**

Firewhirl—The heat generated by a fire produces instability in the lower air level that causes updrafts to form. These updrafts may form into upward spirals containing a mass of flame called firewhirls. Firewhirls may twist off trees or carry embers to other areas, causing hot spots. If a firewhirl moves out of the main fire area, it loses its fire and becomes a normal whirlwind. They frequently occur where heavy concentrations of fuel are burning and a large amount of heat is being released in a small area. Mechanical forces also may be present to cause a firewhirl to form. The favored area for firewhirl formation is the lee side of a ridge where heated air from a fire is sheltered from the general winds. It may also be called a fire devil. *See also* **Blowup Fire; Fire Devil; Firestorm.**

Fire Window Assembly—A window or glass block assembly having a fire resistance or protection rating. The assembly includes all components that support the window installation. Fire windows are tested in accordance with NFPA 80, *Standard for Fire Doors and Fire Windows.* Fire windows were originally designed to protect openings in exterior walls. Therefore, radiant heat transfer was not a significant consideration, since the main function of fire windows was to contain the flames from a fire within the building. *See also* **Fire Barrier; Tempered Glass.**

Fireworks—Incendiary devices or materials used for signaling or entertainment by combustion, deflagration, or detonation. They are sometimes known as pyrotechnics. Fireworks include substances or devices that produce, when ignited or activated, sound, smoke, motion, or a combination of these. Military flares and smoke devices are also considered fireworks. Most fireworks are composed of potassium nitrate (saltpeter), which supplies oxygen, and substances such as charcoal and sulfur that combine with the oxygen, which produce heat and light. After 1800, potassium chlorate was substituted for some or all of the potassium nitrate in the mixture. A large number of flammable substances, such as starch, gums, sugar, shellac, and various petroleum derivatives, are also used in the fireworks mixture in place of charcoal and sulfur. Color is produced in a fireworks display by incorporating compounds of various metals. By the mid-19th century, fireworks became popular in America for the celebration of major events. Numerous injuries associated with fireworks, particularly to children, eventually discouraged their unrestricted use. As a result, most US states and locations in Canada restrict the sale of certain types and sizes of fireworks. Guidelines for fire safety of fireworks are provided in NFPA 1123, *Code for Fireworks Display,* and NFPA 1124, *Code for the Manufacture, Transportation, and Storage of Fireworks. See also* **Pyrotechnics.**

First Aid Firefighting Equipment—Portable fire extinguishers, water pails, small hose lines, and 1.5 in. (3.8 cm) standpipe hoses used in the incipient stage of a fire or in relatively small fire incidents that can be controlled and extinguished by their use.

First Aid Standpipes or **Hoses**—Fire hoses generally 1.5 in. (3.8 cm.) in diameter that are provided for use in the incipient stage of a fire or relatively small fire incidents. They are capable of controlling and extinguishing the fire incident on their own. *See also* **Standpipe System.**

First Alarm—The initial area of a fire alarm or the first signal requesting a first alarm response for a municipal fire department.

First Due—The pumper, ladder truck, and fire officer given the first alarm assignment for a municipal fire department.

First Responders—A group designated by a community as those who will first respond to an incident. It is usually composed of local police, emergency medical service providers, and the fire services.

F

Fixed Fire-Suppression System—A fire-suppression system that provides local application, area coverage, or total flooding protection. It consists of a fixed supply of extinguishing agent, permanently connected distribution piping, and fixed nozzles that are arranged to discharge an extinguishing agent into an enclosure (total flooding), directly onto a hazard (local application), over an entire area (area coverage), or a combination of applications. *See also* **Dry Chemical Fire Extinguishing System.**

Fixed Foam Application Systems—A permanent fire protection system for the control and extinguishment of combustible liquids. It delivers a solution of foam and water that may or may not be aspirated. The system consists of foam concentrate storage, a water delivery system, a foam proportioner system, and application devices that may aspirate the foam. *See also* **Eductor; Foam; Proportioning.**

Fixed Temperature Detector—A device that responds (signals) when its operating element becomes heated to a predetermined temperature level, hence the fixed temperature nomenclature. Fixed temperature detectors are used above the general location of probable fire occurrence and are considered spot-type detectors. They are also used where smoke-type detectors may cause a false activation, such as in kitchens or garages, from particulate or vapors as part of the ongoing operation. The following types are commonly employed:

Bimetallic—A sensing element composed of two different metals having different coefficients of thermal expansion. When heated, it deflects (bends) in one direction. When cooled, it deflects in the opposite direction.

Electrical Conductivity—A line-type or spot-type sensing element. Its electrical resistance varies as a function of temperature.

Fusible Alloy—A sensing element of a special composition (eutectic) metal that melts rapidly at the specified temperature.

Fusible Tubing—Plastic tubing pressurized with an inert gas or compressed air. It melts when a specified temperature is reached, causing a low pressure switch on the system to activate.

Heat-Sensitive Cable—A line-type wire whose sensing element comprises, in one type, two current-carrying wires separated by heat-sensitive insulation. It softens at a specified temperature, thus allowing the wires to make electrical contact. In another type, a single wire is centered in a metallic tube, and the intervening space is filled with a substance that becomes conductive at a critical temperature. During a fire, electrical contact is made between the tube and the wire.

Liquid Expansion—A sensing element consisting of a liquid capable of marked expansion in volume in response to temperature increase.

See also **Fusible Element; Heat Detector; Thermal Element; Thermistor.**

Flambeau—A hand-carried torch used to illuminate city streets for the transport of fire engines and hose reels before the advent of electrical lighting. It consisted of an inverted, cone-shaped tin container packed with tallow and topped by bunched waxed wicks. Flambeaus are now obsolete. They may be found on display in museums or are collected by fire buffs.

Flame—Typically, the glowing gaseous portion of a fire. A flame is a body or stream of gaseous material involved in the combustion process. It emits radiant energy at specific wavelength bands depending on the combustion chemistry of the fuel involved. In most cases of combustion, some portion of the emitted radiant energy is visible to the human eye and may be seen as a flame. Flames generally consist of a mixture of oxygen (air) and another gas, usually such combustible substances as hydrogen, carbon monoxide, or hydrocarbon. The brightest flames are not always the hottest. Hydrogen, which combines with oxygen when burning to form water, has an almost invisible flame even under ordinary circumstances. When hydrogen is absolutely pure and the air around it is completely free of dust, the hydrogen flame cannot be seen even in a dark room.

A typical flame is a burning candle. When the candle is lighted, the heat of the match melts the wax, which is carried up the wick and is then vaporized by the heat. The vaporized wax is then broken down by the heat and, finally, combines with the oxygen of the surrounding air, producing a flame and generating heat and light. The candle flame consists of three zones that are easily distinguished. The innermost zone, a nonluminous cone, is composed of a gas-air mixture at a comparatively low temperature. In the second, luminous cone, hydrogen and carbon monoxide are produced by decomposition and begin to react with oxygen to form water and carbon dioxide (CO_2), respectively. In this cone, the temperature of the flame—about 1,090°F to 1,250°F (590°C to 680°C)—is great enough to dissociate the gases in the flame and produce free particles of carbon. These particles are heated to incandescence and are then consumed. The incandescent carbon produces the characteristic yellow light of this portion of the flame. Outside the luminous cone is a third, invisible cone in which the remaining carbon monoxide and hydrogen are finally consumed. If a cold object is introduced into the outer portions of a flame, the temperature of that part of the flame will be lowered below the point of combustion, and unburned carbon and carbon monoxide will be given off. Thus, if a porcelain dish is passed through a candle flame, it will receive a deposit of carbon in the form of soot. Operation of any kind of flame-producing stove in a room that is unventilated is dangerous because of the production of carbon monoxide, which is poisonous.

All combustible substances require a definite proportion of oxygen for complete burning (a flame can be sustained in an atmosphere of pure chlorine, although combustion is not complete). In the burning of a candle, or of solids such as wood or coal, this oxygen is supplied by the surrounding atmosphere. In gas burners, air or pure oxygen is mixed with the gas at the base of the burner so that the carbon is consumed almost instantaneously at the mouth of the burner. For this reason, such flames are not luminous. They also occupy a smaller volume and are proportionately hotter than a simple candle flame. The hottest portion of the flame of a Bunsen burner has a temperature of about 2,910°F (about 1,600°C). The hottest portion of the oxygen-acetylene flames used for welding metals reaches 6,330°F (3,500°C); such flames have a bluish-green cone in place of the luminous cone. If the oxygen supply is reduced, such flames have four cones: nonluminous, bluish-green, luminous, and invisible.

The blue-green cone of any flame is often called the reducing cone, because it is insufficiently supplied with oxygen and will take up oxygen from substances placed within it. Similarly, the outermost cone, which has an excess of oxygen, is called the oxidizing cone. Intensive studies of the molecular processes taking place in various regions of flames are now possible through the techniques of laser spectroscopy. *See also* **Combustion; Fire.**

F

Flame Arrester—A device used in vents that may contain combustible vapors to prevent flashback (a flame front) into the vent or vent source, if the vapors from the vent were to be ignited. Three types of flame arresters are used in the process industries: thermal-type, water seal-type, and velocity-type. The most common is the thermal type. It consists of an assembly of perforated plates, slots, or screens enclosed in a case and attached to the outlets of the possible combustible vapor releases (breather vents on petroleum storage tanks). The openings allow the passage of vapors or gases but cause heat to be removed from a flame to prevent its propagation past the arrester. When burning occurs within a pipe, some of the heat of combustion is absorbed by the pipe wall. As the pipe diameter decreases, an increasing percentage of the total heat is absorbed by the pipe wall and the flame speed decreases. By using very small diameter (one or two mm) pipes, it is possible to prevent the passage of flame regardless of speed. The Davy miners lamp was the first use of a flame arrester in which a fine mesh screen of high heat absorption properties was placed in front of the flame of the miner lamp to prevent the ignition of methane gas in coal mines. ANSI/UL 255, *Flame Arresters for Use on Vents for Storage Tanks for Petroleum Oil and Gasoline,* specifies requirements for the thermal-type of flame arresters. Thermal-type flame arresters may become clogged with dirt, oily residue or corrosion if made of unsuitable materials. Frequent inspections and cleaning should be performed for flame arrester installations. The water-type flame arrester uses a water seal in which the gas may safely bubble through the liquid. The gas inlet connects to a distribution pipe provided with small openings located below the surface of the liquid of the water seal. The outlet is at the top of the enclosure. Seals of this type are commonly provided at the base of a flare stack. The velocity flame arrester employs a method in which the flow of the process stream is maintained at a higher rate than the velocity of flame propagation through the material. All flame arresters must be designed and installed to the conditions for which they have been tested and approved. May also be referred to as a flame trap or flash arrester. *See also* **Detonation (Flame) Arrester; Safety Lamp** or **Davy Lamp.**

Flame Baffle—A mechanism within a fire shutter assembly that provides a fire-rated barrier along the seam of the shutter hood. It is usually a hinged piece of sheet metal within the hood of a shutter (rolling door), which, when released, closes the space between the top of the interlocking metal slats that form the "curtain" and the hood. *See also* **Fire Shutter.**

Flame Blowout—Use of high velocity air pressure to overcome flame propagation such as blowing out a candle. It is achieved when the ambient air velocity exceeds the flame velocity. Oil well blowout fires are commonly extinguished by the detonation of a high explosive charge that blows out the wellhead's high-pressure flame with a higher-pressure blast wave that separates the flame from the available combustible gas. Well-capping methods are then immediately implemented while keeping the area cool to prevent re-ignition. The use of explosives (dynamite) to blow out well fires originated in 1913 in Taft, California by Karl Kinley. The high thrust output of jet engines have also been employed to blow out wellhead fires in the petroleum industry (Gulf War, Russian oil industry).

Flame Cooling—Removing heat or cooling a flame absorbs the propagating energy of the combustion reaction. When the fuel source temperature is lowered below its ignition temperature, fire extinguishment results. Because heat is continually being released from a combustion process in the form of radiation, conduction, or convection, only a small amount of heat absorption material is necessary for fire extinguishment by cooling.

Flame Curtain—The arrangement of gas burners in a furnace that form a "curtain of fire" and are therefore called curtain burners.

Flame Detection—The verification of a fire through the observance of a flame. Flame detection relies on the detection of the radiant energy from a fire and may be unreliable for smoldering fires or fires obscured by smoke that prevents the observation of radiant energy.

Flame Detector—A radiant energy-sensing fire detector that detects the radiant energy emitted by a flame (infrared, ultraviolet, or visible radiation). These types of detectors are suitable for special applications such as large open or high ceilinged areas (warehouses, lumber yards), or critical areas where a fire may spread very rapidly. Common locations include valves and pumps handling combustible liquids, high air movement areas, etc. It may also be called a flicker detector due to the brief glimpses of radiation energy from a fire event. To prevent false alarms, the sensitivity can be set for the detection of flickering flame for various time intervals, such as 3-, 10-, or 30-second duration, depending on the arrangement of the system. *See also* **Flicker Detector.**

Flame Front—The advancing flame of burning gases in a combustion reaction. *See also* **Flame Speed.**

Flame Front Diverter—A device for diverting a flame front to atmosphere from an enclosure. It senses a pressure wave preceding a flame front of a deflagration and opens a vent to atmosphere.

Flame Front Extinguishing System—A fire extinguishing system that is designed and arranged specifically to suppress propagation of a flame front.

Flame Front Generator—A device that provides a piloted ignition source. Primarily used in the process industries for flare ignition.

Flame Impingement—Flame that directly impacts a surface. Direct flame impingement transfers large quantities of heat. For pressurized vessels containing combustible materials, the occurrence of direct flame impingement may rapidly lead to vessel rupture and boiling liquid expanding vapor explosion (BLEVE) conditions. *See also* **Boiling Liquid Expanding Vapor Explosion (BLEVE).**

Flameout—The condition at which the last portions of flame or glow disappear from the surface of a material. It is commonly noted in material fire testing evaluations for flammability characteristics. It may also refer to a combustion chamber or flaring system where the flame is inadvertently extinguished.

Flameover—The condition where flames only propagate through or across the ceiling layer and target fuel surfaces are not involved. It is also referred to as rollover. In the cotton industry, it is used to describe a fire that spreads rapidly over the exposed linty surface of the cotton bales, where the common term is "flashover," and has the same meaning.

Flameproof—Resistant to the effects of flame exposure (burning), usually by a chemical treatment; however, its use is misleading and the terms flame retardant and flame resistant are preferred instead. The British petrochemical industry uses the term flameproof to classify electrical apparatus for use in atmospheres subject to the release of combustible vapors (hazardous areas) and in this fashion it is used synonymously with the American term "explosionproof" electrical equipment. *See also* **Classified Area.**

Flameproofing—Surface treating or impregnation of wood products and textiles or products with fire-retardant chemicals. It consists of adding metallic salts to delicate fabrics, such as silks, to provide more body. Flameproofing is an archaic term for the description of combustible materials and has generally been replaced by the terms fire resistant or fire retardant to accurately describe the flammability or burning characteristics of the material. In the United States, the Consumer Products Safety Commission (CPSC) is responsible for enforcing the Flammable Fabrics Act (1953), which requires fabrics to meet standards of fire resistance. *See also* **Flammable Fabrics Act.**

Flame Propagation—Flame propagation occurs by two methods: heat conduction and diffusion. In heat conduction, heat flows from the flame front, the area in a flame in which combustion occurs, to the inner cone, the area containing the unburned mixture of fuel and air. When the unburned mixture is heated to its ignition temperature, it combusts in the flame front, and heat from that reaction again flows to the inner cone, thus creating a self-propagating cycle. In diffusion, a similar cycle begins when reactive molecules produced in the flame front diffuse into the inner cone and ignite the mixture. A mixture can support a flame only above some minimum and below some maximum percentage of fuel gas. These percentages are called the lower explosive limit and upper explosive limit (LEL and UEL). A mixture of natural gas and air, for example, will not propagate flame if the proportion of gas is less than about 4 percent or more than about 15 percent.

Flame Propagation Rate—The velocity at which the combustion front travels through a gas or vapor or over the surface of a solid or liquid.

Flame Resistant—A term indicating the ability of a material to withstand the effects of a flame exposure for a limited period. Specific flame-resistant capabilities depend on the properties of the material involved and the duration of flame exposure. Various standardized tests have been developed to denote a flame-resistant material. Draperies, curtains, and other similar furnishings and decorations are generally required to be classified as flame resistant. They are tested in accordance with NFPA 701, *Standard Methods of Fire Tests for Flame-Resistant Textiles and Films.* Many flame-resistant synthetic materials soften and melt when exposed to heat and fire. They also can be subject to twisting, shrinking, dripping, and elongation under fire conditions. *See also* **Fire Resistant; Fire-Resistant Clothing; Fire Resistive.**

Flame Retardant—Material that is constructed or treated so that it will not support or convey flame, or a substance applied to a combustible material to retard burning. Materials rich in oxygen, such as wood or paper, may be treated with compounds of inorganic acids together with a neutralizing substance that vaporizes if fire starts so that the acids produce carbon, steam, and carbon dioxide (CO_2) to smother the fire. Borates, which may also be used, melt at low temperatures to retard oxygen diffusion. Oxygen-poor materials may be treated with bromine compounds or antimony oxide to yield chemicals that block combustion by contributing to the smothering vapors above the fire. *See also* **Flame-Retardant Paint.**

Flame-Retardant Paint—Paint of normal decorative appearance but containing added flame retardant compounds. Antimony oxide and chlorinated compounds are used in oil or emulsion paints and calcium carbonate is used in chlorinated rubber paint. Flame-retardant paints are usually applied to timber and wooden building materials. The exact thickness and substrate to be provided for the application are determined by a certified fire test laboratory. *See also* **Fire-Retardant Paint; Flame Retardant.**

Flame Rod—A flame-sensing instrument primarily used for combustion chambers. It consists of an insulated rod that is electrically conductive and temperature resistant. It is located in the flame area to be supervised and is impressed with voltage between the rod and a ground connected to the fuel nozzle or burner. A current is thereby passed through the flame, which is rectified and detected by a combustion safeguard system. *See also* **Flame, Supervised.**

Flame Speed—The speed of a flame front. It is measured from a fixed reference point and is dependent on flame turbulence, the geometry of the surroundings, and the fundamental burning velocity of the material involved. Flame speed determines the possibility of the generation a blast pressure wave from an explosion. A combustion reaction is classified as a deflagration or detonation, based on the velocity of the flame front propagation through a fuel. Deflagration is designated for flame fronts traveling at less than the speed of sound, and detonation is the term if the flame front travels faster than the speed of sound. *See also* **Burning Velocity; Deflagration; Detonation; Flame Front; Fundamental Burning Velocity.**

Flame Spread—The increase in the perimeter of a fire. Flame spread depends on orientation and the surrounding fluid flow. It can be associated with solids, liquids, forest fuels, smoldering, and gas-phase propagation for premixed systems. Flame spread is influenced by gravity (flame buoyancy) and wind effects. Flame spread on liquid fuel surfaces is enhanced over that of solids due to flows arising from surface tension and natural convection effects. Steady flame spread can be simply expressed as the flame heat flow required to raise the condensed phase from its initial temperature to its ignition temperature.

Relative flame spread speeds are indicated below:

Phenomenon	**(cm/s)**
Smoldering	10^{-3} to 10^{-2}
Downward or horizontal spread (thick solids)	10^{-1}
Upward spread (thick solids)	1 to 10^{2}
Wind-driven spread through forest debris	1 to 30
Horizontal spread on liquids	1 to 10^{2}
Laminar deflagration	10 to 10^{2}
Detonation	~10^{5}

Several empirical flame spread tests are available for comparison of the relative fire risk of materials and are referenced in various fire codes and standards. These include ASTM E-84, *Standard Test Method of Surface Burning Characteristics of Building Materials;* ASTM E-162, *Standard Test Method of Surface Flammability of Materials Using a Radiant Heat Energy Source;* E-648, *Standard Test Method for Critical Radiant Flux of Floor Covering Systems Using a Radiant Heat Source;* and E-1321, *Standard Test Method for Determining Material Ignition and Flame Spread Properties.*

Flame Spread Index (FSI)—A relative performance of fire travel over the surface of a material when tested in accordance with the provisions of NFPA 255, *Standard Method of Test of Surface Burning Characteristics of Building Materials;* ASTM E-84, *Standard Method of Test of Surface Burning Characteristics of Building Materials;* UL 723, *Standard Test Method for Surface Flammability of Materials,* or the Steiner Tunnel Test. The test was developed by A. J. Steiner at Underwriter's Laboratories after World War I. Research on the test was increased after the Coconut Grove Fire in Boston in 1942 in which 492 people were killed and the interior combustible materials were cited as a contributing factor. The method was recognized by ASTM in 1950 and by NFPA in 1955, after three fatal hotel fires (with the loss of 199 lives) in 1946 that cited interior finish as a contributing factor. In the fire test, a building material specimen, about 25 ft. (7.62 m) by 20 in. (0.5 m) is mounted to form the roof of a tunnel about 1 ft. (0.3 m) high. Two gas burners at the front of the tunnel provide a controlled heat source. The flame is developed by burning methane at a rate of 5,000 Btu/min. (5,274 kJ/min.) with an air velocity of 240 ft./min. (72 m/min.). It is designed to provide a moderately severe exposure of approximately 1,400°F (760°C) in the area of flame impingement. The tunnel is calibrated using inorganic reinforced cement board and red oak. The cement board is assigned a flame spread index of zero and red oak of 100. The surface flammability of the test sample is reported in relative terms against the performance of reinforced cement board and red oak. Smoke developed is also measured during this 10-minute test in a relative manner between the performance of reinforced cement board and red oak. The test requires that coatings intended for application to combustible surfaces be tested when applied to that surface. Coatings intended for application to any wood surface may be tested when applied to Douglas fir and coatings for application to noncombustible surfaces are to be tested when applied to 0.25 in. (0.75 cm) of inorganic reinforced cement board.

Building codes have established major performance levels for the surface flammability levels of building materials. These flame spread indexes are 0 to 25 (Class A or Class I), 26 to 75 (Class B or Class II), and 76 to 200 (Class C or Class III). Class D (201 to 500) and Class E (over 500) are also registered by the E-84 test method. Requirements for smoke are typically limited by building codes for materials such as plastics and for air handling systems. It may also be called a flame spread rating. *See also* **Fire Performance; Fuel Contributed Rating; Radiant Panel Test; Smoke Density; Smoke Developed Rating.**

Flame Spread Rating (FSR)—*See* **Flame Spread Index (FSI).**

Flame, Supervised—A flame whose existence is monitored by a flame sensor and is connected to a combustion safeguard system. *See also* **Flame Rod.**

Flame Temperature—Most open flames of any type produce a flame temperature in the region of 2,000°F (1,093°C). The hottest burning substance is carbon subnitride (C_4N_2), which at one atmospheric pressure can produce a flame calculated to reach 9,010°F (4,988°C*). See also* **Adiabatic Flame Temperature.**

Flame Thrower—A device that sprays ignited incendiary fuel for some distance. Normally intended as a weapon.

Flame Trap—*See* **Flame Arrester.**

Flaming Combustion—Combustion in the gaseous phase, usually with the emission of light.

Flaming Debris/Flaming Droplets—Material separating from a burning item during a fire and continuing to flame.

Flammable—In a general sense, it refers to any material that is easily ignited and burns rapidly. Numerous testing methods are available for testing the flammability of materials, therefore one has to know or specify the test standard that should be required or applied to determine the specific flammability of a material. A universal test standard has not been adopted to define the flammability of materials. It is synonymous with the term inflammable, which is now generally considered obsolete due to its prefix "in," which may be incorrectly misunderstood as not flammable (for example, incomplete is not complete). *See also* **Flammable Gas; Inflammable.**

Flammable and Combustible Liquids Code—A fire code published by NFPA that identifies fire safety requirements for the storage, handling, and use of flammable and combustible liquids, including waste liquids. NFPA's Flammable Liquids Committee was founded in 1917. The code was originally written as a model municipal ordinance known as the *Suggested Ordinance for the Storage, Handling, and Use of Flammable Liquids,* and was used in this format from 1913 until 1957. NFPA adopted it as a fire code in 1957. Previous to NFPA's involvement, the National Board of Fire Underwriters undertook technical work of fire hazards of oils and gases.

Flammable Fabrics Act—The Flammable Fabrics Act of 1953 established safety standards with respect to the flammability of fabrics used for wearing apparel. The act was broadened in 1967 to include fabrics used in interior furnishings. In 1973, administration of the act was transferred to the newly created Consumer Product Safety Commission (CPSC). The CPSC is responsible for enforcing the Flammable Fabrics Act (1953), which requires fabrics to meet standards of fire resistance. *See also* **Flameproofing.**

Flammable Gas—A gas that is able to support combustion as a fuel source. All such gases are generically classified as combustible, but identified as "flammable" to emphasize their greater fire risk. *See also* **Combustible; Flammable; Gas, Combustible.**

Flammable Limit or **Flammability Limits**—The range of air-fuel concentrations that can propagate combustion, such as laminar and turbulent flames. Each material has its own flammability limits that are defined by the lower flammable limit (LFL) and upper flammable limits (UFL). The range of flammability limits can be affected by temperatures and pressures above ambient conditions. Flammable limits as tested in the laboratory are also influenced by the surface to volume ratio of the test vessel, direction of air flow, and velocity of the air flow. ASTM E 681, *Standard Test Method for Concentration Limits of Flammability of Chemicals,* is the current standard used in the United States to determine flammability limits. It may also be called concentration limits of flammability or explosive limits. *See also* **Blanketing (or Padding), Gas; Concentration Limits of Flammability; Detonation Limits; Explosive Limits; Flash Point (FP); Lower Flammable Limit (LFL); Upper Flammable Limit (UFL).**

Flammable Liquid—As defined by most fire safety codes (NFPA 30, *Flammable and Combustible Liquids),* generally a flammable liquid is any liquid that has a closed-cup flash point below 100°F (37.8°C). Flash points are determined by procedures and apparatus set forth in ASTM D-56, *Standard Method of Test for Flash Point by the Tag Closed Tester.*

Flammable liquids are classified as Class I per the following criteria:

> *Class I Flammable Liquid*—Any liquid that has a defined closed-cup flash point below 100°F (37.8°C) and a Reid vapor pressure not exceeding 40 psi (2,068.6 mm Hg) at 100°F (37.8°C), as determined by ASTM D-323, *Standard Method of Test for Vapor Pressure of Petroleum Products (Reid Method).*

Class I liquids are further subclassified into A and B, as follows:

> *Class IA Flammable Liquids*—Liquids that have a defined flash point below 73°F (22.8°C) and boiling points below 100°F (37.8°C).
>
> *Class IB Flammable Liquids*—Liquids that have defined flash points below 73°F (22.8°C) and boiling points at or above 100°F (37.8°C).

See also **Combustion; Combustible Liquid; Flash Point.**

Flammable Liquid Storage Locker or **Cabinet**—A cabinet used to safely store incidental amounts of flammable liquids through the incorporation of built-in fire safety features. The cabinet is constructed of metal and is provided with insulation. It is fitted with a three-point latching system for its doors to prevent accidental opening or release, and has a spill lip to prevent liquids from leaking out of the cabinet. Optional vents are available to safely release combustible vapors from the interior. It is normally painted yellow and highlighted in red lettering as a flammable liquid storage location. See Figure F-28. *See also* **Combustible Liquid.**

Flammable Range—*See* **Flammable Limit** or **Flammability Limits.**

Flammability—The capacity or ability of a material to burn with a flame.

Flare—The flame condition of a fire in which burning occurs with an unsteady flame. In the process industries (chemical and petroleum), a flare refers to a primary fire safety system used for the safe remote disposal of gases by burning from normal processes or emergency conditions.

Figure F-28 Flammable liquid storage locker.

Process gases that cannot be safety disposed of may contribute to fire destructiveness or cause a vessel rupture or BLEVE. API RP 521, *Guide for Pressure Relieving and Depressurizing Systems,* provides information on the design and permissible heat radiation levels for flares.

A flare is also a combustible device used to emit a dazzlingly bright light for signaling or illumination on railroads and highways and in military operations; in pyrotechnics the term is applied either to a colored-fire composition burned in a loose heap or to a similar composition rolled into a paper case to ensure longer and more regular burning. The flare in its present form dates from the early part of the 19th century, when the introduction of potassium chlorate permitted the development of compositions to produce colored light. Before this, the only color had been the bluish-white light produced by a mixture of sulfur, saltpeter, and orpiment. These blue lights, as they were called, were and still are often used at sea for signaling and illumination. They were also known as Bengal lights, probably because Bengal was the chief source of supply of saltpeter. *See also* **Burn Pit.**

Flare-Up—A sudden outburst or intensification, such as a sudden bursting (for example, of a smoldering fire) into flame.

Flash—A quick-spreading flame or momentary intense outburst of radiant heat. It may also be used to refer to a spark or intense light of short duration. *See also* **Flash Point (FP).**

Flash Arrester—*See* **Flame Arrester.**

Flashback—Propagation of a flame through a gaseous mixture from the ignition source back to the release point of the flammable material. Also may be used to categorize the re-ignition of a flammable liquid caused by exposure of its vapors to a source of ignition.

Flash Burn—An injury or damage to body tissue caused by the exposure to a flash or sudden intense radiant heat.

Flash Fire—A fire that spreads rapidly through a diffuse fuel, such as dust, gas, or the vapors of an ignitable liquid, without the production of damaging pressure. The fire in the oxygen-enriched NASA *Apollo 1* space capsule, which killed three American astronauts, was considered a flash fire due to the rapidity of its spread and consumption of the contents. Dried grass, leaves, and light branches are considered flash fuels. They ignite readily and fire spreads quickly in them, often generating enough heat to ignite heavier fuels such as tree stumps, heavy limbs, and the matted duff of the forest floor.

Flash Ignition Temperature—This is a test designed for polymeric materials in the Setchkin test (ASTM D-1929, *Standard Test Method for Ignition Properties of Plastics*). A specimen is placed in an electrically heated tubular furnace in a controlled airflow. The temperature at which the gases evolved ignite in the presence of a spark is called the flash ignition temperature. *See also* **Setchkin Test.**

Flashover—A transition phase in the development of a contained fire in which surfaces exposed to thermal radiation reach ignition temperature more or less simultaneously, and fire spreads rapidly throughout the space. A flashover is always a sudden event and may be deadly for occupants or firefighters.

Flashover Simulator—A live fire training exercise to subject firefighters to intensive heat and flashover conditions.

Flash Point (FP)—The flash point of a liquid is the temperature at which the vapor and air mixture lying just above its vaporizing surface is capable of just supporting a momentary flashing propagation of a flame prompted by a quick sweep of a small gas pilot flame near its surface, hence the term flash point. The flash point is mainly applied to a liquid. NFPA 30, *Flammable and Combustible Liquids Code,* determines the flash point of a combustible liquid by one of several ASTM test methods, depending on the viscosity of the fluid. ASTM D 56, *Standard Method of Test for Flash Point by the Tag Closed Cup Tester;* ASTM D 93, *Standard Test Methods for Flash Point by the Pensky-Martens Closed Tester;* ASTM D 3278, *Standard Method of Tests for Flash Point of Liquids by Setaflash Closed Tester;* or ASTM D 3828, *Standard Test Methods for Flash Point by Small Scale Closed Tester,* is specified. The flash point of liquid is one of its characteristics that normally determines the amount of fire safety features required for its handling, storage and transport. *See also* **Closed Cup Tester; Combustible Liquid; Fire Point; Flammable Limit** or **Flammability Limits; Flammable Liquid; Flash.**

Flash Point Index—An index of liquid materials listed by their flash points and trade names that identifies their fire hazards. A flash point index is published by NFPA. Flash point is the minimum temperature that will support the ignition of a premixed flame, while fire point is the minimum temperature that will support the ignition of a diffusion flame.

Flat Load—A method of strong fire hose. It involves laying the hose flat in the storage area. It is commonly used for supply lines and some attack lines.

Flicker—The brief glimpses of radiation energy from a fire event, usually in the visible spectrum, such as a flickering flame.

Flicker Detector (or **Flame Detector)**—A detector that observes ultraviolet, infrared, or visible radiation from a fire event. *See also* **Flame Detector.**

Flooding Factor—A design parameter used in the engineering calculations for the design of carbon dioxide (CO_2) total flooding fire suppression systems based on the volume of the enclosure. It may also be called a volume factor.

Flowswitch—*See* **Waterflow Alarm.**

FM-200—A fire suppression gas developed to replace Halon, technically known as HFC-227. Its chemical name is heptafluoropropane, CF_3CHFCF_3. It has a zero ozone depletion potential (ODP) rating and is rated with a low global warming potential. It requires a design concentration of 7 percent for fire suppression (as compared with 5 percent for Halon applications). It is considered safe for human exposure and does not leave a residue. *See also* **Clean Agent; Halon; Inergen.**

FM Cock—A special approved valve control mechanism that provides a means for positive interlocking of a main fuel safety shutoff valve so that, before the main fuel safety shutoff valve

F

can be opened, all individual burner supervising valves must be in the fully closed position. FM refers to Factory Mutual, as it was developed from its recommendations and design. It may also be called a supervising cock. *See also* **Factory Mutual (FM).**

F

Foam—A fluid aggregate of air-filled bubbles, formed by chemical means, that will float on the surface of flammable liquids, flow over solid surfaces, and cling to vertical and horizontal surfaces. It is made up of three ingredients: water, foam concentrate, and air. When mixed, the ingredients form a stable, homogenous foam blanket that excludes air, seals liquids, and prevents volatile flammable vapors from being emitted. The foam blanket functions to extinguish fires or prevent the ignition of the material. The need for firefighting foams occurs on surfaces on which the cooling effect of water is needed and wherever a continuous foam blanket can provide the benefits of vapor suppression, insulation, delayed wetting, or reflection. Foams resist disruption due to wind and drafts, heat, and flame attack. It is also capable of resealing if an opening is made in the foam blanket. Foam products are commercially available for Class A fuel fires and Class B fuel fires (commonly referred to as Class A foam and Class B foam, respectively). *See also* **Compressed Air Foam System (CAFS); Expansion Ratio; Fixed Foam Application Systems; Foam Concentrate; Foam Maker; Foam Tetrahedron; Overhead Foam Injection; Protein Foam; Subsurface Foam Injection; Twin Agent Unit or Twinned Agent Systems; Water Additive.**

Foam, Alcohol Resistant/Aqueous Film Forming (AR-AFFF)—*See* **Foam Concentrate.**

Foam, Aqueous Film Forming (AFFF)—*See* **Foam Concentrate; Light Water.**

Foam Blanket—A layer of firefighting foam applied to completely cover a combustible liquid or other hazardous liquid surface to prevent or suppress a fire or the release of toxic vapors. Blanket is a common term used to indicate a completely covered surface.

Foam Breakdown—The process of foam collapse and disintegration due to exposure to heat, mechanical stress, and aging as water drains from the foam bubbles.

Foam Cannon—A large foam application nozzle that cannot be hand-held due to its size and reaction forces. See Figure F-29.

Foam Chamber—A foam application device provided for combustible liquid storage tanks for the suppression of surface fires. They are typically used as part of a fixed or semi-fixed surface application fire protection system on the exterior of fixed cone roof tanks or on internal floating roof storage tanks. The chamber is installed just below the roof joint of the tank. Foam solution is piped to the chamber where it is expanded and discharged against a deflector inside the storage

Figure F-29 Foam cannon with pickup tube from portable container.

tank, which diverts the expanded foam back against the inside wall of the tank. This reduces the submergence of the foam and agitation of the liquid fuel surface. The number of foam chambers required for any fixed or semi-fixed storage tank is based on the diameter of the storage tank. A seal is provided in the foam chamber to prevent combustible vapors or liquids from entering the foam system from the storage tank. When the foam system is activated, the seal ruptures from the system solution pressure. See Figure F-30. *See also* **Foam Deflector; Moeller Tube; Type I (Foam) Discharge Outlet; Type II (Foam) Discharge Outlet.**

Foam, Chemical—*See* **Chemical Foam.**

Figure F-30 Foam chamber, top view with lid open showing internal membrane.

Foam Chute—A firefighting foam delivery trough for the protection of combustible liquid storage tanks. It consists of steel sheet formed into a trough securely attached to the inside of the tank wall so that it forms a descending spiral from the top of the tank. Foam is supplied through a foam chamber and is released down the chute until it reaches the combustible liquid surface. It is classed as a Type I outlet by NFPA 11, *Standard for Low Expansion Foam. See also* **Foam Trough; Type I (Foam) Discharge Outlet.**

Foam Concentrate—A fire suppression surfactant material used to seal the vapors from the surface of a combustible liquid. Foam concentrates have two primary classifications recognized by Underwriter's Laboratories (UL): protein or synthetic foams. They include low-, medium-, and high-expansion foam, each designed for a certain fuel exposure and application. Firefighting foams have certain characteristics—they flow freely, they have slow drainage times and the capability of holding water, they produce a stable blanket, they resist liquid exposure, and they resist heat. Modern foam concentrates can be used successfully with either sea, fresh, or brackish water. Foam concentrates for protection of petrochemical locations are divided into five basic types.

1. Protein foam is a hydrolysate base of hoof and horn, fish scales, etc. combined with stabilizers.
2. Fluoroprotein (FP) foam is a protein hydrolysate base combined with fluorochemical surfactants.
3. Film forming fluoroprotein (FFFP) foam is a combination of fluoroprotein surfactants and synthetic foaming agents.
4. Aqueous film forming foam (AFFF) is a combination of fluorocarbon surfactants and

synthetic foam agents. They produce a foam that slides across the surface of hydrocarbon fuels. This is accomplished by the formation of a film that spreads ahead of the foam bubbles.

5. Alcohol resistant/aqueous film forming foam (AR-AFFF) is a foam concentrate suitable for both hydrocarbon and polar solvent exposures and fuels. The concentrate is a combination of synthetic stabilizers, foamers, fluorocarbons, plus proprietary additives.

F

Protein foam (also called regular foam) was the first of the mechanical foams to be developed. It became commercially available in the United States in the early 1940s. The concentrate consists of a protein source (commonly hoof and horn meal or feather meal) which has undergone an alkaline hydrolysis. This protein hyrolysate is then stabilized with metal salts, and has corrosion and bacterial growth inhibitors added. It may also contain freezing point depressants. The concentrate is normally available for 3 or 6 percent proportioning. The foam generated from these concentrates is very thick and viscous and exhibits poor fuel tolerance. Therefore, it spreads slowly over fuel surfaces, resulting in a slow knockdown. Its use has been declining in recent years with its primary use being control and extinguishment of in-depth fuel fires.

AFFF represents the next generation in mechanical foam development. The development of AFFF came about as a result of advances in fluorine surfactant technology. The research and development program was conducted by the Naval Research Labs in the early 1960s principally by Tuve and Peterson, and was referred to as "light water." The objective of the research was to develop an improved foam product that could materially reduce the fire control and extinguishment times for complex Class B fires (aircraft fires), while at the same time being compatible with the dry chemical fire extinguishing agent Purple K. AFFF consists of fluorochemical surfactants, hydrocarbon surfactants, solvents, and foam stabilizers. The concentrate is available for 1, 3, and 6 percent proportioning. A lower percent allows less concentrate to be used, achieving cost and weight or storage space savings. A 1 percent proportioning system may be more susceptible to failure if contamination or particulates are large enough to cause blockages at the proportioner. The foam generated from these concentrates is watery and has a low viscosity. It can be regarded as super a detergent, which in solution has a low surface tension. Because of this, it spreads very quickly over fuel surfaces and around obstructions. A unique feature of AFFFs is that because of fluorosurfactants, they are able to rapidly form a thin water film of the foam solution that actually floats on top of the less dense hydrocarbon fuel below. This very high-speed control of fire is the property that gives AFFF such a great advantage over older protein foam. The effectiveness and durability of the aqueous film is directly influenced by the surface tension of the hydrocarbon. AFFFs are more effective on fuels with higher surface tension coefficients, such as kerosene, diesel fuel, and jet fuels, than they are on fuels with low surface tension coefficients, such as hexane and high octane gasolines. In the United States, AFFFs are almost the only foams used for crash fire-rescue operations. They are also used for in-depth fuel fires, subsurface injection, and in non-aspirating sprinkler systems. AFFF is considered the most versatile of the mechanical foams.

Fluoroprotein foam is an offshoot of the technology developed for AFFF. Chemically, the only difference between fluoroprotein foam and regular protein foam is the addition of a small amount of fluorochemical surfactant. There is generally not enough of this surfactant to allow the formation of an aqueous film as in the AFFFs. The properties of the foam lie in between those of AFFF and protein foam and are dependent upon the ratio of protein to fluorosurfactant. The concentrates available are 3 and 6 percent. Fluoroprotein foam is used on spill fires and in-depth fuel fires. It is also considered by many to be the choice for subsurface injection since it shows good tolerance to fuel contamination and great burnback resistance.

Synthetic foam agents are composed of hydrocarbon surfactants or detergents and are often referred to as medium- or high-expansion foams. Special foam-generating equipment is required to expand these agents to medium and high levels. The concentrates are proportioned between 1 and 6 percent, depending on the desired end use foam properties and the foam-generating equipment limitations. There is little water or dissolved solids in the foam bubble wall structure due to the relatively high expansion ratio. Consequently, they have a lower thermal stability than do low expansion foams. Additionally they have less cooling efficiency and are more easily disrupted by ambient wind conditions and thermal updrafts of a fire. High expansion foams are generally used in the total flooding of enclosures and on liquefied natural gas (LNG) fires and spills where high volume and low water content are necessary because of the cryogenic nature of the fuel. High expansion foams are also used for total flooding of warehouses and other rooms containing Class A materials. Fire extinguishment in these areas amounts to inerting the fire area. There is only a slight cooling effect because of the low water content. Medium expansion foams have not found much application in the United States. Most of the application research for these has been conducted in Europe.

Polar solvent or alcohol-resistant foams are specialized foams that can be used on conventional hydrocarbon fires and additionally may be used on polar solvent or water miscible liquid fuels. The most common alcohol-resistant concentrates available today are based on AFFF concentrates to which a small amount of a high molecular weight polysaccharide (a type of gum) has been added. This polysaccharide is soluble in water but insoluble in polar solvents. As the first portion of foam is applied to the fuel surface and drains rapidly into the fuel, the insoluble polysaccharide is left behind to polymerize and form a gelled barrier between the fuel and the rest of the foam blanket. The concentrates that are commonly available are proportioned at 3 and 6 percent. The early types of alcohol-resistant foams were manufactured from a protein base with various additives. They suffered from severe limitations, such as the amount of time they could be stored in a premixed condition as well as slow transit time in a hose line.

Film forming fluoroproteins (FFFP) are the results of attempts made in the past to combine the burnback resistance of a fluoroprotein foam with the knockdown of an AFFF. The first work on film-forming foams was carried out using protein-based concentrates. However, the result has to be a compromise to some extent, and those foams that are presently on the market usually suffer from poorer burnback resistance than a standard fluoroprotein type. The protein base can also lead to some storage problems due to sedimentation, particularly in hot climates.

Alcohol-resistant foams are suitable for fires on water soluble and certain flammable or combustible liquids and in solvents that are destructive to regular foams, such as alcohols (greater than 15 percent of volume in hydrocarbon, such as gasohol), ketones, etc. This foam has an insoluble barrier in the bubble structure that resists breakdown at the interface between the fuel and the foam blanket.

Aspirating and non-aspirating nozzles can be used for AFFF or FFFP application. A non-aspirated nozzle typically provides a longer reach and quicker control and extinguishment. However, expansion rates and foam drainage times are generally less when AFFF or FFFP is applied with non-aspirating nozzles. It also should be understood that the foam blanket might be less stable and have a lower resistance to burnback than that formed using aspirating nozzles. *See also* **Chemical Foam; Foam; Low Expansion Foam; Mechanical Foam.**

Foam Concentrate Eductor—A device to draw foam concentrate out of a storage location through the principle of a venturi vacuum in the proper percentage for use in firefighting systems. It generally consists of a flexible hose connected to the throat of a venturi fitted with an orifice that is sized to the amount of proportioning specified for the foam concentrate.

Foam Dam—An enclosure for containing applied firefighting foam to a specific location to ensure adequate application to the hazard. Foam dams are normally specified for containing foam application to the rim seal area of a floating roof tank containing a combustible liquid. The foam dam design for floating roof storage tanks is outlined by NFPA 11, *Standard for Low Expansion Foam.* See Figure F-31. *See also* **Overhead Foam Injection; Rim Seal Fire Protection; Seal Protection.**

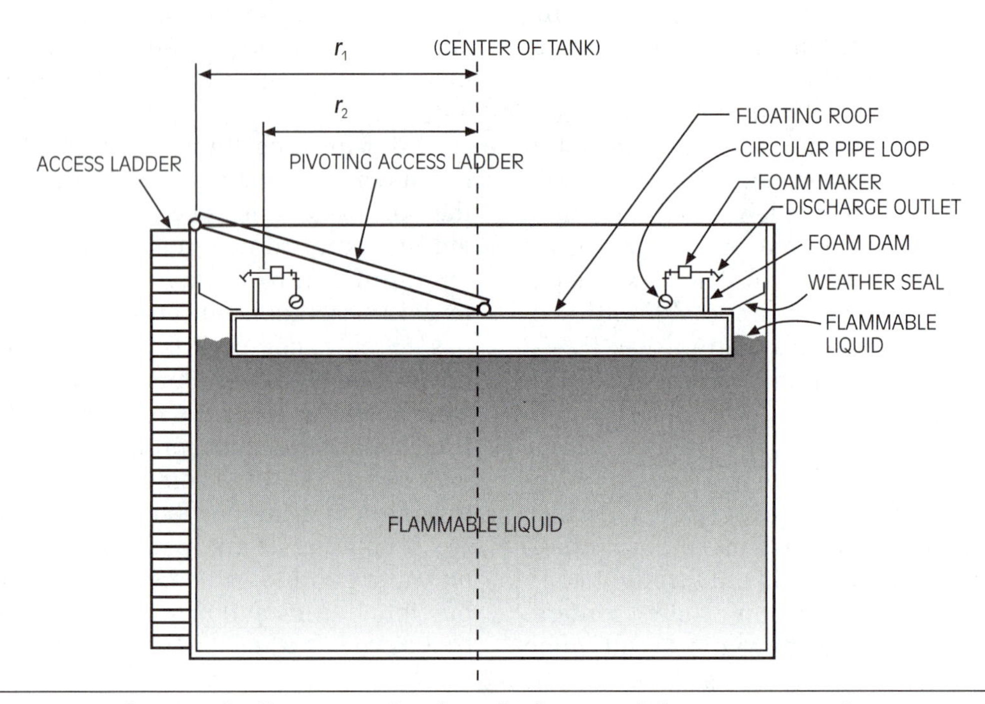

Figure F-31 Sectional view of a floating roof tank with above-seal, low-expansion foam protection using a foam dam.

Foam Deflector—A curved steel plate provided on a foam chamber outlet to direct the supplied foam onto the surface of the combustible liquid. It is normally provided to the foam chamber for a topside foam application of a combustible liquid storage tank for dispersal into the seal area of the floating roof tank. *See also* **Foam Chamber.**

Foam, Film Forming Fluoroprotein (FFFP)—*See* **Foam Concentrate.**

Foam Fire Extinguishing System—A foam fire extinguishing system designed and installed in accordance with a recognized standard. It can be a fixed discharge outlet system utilizing fixed storage and piping connected to fixed outlets or monitor nozzles, or a local application foam water hose line system connected to fixed storage tank supply system.

Foam, Fluoroprotein (FP)—*See* **Foam Concentrate.**

Foam Maker—A device designed to introduce and mix air into a pressurized foam solution stream (low/medium expansion nozzle, high expansion nozzle, or compressed air foam system) for the purpose of aspirating the firefighting foam prior to its delivery to an application device. It consists of an orifice plate (sized for the required flow rate at a given pressure), an air inlet section, and a mixing barrel. A foam maker is normally installed in the piping of a fixed or semi-fixed foam system for open-top floating roof combustible liquid storage tanks and dike protection systems. *See also* **Foam.**

Foam Maker, Floating Roof—A device designed to introduce and mix air into a pressurized foam solution stream to protect open top floating roof combustible liquid storage tanks. It supplies expanded foam to the floating roof seal area where a fire normally occurs.

Foam Maker, High Back-Pressure (HBPFM)—A device designed to introduce and mix air into a pressurized foam solution stream (low/medium expansion nozzle, high expansion nozzle, or compressed air foam system) for the purpose of subsurface injection into a combustible liquid storage tank. It consists of an orifice plate (sized for the required flow rate at a given pressure), an air inlet section, and a mixing barrel. It produces expanded foam with ratios of between 2 to 1 and 4 to 1. Foam ratio refers to the volume of foam solution (concentrate and water) to the volume of the mixed foam solution and air that is produced. It is capable of being used with backpressures (storage tank head pressure and friction losses in the application pipe from the foam maker) that are as much as 40 percent of the operating inlet pressures to the device. It injects expanded foam through the wall of the tank near the bottom. The foam tends to rise to the surface to form a fire suppression blanket. The foam can be supplied through a dedicated foam delivery line or through a connection on existing piping to the tank.

Foam Monitor—A large capacity foam solution application device that usually air aspirates the foam solution in large discharge streams. A foam monitor can be manually or remotely controlled or can oscillate automatically. The nozzle may be portable or fixed.

Foam Nozzle—A foam solution application device that usually air aspirates the foam solution. A foam nozzle may be portable or fixed. See Figure F-32.

Foam Proportioner—A device for mixing foam concentrate into firewater in a precise percentage to create a foam solution. *See also* **Foam Proportioning; Foam Solution.**

Foam Proportioning—A method of foam concentrates and water to form a foam solution.

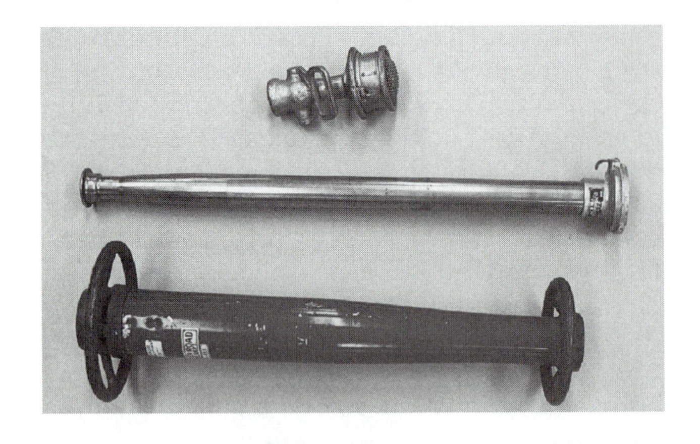

Figure F-32 Typical foam nozzles.

There are many different methods of mixing the foam concentrate with water in a foam system installation. These include premix, in-line eductor, bladder tank with ratio controller (proportioner), foam pump with in-line balanced pressure proportioning unit, and balanced pressure pump proportioning skid.

A premix system is used for small systems up to a maximum requirement of 500 gallons (1,892 liters) or less. The premix solution is stored in either an atmospheric-type storage container or a pressure vessel. If stored in an atmospheric container, a pump must be used to push the foam solution to the discharge device. In the pressure vessel system, normally a bank of compressed air or nitrogen cylinders is used as the method of expelling the foam solution out of the vessel to the discharge devices.

An in-line eductor is a simple method of supplying foam concentrate into the water stream. It is highly efficient when water pressure and flow are sufficient through the unit. All in-line eductors are sized for a particular flow, and the discharge device must pass the same or greater amount to ensure that the correct proportioning occurs. All eductors lose approximately 35 percent of the incoming pressure through friction loss through the unit. In-line eductors are used with atmospheric foam concentrate storage tanks.

Bladder tanks with ratio controllers are a very reliable method for mixing foam concentrate with water. They require no external power other than water pressure and flow to ensure correct operation. The bladder tank is a coded pressure vessel with an internal bladder that holds the concentrate. When water is discharged into the tank between the bladder and the tank shell, it subjects the bladder to pressure that then forces the foam concentrate to migrate out of the bladder through interconnections to the ratio controller. Bladder tanks can be vertical or horizontal in orientation. The ratio controller is the point in a fixed foam fire protection system at which the foam concentrate and the water meet and the concentrate is metered through a fixed orifice into the water stream at the correct mix ratio. Three features must be considered when specifying a ratio controller: system flow requirements, type of foam concentrate to be used, and residual system water pressure available.

A foam pump with an in-line balance proportioner (ILBP) is an accurate method of metering foam concentrate into the water supply, especially if the ratio controller is installed some distance from the foam concentrate supply. With the ILBP, the foam concentrate is supplied to the ratio controller from a positive displacement rotary gear pump driven by either an electric, diesel, or water turbine motor. The ILBP is installed in the system riser and senses the water and concentrate pressures. It self-adjusts a diaphragm valve that then meters the correct quantities of foam concentrate into the water stream. The foam concentrate is stored in an atmospheric storage tank.

A foam balanced pressure pump proportioning skid is similar in operation to the foam pump/ILBP system with the exception that the mixing of the foam concentrate and water takes place on the skid. The skid typically contains the foam pump, controller, balancing valve, duplex gauge, and ratio controller.

In a pressure proportioning tank, the foam concentrate is stored in a pressure-rated vessel. It employs the water pressure as the source of power. Water enters the tank and pressurizes the concentrate, and at the same time, water flowing through an adjacent venturi or orifice mounted on the tank creates a pressure differential while the water pressure simultaneously forces the concentrate up a pickup tube and into the discharging water stream. This type of system is only suitable for use with protein-based foam concentrates or concentrates that have densities greater than 1.10. It is not suitable for use with any AFFF concentrates.

In direct injection, foam concentrate is pumped into a riser through an orifice directly mounted on the riser. The orifice is sized to pass the correct quantity of foam concentrate at a given pressure. This is also based on the water supply being of a constant flow and pressure. This type of system is only suitable where pressures and flows are constant in an open-type discharge system. *See also* **Foam Proportioner; Foam Solution; Proportioner.**

Foam, Protein—*See* **Foam Concentrate.**

Foam Pump—A pump that is used to deliver foam concentrate to foam proportioning systems for enhanced fire control and suppression by a foam solution or an aspirated foam solution. Foam pumps are generally of the positive displacement type because of the high viscosities of foam concentrations. If centrifugal pumps are used for pump foam concentrates, they may experience "slippage" because of the high viscosities of the foam concentrates. Positive displacement, rotary gear pumps that are driven by diesel engines, electric motors, or a water turbine motor are also commonly used. Rotary gear pumps consist of two gears that rotate within the walls of a circular housing. The inlet and outlet ports are on opposite sides of where the gear teeth mesh together. Fluid is drawn into the clear space between the meshing gears and its pressure is subsequently increased by the rotation of the gears acting together. *See also* **Rotary Pump.**

Foam Solution—Fire suppression foam concentrates mixed in proper proportion with water as required by the specification of the foam concentrate (0.5, 1, 3, or 6 percent by water solution mix). Foam solution is created in a foam proportioner that inserts the proper percentage of foam concentrate into the firewater stream for delivery to the foam application devices. *See also* **Foam Proportioner; Foam Proportioning.**

Foam Stability—The relative ability of a firefighting foam to withstand collapse or breakdown from external causes such as heat or chemical reaction. Foams with a high stability are favorable because they prolong the protection afforded before additional quantities of foam need to be reintroduced.

Foam Tetrahedron—A graphical, symbolic representation of the four factors needed for the production of firefighting foam. Each side of the tetrahedron is representative of a factor. The four factors include foam concentrate, water, air, and foam solution proportioning/aspiration. Partial provision or lack of one of the elements prevents an adequate foam from occurring or continuing. *See also* **Fire (Safety) Symbols; Foam.**

Foam Trough—A foam delivery system for combustible liquid storage tanks for gently delivering a foam to the surface of a combustible liquid. It consists of sections of steel sheet formed into a chute attached to the inside of the tank wall so that it forms a descending spiral from the top to close to the bottom of the tank. It is classed as a Type I outlet by NFPA 11, *Standard for Low Expansion Foam. See also* **Foam Chute; Type I (Foam) Discharge Outlet.**

F

Foam-Water Deluge System—A system having a pipe connected to and including a source of foam concentrate and a water supply. Water and foam concentrate, such as protein, fluoroprotein, or aqueous film forming foam (AFFF), are delivered to open discharge devices for extinguishing agent discharge and for distribution over the area to be protected. The piping is connected to the water supply through an automatic valve that is actuated by the operation of a detection system installed in the same areas as the discharge devices. When this valve opens, water flows into the piping system, foam concentrate is injected into the water, and the resulting discharge of foam solution through the foam-water discharge devices generates and distributes foam. Upon exhaustion of the foam concentrate supply, water discharge follows the foam and continues until it is shut off manually.

Foam Weep—The portion of foam that is separated from the principal foam stream during discharge, falls at short range, and fails to meet the desired application point.

Fog Foam—Firefighting foam applied in the form of a spray.

Foolproof—So plain, simple, obvious, and reliable as to leave no opportunity for error, misuse, or failure to implement the correct action. Some fire protection systems or devices are designed to be foolproof to reduce the possibility of failure or error.

Forcible Entry—Access techniques used by the fire service to obtain entry to locked or blocked enclosures (buildings, houses, vehicles) for emergency purposes with minimum damage to the enclosure. Specilized tools have been devised to assits firefighters in forcible entry operations. *See also* **Claw Tool; Halligan Tool; Hux Bar; Kelly Tool.**

Forensic Fire Engineering—The study and investigation of a fire origin and determination, failure analysis, and legal implications. Specialists in forensic science apply their knowledge to perform these fire investigations and studies.

Forest Fire—Natural or human-caused fires that unnecessarily burn forest vegetation. Foresters usually distinguish three types of forest fires: ground fires, which burn the humus layer of the forest floor but do not burn appreciably above the surface; surface fires, which burn forest undergrowth and surface litter; and crown fires, which advance through the tops of trees or shrubs. It is not uncommon for two or three types of fires to occur simultaneously. These fires often reach the proportions of a major fire. Approximately 95 percent of all forest fires are caused by people, while lightning strikes are responsible for 1 to 2 percent. A big forest fire may crown, that is, spread rapidly through the topmost branches of the trees before involving undergrowth or the forest floor. Consequently, violent blowups are common in forest fires, and they may assume the characteristics of a firestorm.

Forest fires are often called wildland fires. They are spread by the transfer of heat to grass, brush, shrubs, and trees. Because it is frequently difficult to extinguish a forest fire by attacking it directly, the principal effort of forest firefighters is often directed toward controlling its spread by creating a gap, or firebreak, across which fire cannot move. After the firebreaks are made, the fire crews attempt to stop the fire by several methods: trenching, direct attack with hose streams, aerial bombing, spraying fire-retarding chemicals, and controlled backburning. As much as possible, advantage is taken of streams, open areas, and other natural obstacles when establishing a firebreak. Wide firebreaks may be dug with plows and bulldozers. The sides of the firebreaks are soaked with water or chemicals to slow the combustion process. Some parts of the fire may be allowed to burn themselves out. Firefighting crews must be alert to prevent outbreaks of fire on the unburned side of the firebreaks. Firefighting crews are trained and organized to handle fires covering large areas. They establish incident command posts, commissaries, and supply depots. Two-way radios are used to control operations, and airplanes drop supplies as well as chemicals. Helicopters can be used to serve as command posts and transport firefighters and their equipment to areas that cannot be reached quickly on the ground. Some severe wildfires have required more than 10,000 firefighters to be engaged at the same time.

The US Forest Service maintains research laboratories that develop improved firefighting equipment and techniques, and maintains a school that trains firefighters in the latest firefighting techniques. International conferences on wildland fire prevention and firefighting have been held with greater frequency in recent years.

Fire management programs are extensive both in the United States and other countries. Programs include fire prevention, firefighting, and the use of fire in land management.

Most forest fires result from human carelessness or deliberate arson. Fewer fires are started by lightning. Weather conditions influence the susceptibility of an area to fire; such factors as temperature, humidity, and rainfall determine the rate and extent to which flammable material dries and, therefore, the combustibility of the forest. Wind movement tends to accelerate drying and to increase the severity of fires by speeding up combustion. By correlating the various climatic elements with the flammability of branch and leaf litter, the degree of fire hazard may be predicted for any particular day in any locality. Under conditions of extreme hazard, forests are closed to public use.

Although organizations involved with fire control have traditionally fought all fires, certain fires are a natural part of the ecosystem. Complete fire exclusion may bring about undesirable changes in vegetation patterns and may allow accumulation of fuel, with increased potential for feeding catastrophic fires. In some parks and wilderness areas, where the goal is to maintain natural conditions, lightning-caused fires may be allowed to burn under close surveillance.

One of the most important aspects of forest fire control is a system of locating fires before they are able to spread. Land-based forest patrols and lookouts have been largely replaced by surveillance aircraft that detect fires, map their locations, and monitor their growth.

Ground fires, once established, are difficult to extinguish. When the humus layer is not very deep, a ground fire may be extinguished with water or sand. Most ground fires, however, are controlled by digging trenches around the burning area and allowing the fire to burn itself out. Surface fires are limited by clearing the surrounding area of low vegetation and litter, or digging emergency furrows to confine the area. Crown fires are difficult to extinguish. They may be allowed to burn themselves out, they may be halted by streams, or they may be limited by backfired areas. Backfiring consists of carefully controlled burning of a strip of forest on the leeward side of the blaze, so that when the fire reaches the burned area it can go no farther.

Foresters may purposely ignite prescribed fires under carefully controlled conditions to remove unwanted debris following logging, to favor tree seedlings, or to keep fuels from accumulating. Since most grasses and shrubs grow well after fires, and animals are attracted to the tender and nutritious new growth, prescribed fires often benefit both wildlife and livestock. The mosaic of vegetation of different ages that results from frequent fires favors a rich diversity of plant and animal life. *See also* **Bushfire; Fuels Management; Unburned Island; Wildland Fire.**

Forward Lay—The laying of fire hose in a direction from a water source (fire hydrant) to the fire incident from a hose cart or fire engine. Whenever a forward lay is conducted, it places the fire pumper near the fire incident. Consequently, the fire pumper may not be able to use its pump because of the long distance of a supply hose. This type of operation was commonly practiced before the introduction of pumper trucks. The manner in which a fire hose is stored for use should account for the method of hose lays to be used and type of hose couplings (male or female) available at each end of the hose. It may also be called hydrant-to-fire lay or straight lay. See Figure F-33. *See also* **Hose Lay; Reverse Lay.**

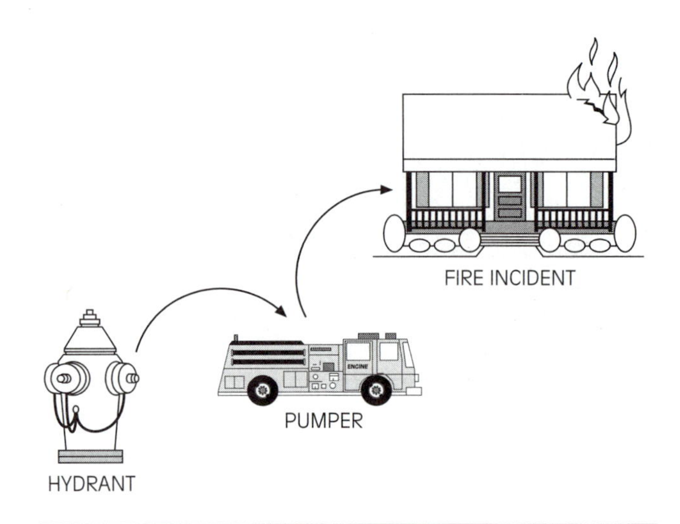

Figure F-33 Forward lay.

Four-Way Valve—A hydrant valve used by the fire service that allows connection of a fire pumper without the shutdown of outlets in use for hose lines from the subject hydrant.

Frangible Bulb—A small sealed glass bulb filled with liquid that is used to detect a fire and actuate protective devices. Exposure to heat causes the liquid in the bulb to expand (boil) resulting in failure (bursting) of the bulb. Frangible bulbs are most commonly utilized in automatic sprinklers, whereupon their bursting from heat exposure, they release the system water supply for distribution by the individual sprinkler head deflector.

Free-Burning—The burning of materials in the open air, where overventilation of the combustion process usually occurs. Also used to describe an unwanted fire that is not under control (free-burning forest fire). It may also be called open burning and is sometimes used to describe the second stage of a fire, that is, incipient, then free-burning, then smoldering. *See also* **Blaze; Burn; Open Burning.**

Frictional Ignition—Ignition caused by rubbing abrasive surfaces, causing sparks, or causing a hot surface to occur that has sufficient energy to cause a combustible material to ignite. *See also* **Ignition Energy; Ignition Source.**

Friendly Fire—An insurance industry term for a fire that is a deliberately ignited flame or glow that stays within its intended confines (furnace, fireplace, stove, lantern, etc.), and is commonly used for heating, cooking, lighting, or aesthetic purposes. A friendly fire is in contrast to a hostile fire, one that escapes from its intended confines. A standard fire insurance policy covers only hostile fires. *See also* **Fire; Hostile Fire.**

Frost Valve—A valve incorporated into a fire hydrant that opens when the hydrant valve is closed to allow drainage and prevent water freezing in the hydrant.

Fuel—A material that yields heat through combustion. Fuels can be in a solid, liquid, or gaseous state. Solid fuels are generally more stable and therefore easily transportable and stored (wood, coal), whereas liquid and gaseous fuels are volatile materials and require a specific enclosed container (tank or pressure vessel) for transport and storage.

Fuel Break—*See* **Firebreak.**

Fuel Contributed Rating—A relative index measured by NFPA 255, *Standard Method of Test of Surface Burning Characteristics of Building Materials,* for the purpose of comparative examination of a materials fire hazard when used as a material of construction. *See also* **Flame Spread Index (FSI); Smoke Developed Rating.**

Fuel-Controlled or **Fuel-Limited Fire**—A fire in which the heat release rate and the growth rate are controlled by the characteristics of the fuel (quantity and geometry), and adequate air for combustion is available. *See also* **Ventilation-Limited Fire** or **Ventilation-Controlled Fire.**

Fuel Loading—The amount of combustible material per a specified area. It is normally estimated as the average weight in pounds of combustibles per square foot of area. The larger the amount of combustibles, the greater the fuel loading and the greater the potential fire hazard. The amount of combustibles may also be referred to as the fire loading of a building. *See also* **Crib; Fire Load** or **Fire Loading; Heavy Content Fire Loading.**

Fuels Management—A program to control the growth of natural vegetation to lessen the potential fire intensity of a wildfire. Generally used in forested or highly vegetative areas where the risk of a fire spreading is great. *See also* **Burnout; Forest Fire.**

Fully Developed Fire—The state of a fire in an enclosure in which the fire has reached its maximum possible size. The maximum fire is controlled by the quantity of readily available fuel or by the ventilation into the enclosure.

Fully Involved Fire—A building or structure fire in which the interior is completely on fire and its condition (exposure to flame, heat, or smoke) is such that immediate interior access is not possible until firefighting efforts are applied to it.

Fundamental Burning Velocity—The burning velocity of a laminar flame under specific conditions of composition, temperature, and pressure of a combustible gas. *See also* **Burning Velocity; Flame Speed.**

Fuse—A safety device that protects electrical circuits from the effects of excessive current to prevent overheating and the occurrence of a fire. A fuse commonly consists of a current-conducting strip or wire of easily fusible metal that melts, interrupting the circuit whenever that circuit is made to carry a current larger than that for which it is intended. The screw-plug fuse was once commonly used in domestic electrical systems. It contains a short length of wire (the fusible element) enclosed in a fire-resistive container that has a screw-threaded base. The wire is connected to metal terminals at both the screw base and at the side, and the whole is covered with a transparent glass or mica window to see whether the fuse has melted. The cartridge fuse, a type of fuse widely used in industry where high currents are involved, has a fusible element connected between metal terminals at either end of a cylindrical insulating tube. A fuse is a non-resetable device that must be replaced with a new fuse once it has been activated. *See also* **Circuit Breaker.**

Fusible Alloy—A low melting alloy, usually 140°F to 284°F (60°C to 180°C), commonly made of bismuth, tin, lead, and other metals. They are used in electrical fuses to automatically open the circuit (melt the fuse) from the flow of excessive current. Fusible alloys have to be replaced after they have been activated, that is, they are non-resetable. *See also* **Fusible Element; Fusible Metal.**

Fusible Element—A component (link, plug, pellet, etc.) that is manufactured to melt at a specific temperature. In fire protection applications, fusible links, plugs, pellets, and similar devices are provided with a fusible element that melts once the heat effects of a fire are met. Once the element melts, it causes the sprinkler heads, fire alarms, fire dampers, or doors to activate as required by the fire condition to afford protection. *See also* **Eutectic Solder; Fixed Temperature Detector; Fusible Alloy; Fusible Metal; Heat Detector; Thermal Element.**

Fusible Link—A release device activated by the heat effects of a fire. It usually consists of two pieces of metal joined by solder that is specified to melt at a certain temperature. Fusible links are named for the "linking together" of two separate components that are released upon heat detection. They are manufactured at various temperature ratings and are subject to varying normal maximum tension. Fusible links are available in temperature ratings of 125°F to 500°F (51.6°C to 260°C) and in various load ratings. Temperature ratings for a particular installation are usually chosen that are at least 50°F (10°C) above maximum expected ambient conditions. When installed and the specified fixed temperature of the device is reached, the solder melts and the two metal parts separate due to the applied tension. As a result, specific actions are initiated, such as signaling the occurrence of a fire; closing fire doors, windows, or dampers; actuating fire extinguishing systems; or releasing water from a sprinkler head. The invention of the fusible link is attributed to Edward Atkinson, President of the Boston Manufacturers Mutual Fire Insurance Company for application in the use of self-closing hatches, doors, and shutters in the early 1900s.

Fusible Metal—A metal alloy having a specific melting point, used as solder and for safety plugs and fuses. Cadmium lowers the melting point of metals with which it is alloyed; it is commonly used with lead, tin, and bismuth in the manufacture of fusible metals for automatic sprinkler systems, fire alarms, and electric fuses. The melting point chosen for fire protection systems is normally 50°F (10°C) above the maximum expected ambient temperatures as a safety factor. They are generally available in similar temperature ratings to those of automatic sprinklers and have load ratings (maximum tension) of 5, 8, 10, 15, and 20 pounds (2.3, 3.6, 4.5, 6.8, and 9 kg). Drop weight releases are commonly designed for the support of loads of 50 to 100 pounds (23 to 45 kg). Fusible solders were first discovered by Sir Isaac Newton in 1699. He found that certain alloys have a lower melting point than any of their constituents. *See also* **Fusible Alloy; Fusible Element; Heat Detector; Thermal Element.**

Fusible Plug or **Pellet**—A heat-activated solder provided in an outlet of a pneumatic fire detection system or sprinkler system to indicate the presence of a fire condition and to disperse the gas or fluid in the contained system. The melted solder allows release of the gases contained in the pneumatic system for fire alarm activation, device activation, or water release from a sprinkler head. *See also* **Heat Detector.**

Fusible Solder—*See* **Fusible Metal.**

Fusible Valve—A shutoff valve provided with a fusible element that causes it to automatically close when subjected to elevated temperatures. They are usually provided on a fuel line at a location where a fire may occur (heater or burner). The valve is normally kept open until the fusible element melts, causing the valve to close and preventing additional fuel to feed the fire incident. Fusible valves commonly have a temperature rating of 165°F (74°C), but may be supplied with a higher rating for special circumstances. *See also* **Fire Valve.**

Gas Blanketing or **Inerting**—The provision of an oxygen deficient gas into an enclosed area for the purpose of excluding oxygen from the enclosure. It is normally instituted to prevent a combustible air-gas mixture from forming in the enclosure. The process industry sometimes uses gas blanketing for storage tanks as a safety feature when a gas supply is readily available. Large oceangoing tanker vessels are equipped with a continuous gas inerting system that collects expelled combustion gases from the vessel's engines to blanket the liquid cargo holds or tanks. *See also* **Inerting.**

Gas Cartridge or **Cylinder Extinguisher**—A portable fire extinguisher in which the firefighting agent and expelling medium are stored together but in separate pressure vessels until required, at which time the agent container is pressurized by the expelling medium. The cartridge is a small replaceable cylinder normally filled with carbon dioxide (CO_2). Nitrogen-filled cartridges are used where low temperatures may be encountered. The fire extinguisher is operated by a lever that punctures a sealed orifice in the cartridge, which then pressurizes the large container of fire extinguishing agent from the cartridge. A separate nozzle valve is provided on the discharge hose attached to the large container to release and control agent flow onto the fire. A cartridge operated multipurpose dry chemical portable fire extinguisher is an example of a gas cartridge extinguisher. *See also* **Fire Extinguisher, Portable.**

Gas, Combustible—Gas that is capable of being ignited and burned, such as hydrogen, methane, propane, etc. *See also* **Flammable Gas.**

Gas Detector, Combustible—*See* **Combustible Gas Detector.**

Gas, Sewage—The fermentation or decomposition of organic matter results in sewage gas. These gases can be produced when organic matter has settled as a solid in sewer lines. This can occur as a result of flat grades, crevices, sumps, or obstructions where consistent flow of sewage is lacking, or as a result of bacterial action on wood or other organic material immersed in water. These flammable gases are principally methane, hydrogen sulfide, and hydrogen. Historical evidence suggests sewage gases seldom reach explosive concentrations in sewers and drains. When sewer gases are mixed with other flammable liquids and gases present in sewers, explosive conditions might exist.

General Staff—For an incident command these include operations, planning, logistics and administration.

Glow—The visible light emitted by a substance because of its high temperature.

Glowing Combustion—Luminous burning of solid material without a visible flame. A stage in the ignition of a solid material that occurs before sufficient volatile fuel has evolved to sustain a gas-phase flame. *See also* **Smoldering.**

Goad Map—*See* **Sanborn Map.**

Grading Schedule—*See* **Municipal Grading Schedule.**

Grass Fire—A generic term for a fire occurring in vegetative ground cover (grass, brush, etc.).

Greek Fire—A gelatinous, incendiary mixture used in warfare before gunpowder was invented. Flammable liquids had long been in use, but it was not until the 7th century that Greek fire was invented, presumably by Callinicus (circa 620–673), an Egyptian architect who had fled from Syria during the Muslim invasions. The formula was closely guarded as a state secret for many centuries by the Byzantine Empire. The exact composition of Greek fire is still disputed, but it was probably composed of a mixture of flammable materials such as sulfur and pitch on a petroleum base, allowing it to float and continue burning on water. This jelly-like mixture was sprayed on the enemy from tubes through which it was forced under pressure by pumps.

Ground Fault Interruptor—Device installed on an electrical circuit to automatically detect and trip the circuit should a grounding condition occur.

G

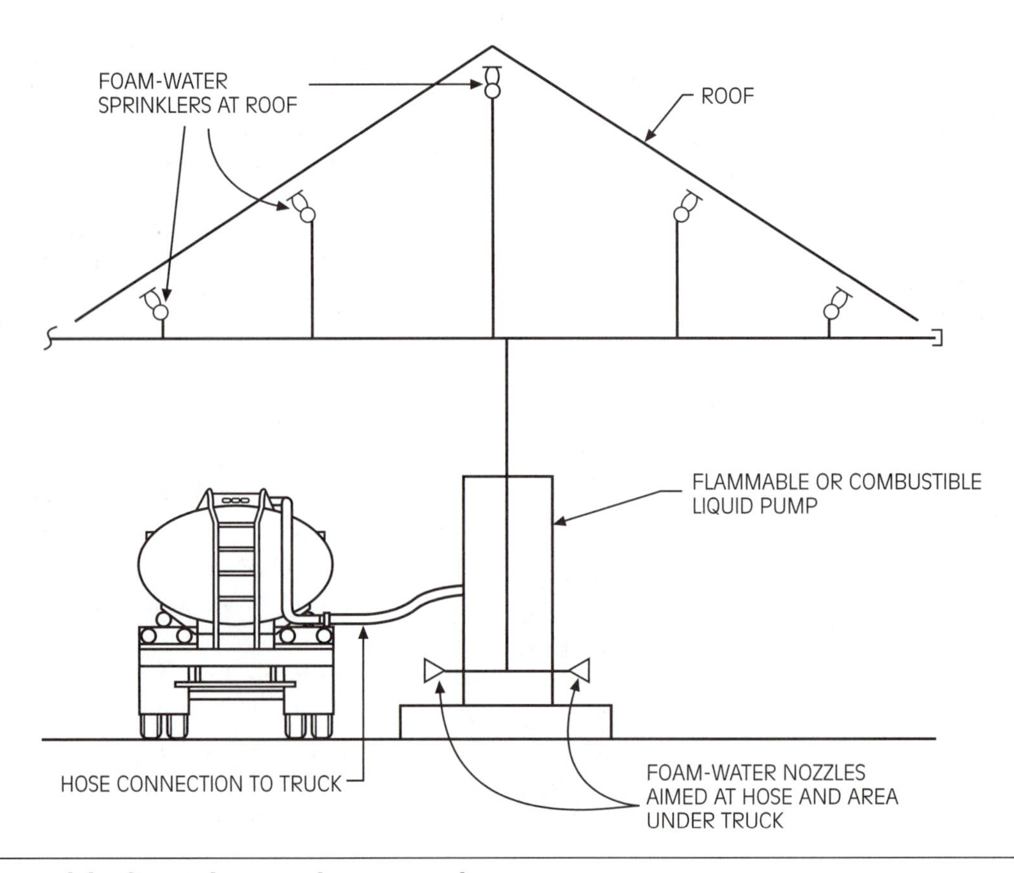

Figure G-1 Truck loading rack—ground sweep nozzles.

Used to prevent electrical shocks to individuals should they be in contact with the circuit.

Ground Fire—*See* **Forest Fire.**

Ground Sweep Nozzles—Water spray nozzles mounted at ground level to provide protection against combustible liquid spill fires. Commonly provided on the underside of a fire vehicle to disperse flames from the underside of the vehicle or at truck loading racks for petroleum products. See Figure G-1.

Gusset Plate—Flat sheet metal plate used to connect wood truss members. Fails quickly under heat and fire conditions that lead to structural collapse.

Gutter Line—Slang terminology for a fire hose line that is placed or laid in a road drainage trough (gutter) or as close to the side curbing as possible. The placement of hose in the gutter allows other vehicles to enter or maneuver at a fire scene without having to run over the fire hose. See Figure G-2.

Figure G-2 Gutter lines.

Haines Index—An atmospheric index used to indicate the potential for wildfire growth by measuring the stability and dryness of air over a fire.

Half Hitch—A type of knot used in the fire service, usually in conjunction with another type of knot. Basically is a loop of rope around an object. It is typically used to maintain the orientation of an object be hoisted. See Figure H-1.

Halligan Tool—A forcible entry tool used by the fire service. It consists of a long bar with a claw at one end and one or two spikes at right angles at the opposite end. It is generally used to pry open locked or stuck doors and windows to gain access for firefighting operations or rescue activities. It was designed by Chief Halligan of the New York City Fire Department. It may also be called a Hooligan Tool. See Figure H-2. *See also* **Claw Tool; Forcible Entry; Hux Bar; Kelly Tool.**

Halogenated Agents—Chemical compounds (halogenated hydrocarbons) that consist of a carbon atom plus one or more elements from the halogen series of elements (fluorine, chlorine, bromine, iodine, and astatine group). Halon 1201 and 1301 are common halogenated agents that have been used for fire suppression applications. *See also* **Halon.**

Halon—Halon is a contraction of halogenated hydrocarbons, commonly bromo-trifluoro-methane (Halon 1301, $CBRF_3$) and bromo-chloro-difluoro-methane (Halon 1211, $CBrCLF_2$). A Halon may be any of a group of halogenated aliphatic hydrocarbons, most of which are derived from methane or ethane by replacing

Figure H-1 Half hitch.

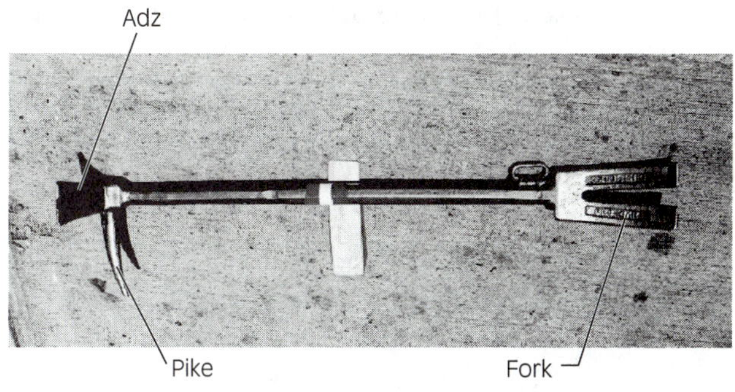

Figure H-2 Halligan tool.

some or all of the hydrogen atoms by atoms of the halogen series (fluorine, bromine, chlorine, or iodine). The effectiveness of Halons in extinguishing fires arises from their action in interrupting chain reactions that propagate the combustion process. Halon 1301 is considered three to ten times more effective than carbon dioxide (CO_2) as a fire extinguishant. Halogenated hydrocarbons take the form of liquefied gas or vaporizing liquids at room temperature. Halons are nonconductors of electricity and can be used in fighting fires in flammable liquids and most solid combustible materials, including those in electrical equipment. They are ineffective on fuels containing their own oxidizing agent or highly reactive metals, such as sodium or potassium. Halon 1301 (bromotrifluoromethane, $CBRF_3$) is especially favored for extinguishing fires involving electronic equipment because it leaves no residue and does not cause electrical short circuits or damaging corrosion of the equipment.

Because Halon is a compound considered destructive of the Earth's ozone shield, environmentalists have urged restriction of its use. It is generally considered obsolete for fire protection purposes and has been eliminated, phased out, or replaced with agents that are nondestructive to the ozone layer. *See also* **Fire Suppressant Agent; FM-200; Halongenated Agents; Inergen.**

Halyard—Rope used on a extension ladder by the fire service to extend the fly sections. May also be called a fly rope.

Handline—A small fire hose, usually 2.5 in. (6.5 cm) or smaller, that is manually handled by firefighters and can be maneuvered without mechanical assistance. *See also* **Booster Hose.**

Hand-Propelled Extinguisher—A fire extinguishing agent that is manually applied by individuals using a bucket, pail, scoop, shovel, etc.

Hand Pump—A small pump that is worked by hand. It usually delivers between 1 and 3 gpm (3.8 and 11.3 liters/min.). It is useful for very small or incipient fires with a nearby water supply.

Hard Suction Hose—*See* **Hose, Hard Suction.**

Hardy Cross Method—A mathematical calculation method of reiterative analysis to determine the flow of water in a network of gridded or looped pipes. It is the oldest and most widely used method for analyzing flow in pipe networks. It was developed by Professor Hardy Cross (1885–1959) at the University of Illinois, in 1936. The Hardy Cross method is an adaptation of the Newton-Raphson method, which solves one equation at a time before proceeding to the next equation during each iteration, instead of solving all equations simultaneously. *See also* **Newton-Raphson Method.**

Hazardous Fire Area—Terminology used in the uniform fire code to describe an area of grass, brush, or forest with limited accessibility such that a fire originating in such an area would be unusually difficult to extinguish and would lead to significant damage and potential erosion. Similar to the term fire hazardous zone used in the petroleum processing industry to describe an area of high risk for petroleum fires particularly at pumps, compressors, and fired heaters where extra fire precautions are implemented, such as fireproofing of structural members. *See also* **Fire Hazardous Zone (FHZ).**

Hazardous Material—A material that presents a potential danger to the life, health, property, or the environment due to its fire, explosion, toxicity, or similar effects. May be referred to as HAZMAT. *See also* **Material Data Safety Sheet.**

Hazardous Materials Identification System (HMIS)—A simple identification system (placard) used to rapidly determine the relative toxicity, combustibility, and reactivity hazard of a substance during emergency situations. It uses a three-place format with numerical indexes from 0 to 4. The first place is for toxic properties, the second place is for flammability, and the third place is for reactivity with other chemicals. Most portable fire extinguishers have a zero numerical index in the second and third places because they are nonflammable and relatively inert. NFPA 704 provides guidance in the application of the hazardous material identification system. *See also* **Fire Hazard Identification; Material Safety Data Sheet (MSDS).**

Hazen-Williams Formula—A mathematical friction loss flow formula showing the relationship among various factors governing the flow of water in pipes. It was developed by Gardner S. Williams and Allen Hazen in 1905. The formula is based on the work of Antoine de Chezy (1718–1798). The Hazen-Williams formula is widely applied in the design of water-based fire protection systems and in the hydraulic evaluation of water distribution systems.

The Hazen-Williams formula for friction loss is as follows:

$$P = [(4.52) \times (Q^{1.85})]/[(C^{1.85}) \times (D^{4.87})]$$

Where:

P = friction loss, psi per foot
Q = water flow in gpm
C = Hazen-Williams pipe internal roughness coefficient (C factor)
D = internal diameter of pipe, in inches

See also **C Factor; Hydraulic Calculation; Moody Diagram.**

HAZMAT—*See* **Hazardous Material.**

HAZOP—An acronym for hazard and operability study in which the hazards and operability of a system are identified and analyzed in a systematic manner to determine if adequate safeguards are in place. HAZOP studies form a qualitative risk analysis and are primarily used in the process industries to evaluate unforeseen operations resulting in hazards affecting the facility. It forms one of the tools for effective process safety management and can identify potential sources of a fire incident. The HAZOP technique originated with the chemical industry in England in the 1960s.

Head—The energy of a fluid derived from its height, velocity, or pressure. The static head for water is about 0.434 psi for every 1 ft. of elevation (9.82 kPa for every 1.0 meter). It is also sometimes called head pressure when specifically referring to the pressure of the vertical height of a liquid column. It may also refer to the advancing front of a forest fire. *See also* **Static Pressure.**

Head (Sprinkler)—*See* **Sprinkler Head.**

Heat—A form of (kinetic) energy associated with the motion of atoms or molecules and capable of being transmitted through solid and fluid media by conduction, through fluid media by convection, and through empty space by radiation. Excessive heat exposure may cause burns to individuals and ignition and burning to materials. Limits of heat exposure or clearance distances are imposed to prevent injury or damage where heat sources may be constantly encountered (heat radiation levels, exposed equipment operating temperature limits, exhaust gases, etc.). *See also* **Heat Conduction.**

Heat Actuated Device (HAD)—A pneumatic rate-of-rise fire detector based on heat exposure. A hollow enclosure expands due to the presence of heat rise and activates an alarm device through a pressure switch. An HAD is a spot-type detector, and area coverage is provided by providing several detectors that are arranged and

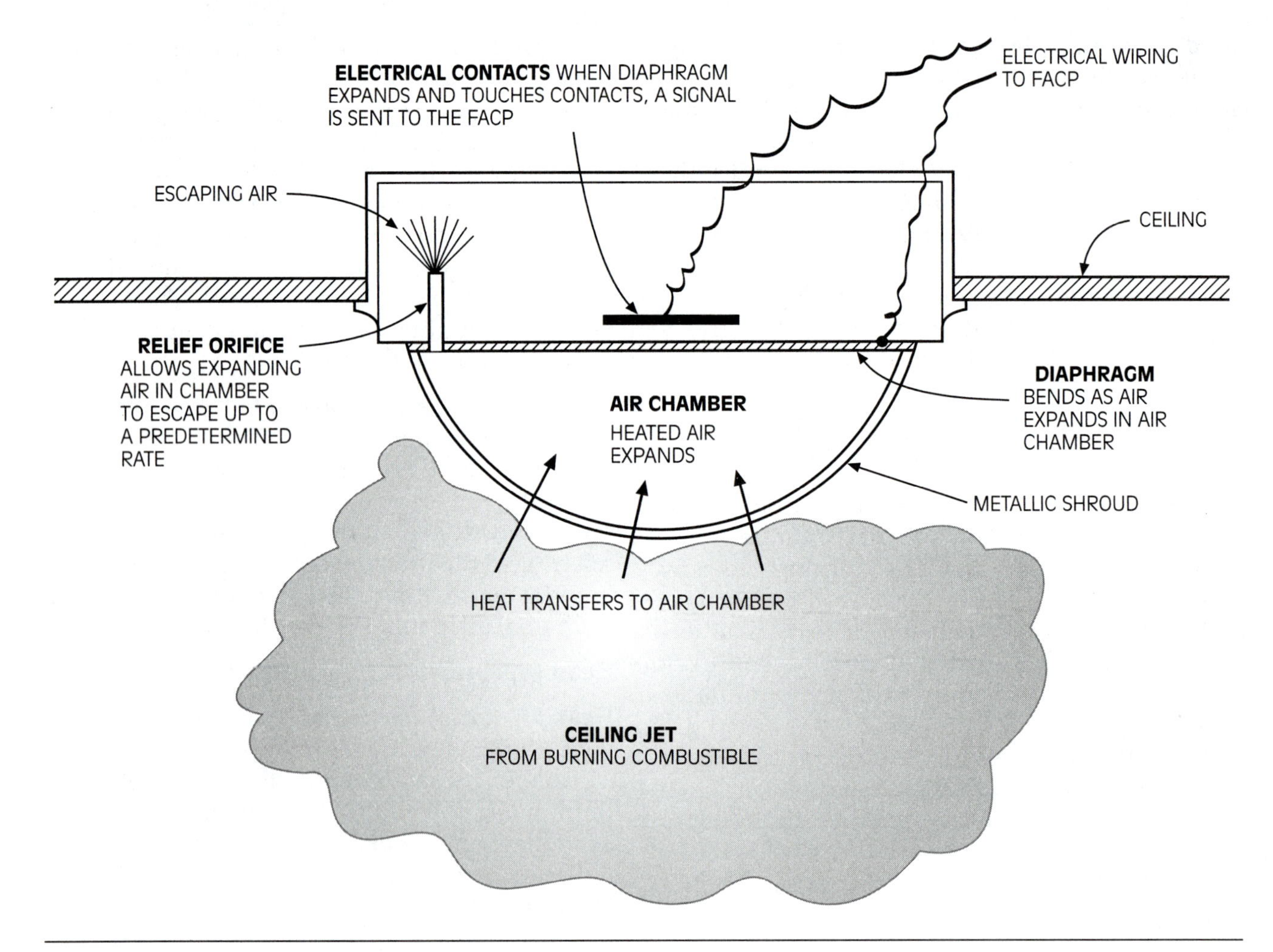

Figure H-3 Spot-type pneumatic rate-of-rise heat detector.

spaced according to specific requirements listed in fire codes (NFPA 72, *National Fire Alarm Code*). See Figure H-3. *See also* **Heat Detector.**

Heat Collector Plate or **Canopy**—A covering provided over a heat detector or automatic sprinkler placed in the open, to trap and collect updrafts of heat from a fire incident to aid in its detection or sprinkler activation. They commonly consist of a sheet of steel formed into a 12 in. by 12 in. (30.5 cm by 30.5 cm) canopy.

Heat Conduction—The passage of heat energy through a material. Heat conduction involves transfer of energy and entropy between adjacent molecules. In fires, heat conduction is usually a slow process compared to other methods of heat transmission (convection or radiation). *See also* **Conduction; Heat.**

Heat Convection—The process in which heat is transferred from one place to another by the movement of matter, that is, by liquid or gas. Fires usually produce heat convection currents that rise upward, unless affected by other influences (wind, obstacles, etc.). *See also* **Convection.**

Heat Detection—The sensing and signaling of elevated temperatures. The detection of heat can be made by sensing a specific level of heat by temperature set points or by the rate of temperature rise. Area heat detection can be accomplished by installing spot devices or linear detection wires made of dissimilar metals. Heat detection set points are usually set above maximum expected ambient temperatures with a safety factor included to avoid false alarms or false activation circumstances. *See also* **Pneumatic Fire Detection System; Thermal Detection.**

Heat Detector—A device that senses elevated ambient temperature based upon a specific set point (fixed temperature) or the rate of heat (temperature) rise. Heat detectors are most effective for fires that evolve heat rapidly with very little smoke. They are not suitable where large losses could be caused by just a small fire (for example, in computer rooms). In general, heat detectors are slower to respond than smoke detectors but less likely to cause false alarms. A fixed temperature heat detector responds when the temperature reaches a set level. They are available to cover a wide range of operating temperatures from 135°F (57°C) and up. Detectors sensitive to various ranges can be installed in a zoned fire detection system. A rate compensated-type heat detector responds to the surrounding air temperature rather than to the detector element temperature, thereby reducing thermal lag. The rate-of-rise heat detector activates when the rate of temperature increase exceeds a predetermined value, typically 12°F to 15°F (7°C to 8°C) per minute. The rate-of-rise detectors are designed to compensate for the normal changes in ambient temperature caused by work conditions or processes. Very few heat detectors still use bimetallic strips that sense a change in electromotive force due to heat absorption. Most heat detectors use thermistors and electronics to provide better performance at a lower cost. The mining and petroleum industries sometimes employ fusible plastic tubing (or metal tubing with strategically located fusible plugs), pressurized with inert gas or compressed air, that is draped around likely fire hazards. Heat from the fire causes the tube (or fusible plugs) to melt, releasing the gas pressure and activating a mechanical pneumatic actuator or pressure switch and sending an alarm. *See also* **Fire Alarm Signal; Fixed Temperature Detector; Fusible Element; Fusible Metal; Fusible Plug** or **Pellet; Heat Actuated Device (HAD); Rate Compensated (Heat) Detector; Rate-of-Rise (Heat) Detector.**

Heat Exhaustion—An illness characterized by fatigue, nausea, headache, dizziness, pallor, weak pulse, thirst, rapid and shallow breathing, and profuse sweating from exposure to excessive heat. Firefighting activities can lead to heat exhaustion if adequate prevention measures are not taken. Factors that may predispose a person to heat exhaustion include sustained exertion in a heated environment, failure to replace water lost in sweating, and lack of acclimatization. Heat exhaustion treatments include moving to a cooler environment, resting in a recumbent

H

position, and administering fluids by mouth. In severe cases, heat exhaustion can lead to collapse, coma, or death. It may also be called heat prostration. *See also* **Fire Injury; Heat Prostration; Heat Stress (Temperature Stress)** or **Hyperthermia; Heat Stroke.**

Heat Flux—The rate of heat transfer per unit area that is normal to the direction of heat flow. It is a total of heat transmitted by radiation, conduction, and convection. A radiant heat flux of 312.5 Btu ft^{-2} h^{-1} (1 kW m^{-2}) (that is, direct sunlight) will be felt as pain to exposed skin. A radiant heat flux of 1,250 Btu ft^{-2} h^{-1} (4 kW m^{-2}) will cause a burn on exposed skin. A flux density of 3,125 to 6,250 Btu ft^{-2} h^{-1} (10 to 20 kW m^{-2}) may cause objects to ignite, and a flux density of 11,813 Btu ft^{-2} h^{-1} (37.8 kW m^{-2}) will cause major fire damage. Jet fires may have heat fluxes of 62,500 to 93,750 Btu ft^{-2} h^{-1} (200 to 300 kW m^{-2}), and pool fires may have values of 9,375 to 15,625 Btu ft^{-2} h^{-1} (30 to 50 kW m^{-2}). Recent research indicates that heat flux is a more realistic method to determine heat transmission into fire barriers. A heat flux of 3,125 to 6,250 Btu ft^{-2} h^{-1} (10–20 kW m^{-2}) in an enclosed compartment usually corresponds to the smoke layer temperature of 932°F (500°C) and is taken as the point of flashover initiation.[3] Heat flux may also be called heat flow rate.

Heat Flux, Critical—A heat flow rate threshold level below which ignition is not possible.

Heat of Combustion—The quantity of heat given off by a particular substance during the combustion process is its heat of combustion. All combustion reactions release heat and therefore are exothermic in nature. The quantity of heat can be calculated; usually it is expressed in joules or calories per gram and for a particular fuel. It is known as the calorific value (1 calorie = 42 joules). The heat of combustion provides a measure of fuel efficiency. *See also* **Calorimetric Bomb Test.**

Heat of Decomposition—The amount of heat released during a chemical decomposition reaction.

Heat of Ignition—The heat energy that is sufficient to act as an ignition source. Heat energy comes in various forms and usually from a specific object or source. The heat of ignition is divided into two parts: equipment involved in ignition and form of heat of ignition. Heat of ignition can be an open flame, a hot surface, an arc, spark, or some other similar form. *See also* **Ignition Energy.**

Heat Prostration—A form of heat exhaustion. It can be distinguished readily from heat stroke by the moderate or absent elevation of body temperature and by the persistence of heavy sweating. *See also* **Heat Exhaustion.**

Heat Protective Shield—*See* **Heat Shield.**

Heat Radiation—The amount of heat transmitted by radiation. Radiant heat can be expressed mathematically as $E = f(e, A, T^4)$ where E represents the radiative power of a flame. It is a function of the flame emissivity e, the exposed area A, and the flame temperature T to the fourth power. Flame temperature is the dominant factor in the function. Individuals with appropriate clothing can tolerate a radiant heat intensity of about 1,500 Btu/h ft^2 (4.73 kW/m^2) for several minutes. *See also* **Heat Transfer.**

Heat Release Rate (HRR)—The rate at which heat energy is generated by burning. The heat release rate of a fuel is related to its chemistry, physical form, and availability of oxidant, and is ordinarily expressed as kilowatts (kW) or

[3] Quintiere, J. G., *Principles of Fire Behavior,* Thomson Delmar Learning, Clifton Park, NY, 1998.

Btu/second. Currently, the cone calorimeter, first announced in 1982, is typically used (or the furniture calorimeter) to determine heat release rates (Ref. ASTM E-1474-92 and ANSI/NFPA 264A).

Heat Resistant—The ability to withstand the effects of high temperature. The most heat-resistant substance was invented in 1993 and is known as NFAARr or Ultra Hightech Starlite. It can temporarily resist temperatures of 18,032°F (10,000°C). *See also* **Fireproofing.**

Heat-Resistant Glass—A glass with a low coefficient of expansion, such as borosilicate glass. It is used as a wire-free, fire-resistant glass. *See also* **Wired Glass.**

Heat Shield—A protective barrier, coating, or system against heat effects. A heat shield does not prevent the spread of fire or act as a smoke barrier, but only limits the temperature of the unexposed side of the heat shield to a predetermined amount. Heat shields are not tested to a recognized standard but are required to perform to a specified heat radiant level and unexposed side temperature limitation. They may be as simple as a noncombustible reflective material against the effects of radiated heat. In the process industries, a double-skinned membrane with an interior insulating layer has been used against the heat effects from flares. Spacecraft generally are provided with a heat shield to resist the effects of air friction upon reentry into the Earth's atmosphere, where temperatures of 2,700°F (1,500°C) can be achieved. American Petroleum Institute (API) Recommended Practice (RP) 521, *Guide for Pressure-Relieving and Depressuring Systems,* provides guidance in the level of heat exposure pain thresholds and tolerance for normal work or emergency activities. Firefighter protective clothing is classified by various fire exposure performance standards mentioned in several NFPA standards: 1971, *Standard on Protective Clothing for Structural Fire Fighting;* 1972, *Standard for Helmets for Structural Fire Fighting;* 1973, *Standard on Gloves for Structural Fire Fighting;* 1974, *Standard on Protective Footwear for Structural Fire Fighting;* 1976, *Standard on Protective Clothing for Proximity Fire Fighting;* 1977, *Standard on Protective Clothing and Equipment for Wildland Fire Fighting.*

Heat Stress (Temperature Stress) or **Hyperthermia**—Physiological stress induced on the body by excessive temperature (heat). It can impair functioning and cause injury or death. Exposure to intense heat increases body temperature and pulse rate. Heat stress can be imposed by excessive air temperature, water vapor pressure, radiant heat, air circulation, level of physical activity, or type of clothing. If body temperature is sufficiently high, sweating may cease, the skin may become dry, and deeper and faster breathing may follow. Headaches, nausea, disorientation, fainting, and unconsciousness also may occur. Heat stress occurs when the human body's core temperature reaches 105.8°F (41°C); the normal temperature is 98.6°F (37°C). *See also* **Heat Exhaustion; Heat Stroke.**

Table H-1 Heat Stress Tolerance[4]

Exposure Temperature °F (°C)	Relative Humidity %	Tolerance Time
120 (49)	10	~ 10 days
120 (49)	50	~ 2 hours
120 (49)	100	~ 10 minutes
212 (100)	0–100	~ 10 minutes

[4] Dinneno, P. J. ed., *SFPE Handbook of Fire Protection Engineering,* Second Edition, D. A. Puser, Chapter 2, Section 8, National Fire Protection Association, Boston, MA, 1995.

Heat Stroke—A heat stress illness caused by heat exposure that is considered a medical emergency. Firefighting activities can lead to a heat stroke if adequate prevention measures are not taken. Symptoms include hot, red, dry skin, a rectal temperature of 104°F (40°C) or higher, rapid strong pulse, heavy breathing, confusion, possible convulsions, and/or loss of consciousness. Factors that may predispose a person to heat stroke include sustained exertion in the heat by un-acclimatized workers, lack of physical fitness, obesity, recent alcohol intake, dehydration, individual susceptibility, and chronic cardiovascular disease. Immediate treatment of suspected heat stroke is required. Treatments to rapidly reduce body temperature include immersing in chilled water, rinsing with alcohol, wrapping in a wet sheet, and fanning with cool, dry air. A physician's care is required to treat possible secondary disorders such as shock or kidney failure. If left untreated, heat stroke can lead to delirium, convulsions, coma, and even death. Heat stroke is a more serious condition than heat exhaustion. *See also* **Fire Injury; Heat Exhaustion; Heat Stress (Temperature Stress)** or **Hyperthermia.**

Heat Transfer—Any or all of several kinds of phenomena, considered mechanisms, that convey energy and entropy from one location to another. The specific mechanisms are usually referred to as conduction, convection, and thermal radiation. Conduction involves the transfer of energy and entropy between adjacent molecules, usually a slow process. Convection involves movement of a heated fluid, such as air, usually a rapid process. Thermal radiation refers to the transmission of energy as electromagnetic radiation from its emission at a heated surface to its absorption on another surface, a process requiring no medium to convey the energy. *See also* **Conduction; Convection; Heat Radiation; Thermal Convection.**

Heat Wave—For fire protection applications, a heat wave is the transmission of heat downward in crude or heavy oil in a confined space or storage tank into the bulk of the unheated oil. This occurs because of the relative different densities of the liquids that make up these oils. As the lower densities or fractions burn at the surface of the oil, the heavier fractions with a higher boiling point do not heat up and burn, but sink to the bottom of the container. Because the oil in the tank is above the boiling point of water, an explosive expulsion (frothing or boilover) of the tank over may occur once the heat wave reaches near the bottom of the tank where the water has settled (causing it to turn to steam). The heat wave for heavy crude oil or fuel oil that has been burning for some time generally travels at 9 to 15 in. (23 to 38 cm) per hour, depending on the fuel type. *See also* **Boilover.**

Heavy Content Fire Loading—Terminology for the storage of combustible materials in high stacks in close proximity to each other. The arrangement causes an increase in the amount of fuel loading above average levels. *See also* **Fire Load** or **Fire Loading; Fuel Loading.**

Heavy Stream—A large diameter stream of water for fire protection applications. It consists of a flow of water generally greater than 400 gpm (25.2 liters per second). A heavy stream is usually too difficult to control with manual methods and is directed by monitor nozzles, deluge guns, water cannons, ladder pipes, or turret pipes. *See also* **Hose Stream; Master Stream; Monitor Wagon; Water Cannon.**

Heavy Timber Construction—*See* **Building Construction Types.**

Helmet—*See* **Fire Helmet.**

Hephaestus—The Greek god of fire and metalworking. Representative of fire worship in Greek civilization. *See also* **Vulcan.**

Higbee Indicators or **Cut**—Notches or grooves provided in fire hose couplings to indicate by sight or touch the beginning of the thread. They are provided and used to prevent damage to the coupling by mistakenly cross-threading the coupling. *See also* **Blunt Start.**

High Backpressure Foam Maker—A foam maker utilizing the venturi principle for aspirating air into a stream of foam solution to form foam under pressure. Sufficient velocity energy is conserved in this device that the resulting foam can be conducted through piping or hoses to the hazard being protected. It is used where considerable backpressure exists from the hazard being protected (combustible liquid storage tank). See Figure H-4. *See also* **Subsurface Foam Injection.**

High Challenge Fire Hazard—Generally understood as fire hazards posed by high-piled storage.

Higher Education Project—A program of the Emergency Management Institute to promote emergency management degree programs nationwide.

High Expansion Foam—An aggregation of bubbles that are mechanically generated by the passage of air or other gases through a net, screen, or other porous medium that is wetted by an aqueous solution of surface active foaming agents. Under proper conditions, firefighting foams with expansion ratios from 20:1 to 1,000:1 can be generated. Expansion ratios for high expansion foams range from 200 and above. These foams provide a unique agent for transporting water to inaccessible places; for total flooding of confined spaces; and for volumetric displacement of vapor, heat, and smoke. Tests have shown that under certain circumstances, high expansion foam, when used in conjunction with water sprinklers, provides more positive control and extinguishment than either form of extinguishment system by itself. High-piled storage of rolled paper stock is an example. Optimum efficiency in any one type of hazard depends to some extent on the rate of application and the foam expansion and stability. High expansion foam is an agent for control and extinguishment of Class A and Class B fires and is particularly suited as a flooding agent in confined spaces. High expansion foam can also be used on solid and liquid fuel fires, but the in-depth coverage it provides is greater than for medium expansion foam. Therefore, it is most suitable for filling volumes in which fires exist at various levels. For example, experiments have shown that high expansion foam can be used effectively against high-rack storage fires, provided that the foam application is started early and the depth of foam is rapidly increased. It also can be used to extinguish fires in enclosures where it might be dangerous to send personnel (in basement and underground passages). It can be used to control fires involving liquefied natural gases (LNG) and liquefied petroleum gases (LPG), and to provide vapor dispersion control for LNG and ammonia spills. High expansion foam is particularly suited for indoor fires in confined spaces. Its use outdoors can be limited because of the effects of wind and lack of confinement.

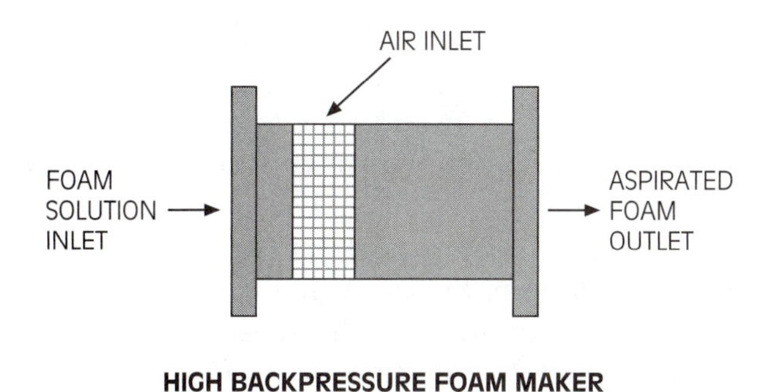

Figure H-4 High backpressure foam maker for subsurface injection.

High expansion foams have the following effects on fires:

1. High-expansion foams generated in large volumes can prevent the free movement of air, which is necessary for a combustion process to continue.
2. When they are forced into a fire, the water in the foam is converted to steam, thus reducing the oxygen concentration by dilution of the air.
3. The conversion of the water to steam absorbs heat from the burning fuel. Any hot object exposed to the foam continues the process of breaking the foam, converting the water to steam, and cooling.
4. Because of their relatively low surface tension, solution from the foams that is not converted to steam will tend to penetrate Class A materials. However, deep-seated fires might require overhaul.
5. Where it is applied and reaches sufficient depth, medium and high expansion foams can provide an insulating barrier for the protection of exposed materials or structures not involved in a fire and can thus prevent fire spread.
6. For LNG fires, high expansion foam will not normally extinguish a fire, but it will reduce the fire intensity by blocking radiation feedback to the fuel.
7. Class A fires are controlled when the foam completely covers the fire and burning material. If the foam is sufficiently wet and is maintained long enough, the fire can be extinguished.
8. Class B fires involving high flash point liquids can be extinguished when the surface is cooled below the flash point. Class B fires involving low flash point liquids can be extinguished when a foam blanket of sufficient depth is established over the liquid surface.

The development of high expansion foams for firefighting purposes started with the work of the Safety in Mines Research Establishment of Buxton, England, which was working on the difficult problem of fires in coal mines. It was found that by expanding an aqueous surface active agent solution to a semi-stable foam of about 1,000 times the volume of the original solution, it was possible to force the foam down relatively long corridors, thus providing a means for transporting water to a fire inaccessible to ordinary hose streams. This work led to the development of specialized high expansion foam-generating equipment for fighting fires in mines, for application in municipal industrial firefighting, and for the protection of special hazard occupancies. Medium expansion foam was developed to meet the need for a foam that was more wind resistant than high expansion foam for outdoor applications. *See also* **Expansion Ratio; High Expansion Foam Generator; Water Additive.**

High Expansion Foam Generator—A device for the formation of high expansion foam. It normally is powered by the pressure of the incoming foam solution driving a hydraulic motor. The foam solution is sprayed on a screen where the air stream created by a fan attached to the hydraulic motor creates foam bubbles. The continuous flow of solution and air produces a high volume of foam. See Figure H-5. *See also* **High Expansion Foam.**

High Explosive—A material that is capable of sustaining a reaction front that moves through the unreacted material at a speed equal to or greater than that of sound in that medium, typically 3,300 ft./sec. (1,000 meters/sec.). A material that is capable of sustaining a detonation. *See also* Detonation; Explosive.

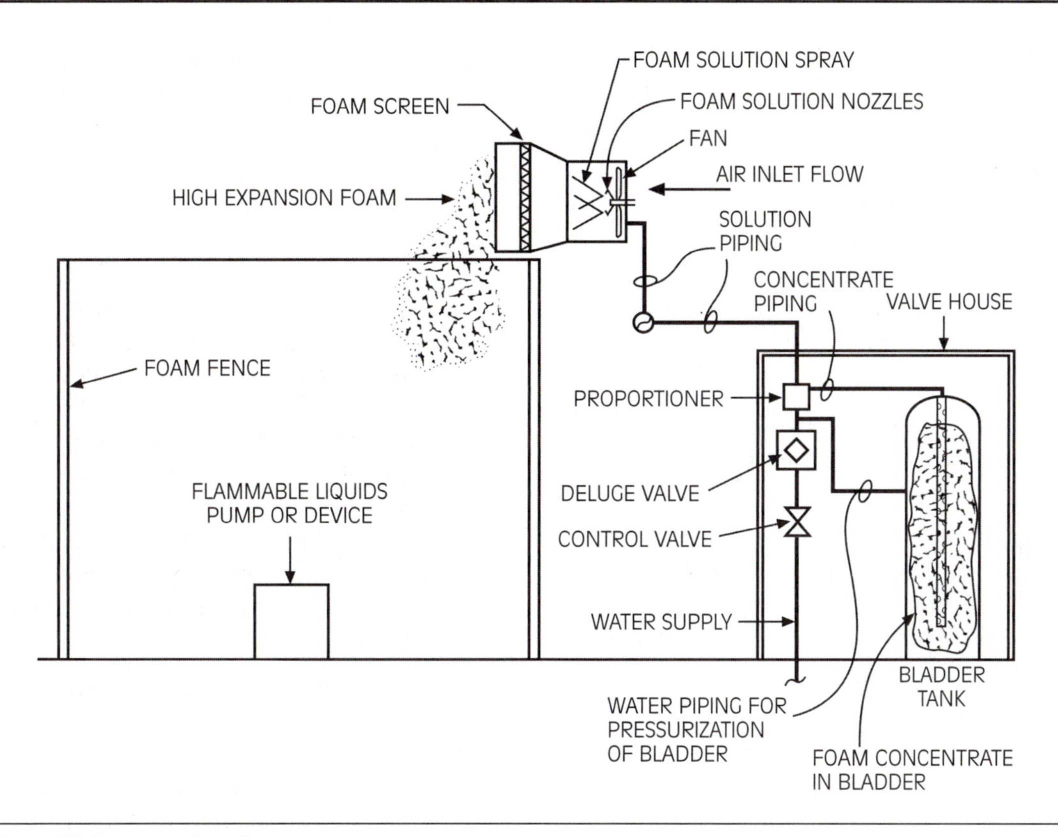

Figure H-5 High expansion foam generator.

Highly Protected Risk (HPR)—Term used within the insurance industry to describe a property risk that has sprinkler protection and is considered a superior facility from a fire protection viewpoint (low probability of loss). It therefore has a very low insurance rate compared to other industrial risks.

As defined by the insurance industry, a highly protected risk (HPR) occupancy also has the following features in addition to a loss-prevention-responsive management:

1. The occupancy is protected by automatic sprinklers with adequate water supply.
2. The sprinkler system is capable of controlling a fire incident without human intervention.
3. The sprinkler system alarms are supervised to give early notification to the fire department.
4. The early detection and alarm system provides the responding fire department with a greater opportunity to extinguish the fire.

High Order Explosion—A rapid pressure rise or high-force explosion characterized by a shattering effect upon the confining structure or container and long missile distances.

High Rise Intensity Fire—*See* **Time Temperature Curve, Hydrocarbon Fire.**

High Sensitivity Smoke Detection (HSSD)—*See* **Aspirating Smoke Detection (ASD).**

H

Holdover Fire—A fire that remains dormant (smoldering) for a long time. It may also be called a hangover fire or a sleeper fire. *See also* **Smoldering.**

Holocaust—Great destruction resulting in extensive loss of life and property, especially referred to when fire is involved. The word "holocaust" is derived from the Greek holokauston, which originally meant a sacrifice totally burned by fire.

Homeland Security—A department of the U.S. government to administer and coordinate national threats and emergencies.

Hook and Ladder—An obsolete term used to describe an aerial ladder truck. It was used in the fire service when hooks were commonly employed in firefighting efforts to pull down buildings to create a firebreak.

Horseshoe Load or **Fold**—*See* **Accordion Horseshoe Load** or **Fold.**

Hose Bed—Portion of a fire apparatus that carries firefighting hoses.

Hose, Booster—A hose intended for use on a fire apparatus for suppression activities. It has a rubber tube, a braided or spiraled reinforcement, and an outer protective cover. The hose is available in sizes up to 1.5 in. (3.8 cm). *See also* **Hose Reel.**

Hose, Braided—Nonwoven rubber fire hose made by braiding one or several layers of yarn with interlaying layers of rubber over a rubber tube that is encased in a rubber cover.

Hose Bridge—A temporary ramp placed over a fire hose to allow vehicles to pass over it without damaging it. It may also be called a hose jumper. *See also* **Hose Ramp.**

Hose Cabinet—A specifically designated cabinet that is provided for a folded fire hose, nozzle, and control valve that may be wall mounted or partially recessed. The cabinet fire hose is normally part of the standpipe system for the facility.

Hose Cart—A cart for carrying fire hose or equipment in streets or large buildings. The term originated at the turn of the 19th century. The carts were originally called turtlebacks because of their broad curvature in profile. Because of the limited capacity of the turtleback cart, the jumper hose cart was invented in 1810. This was a large capacity hose cart that could be "jumped" over curbs (hence its name). About 1812, a four-wheeled version of the jumper cart was developed and was called a hose carriage. In the 1870s, the hose carriage was revised into a hose wagon that was pulled by horses and driven by firefighters. Sled-equipped hose wagons were called pungs. *See also* **Hose Truck; Pung.**

Hose Clamp—A manual device placed on a fire hose to control the flow of water in the hose. Hose clamps are generally used to immediately stop the flow of water in a section of hose line due to a burst hose or massive coupling leakage that requires immediate replacement. They are also used to insert a length of hose, change a nozzle installation, or drain a section of hose. Hose clamps can be manual, press, or screw types, or they can be hydraulically operated. The use of hose clamps has to be undertaken carefully to avoid damage to the hose.

Hose, Discharge—A fire hose connected to a fire pump outlet for transferring water to use in firefighting operations.

Hose, Drafting—Semi-rigid hose used to transfer a supply of water to a fire pumper truck or to suction water by the truck from an open body of water. *See also* **Hose, Hard Suction.**

Hose, Fire—A flexible conduit for channeling water to fire control and suppression devices (nozzles). Fire hoses enable firefighters to work closer to the fire without endangering their engines and to increase the accuracy of water placement. It also makes it possible to draw water from rivers and ponds.

The earliest notation of a type of fire hose is from ancient times. The Greek engineer and architect Apollodorus (fl. early 2nd century CE), who worked primarily for the Roman emperor Trajan (reigned 98–117), stated that to convey water to a high place exposed to "fiery darts," the gut of an ox having a bag filled with water attached to it might be employed, for on compressing the bag, the water would be forced up through the gut to the place of destination.

In 1673, Jan (then the Superintendent of the Amsterdam Fire Brigade in Holland) and Nicholas Van Der Heiden (father and son), are credited with creating and using the first reliable flexible fire hose by sewing strips of leather together longitudinally. They used strong linen thread to close the seams of the hose. The English referred to their invention as "hose" after the popular word for stockings. The Van Der Heiden's hose was notorious for leaks, but was found more useful than a bucket brigade, which was used at the time. Seamless woven hose made of hemp and later of linen was produced in Europe in the 1720s, but it was less durable than the leather hose, failing after short usage time. In 1808, the Philadelphia firm of Sellers and Pennock developed leather hose that replaced linen seaming thread with iron rivets (later copper, circa 1819) that provided a tight, leakproof seam. The 50-ft. (15 m) lengths coupled with brass fittings enabled firefighters to convey water through narrow passages, up stairways, and into buildings, while the pumps operated in the street.

The first fire hose of rubber lined cotton weave to replace riveted leather hose was invented by James Boyd of Boston, Massachusetts. He obtained a patent on May 30, 1821 on a "new and useful improvement in the mode of manufacturing fire engine hose." In 1819, he established James Boyd & Sons in Boston and manufactured Boyd's Patent Double Fire Engine Hose. This type of hose was used until 1870 when rubber lined cotton hose took over and became the standard type. Standards on fire hoses for both mill and fire department use were among the earliest standards issued by the National Fire Protection Association (NFPA). It issued its first standard on fire hose in 1898. Luminous fire hoses are now available that assist firefighters in a darkened smoke-filled environment. *See also* **Attack Line; Linen Hose.**

Hose, Forestry—A fire hose designed and constructed to meet specialized requirements for fighting wildland fires.

Hose, Hard Suction—A noncollapsible hose primarily used by mobile firewater pumps for drafting large volumes of water from supplies (lakes, rivers, wells, etc.) without any head conditions, that is, they need to be pumped (or lifted) to be utilized. It may also be used for supplying pumps on fire apparatus from hydrants if specified for that purpose. The hose contains an integral semi-rigid or rigid reinforcement designed to prevent collapse of the hose under partial vacuum or vacuum conditions created from the suction of a pump. Common drafting hoses used with pumpers in the United States have an internal diameter of 4.5 in. (11.4 cm), but can range from 2.5 to 6 in. (6.4 to 15.2 cm).

In the early 1670s, the first fire engine drafting hoses were invented by Jan of Holland and were called "water snakes." They were made of sailcloth and stiffened with cement. Metal rings were provided on the ends to maintain the shape of the hose and were used with manually operated fire engines containing piston pumps. Later hoses were made of metallic cylinders placed

Figure H-6 Hard suction hoses, 2.5-in. and 4-in. (6.3-cm and 10.2-cm) diameter.

end to end, and covered with leather, or stout spiral wire covered with leather in order to prevent collapse. Prior to the invention of the suction hose, leather water buckets were used in a bucket brigade to fill the hand pumper or to fight a fire. See Figure H-6. *See also* **Draft; Hose, Drafting; Hose, Soft Suction; Steamer Connection.**

Hose House—An enclosure located over or adjacent to a fire hydrant or other water supply designed to contain the necessary hose nozzles, wrenches, gaskets, and spanners to be used in firefighting in conjunction with and to provide aid to the local fire department or fire brigade. The hose house affords protection for the hose, hydrant, and equipment from ambient weather conditions and allows a convenient pre-staging point for emergency firefighting equipment. Care must be taken in the construction of the hose house to avoid corrosion to equipment and deterioration of the house from periodic water leakages.

Hose Jacket—A portable device used to enclose a leaking portion of hose in order to maintain the hose in use with minimal water and pressure loss.

Hose Lay—The method of laying out fire hose and accessories at the scene of a fire for the purpose of firefighting. Either a forward or reverse lay may be used. *See also* **Forward Lay; Reverse Lay.**

Hose Layout—The evolution, positioning, and arrangement of fire hoses for a fire incident. *See also* **Evolution.**

Hose, Linen—Unlined fire hose, formerly used for first aid standpipes, consisting of a line or flax fabric without a rubber lining. In present practice, cotton or synthetic fiber is more commonly used.

Hose Load—The amount or arrangement of firefighting hose stored on a fire truck. *See also* **Accordion (Hose) Load** or **Fold.**

Hose Pipe—A British term for a hose line.

Hose Rack—A device for compactly storing collapsible fire hose. The hose is normally permanently connected to a water supply source at one end and a nozzle provided at the free end. A shutoff valve is provided on the pipe supplying water to the hose rack. The hose is stored uncharged in a collapsed accordion style arrangement. Each fold of the hose is loosely held in the rack except for the last fold, which has a more rigid holding assembly. When needed by a single individual, the water isolation valve is opened, but the water is held back until all the hose held in the rack has been pulled out, and the last kink forcibly released by a tug on the hose. Therefore the hose cannot be used until all of it is completely unstrung from the rack. It is primarily installed inside buildings within a protective and aesthetic enclosure. The term hose rack is also used to denote a structure used to support hose that is to be dried. *See also* **Accordion (Hose) Load** or **Fold; Kink; Standpipe Hose Cabinet.**

Hose Ramp—A protective cover provided over a fire hose to allow the crossing of automotive vehicles and prevent damage to the hose. *See also* **Hose Bridge.**

Hose Reel—A device for compactly storing hard rubber fire hose (usually booster hose). It also allows the firewater nozzle to be readily used as the hose is being unwound from the reel. The hose is permanently connected to a water supply source at one end and a nozzle at the free end. A shutoff valve is provided on the pipe supplying water to the hose reel. It is primarily installed on mobile fire apparatus and in industrial facilities. They generally are provided with 0.75, 1.0, or 1.5 in. (1.9, 2.54, or 3.8 cm) hoses that can be up to 200 ft. (60 m) in length and can be adapted for the application of foam water. See Figure H-7. *See also* **Hose, Booster.**

Hose Roller—A device used at the edge of buildings or windows to prevent the chafe occurring on a fire hose. It contains rollers to allow the hose to freely move without damage.

Hose, Soft Suction—A collapsible hose, either woven jacket or rubber covered, used to provide water to the suction of fire pumpers. It usually consists of short sections of large diameter hose that is connected to pressurized water supply sources (hydrants). It may also be called a soft sleeve. *See also* **Hose, Hard Suction.**

Hose Station—A location provided with a permanent fire hose for fire protection applications. It consists of a hose connection, hose, nozzle, and method of storing the hose.

Hose Stream—A flow of water in the air from an application nozzle onto a fire event for control and extinguishment. A stream of water is generally narrow and concentrated as opposed to a broad spray that radiates the water at an angle outward from the nozzle in droplet form. It may also be called a fire stream. *See also* **Heavy Stream; Master Stream; Water Spray.**

Hose Stream Test—Part of a fire testing procedure whereby a fire hose stream of water is directed onto the surface of a building material assembly (walls, doors, windows, etc.) after the period of fire exposure. The assembly is subjected to the erosion, cooling, and impact effects of the hose stream. The assembly fails the test if

H

Figure H-7 Hose reel with hard rubber line on mobile apparatus.

an opening develops that allows a water stream to project from the opposite side of the assembly.

Hose, Suction—*See* **Hose, Hard Suction; Hose, Soft Suction.**

Hose Tower—A tower used to hang fire hose vertically after it has been used to allow it to dry by means of gravity runoff and natural evaporation. The hose tower is usually located as a part of or near the fire station. After the hose has dried out, it is removed from the tower and stored normally. Fire hose should not be dried on hot pavements or under intense sunlight, which may cause premature aging of the hose. The hose tower must also be properly ventilated and the temperature controlled so the hose will not be damaged by excessive heat. To avoid damage to the hose connection, the hose should not be suspended from its couplings when drying in the tower. NFPA 1962, *Standard for the Care, Use, and Service Testing of Fire Hose Including Couplings and Nozzles,* provides guidance for the care of fire hose.

Hose Tray—A platform used to carry fire hose loads. It is primarily used in forest service firefighting operations and is airlifted by helicopters to the fire site. One end of the hose may be connected to a water source, and the hose is laid while it is being transported in flight.

Hose Truck—A fire service vehicle for the supply of fire hose at a fire incident. *See also* **Hose Cart.**

Hose, Unlined Fire—A fire hose that consists of a woven jacket, usually of linen yarns, that is manufactured so that the yarn swells when wet, tending to seal the hose. It is not provided with an internal lining of rubber. It has about twice the friction loss of lined hose for a given flow and diameter. It has been commonly provided for building first aid hoses and firefighting.

Hose Wagon—*See* **Hose Cart.**

Hostile Fire—An insurance industry term for a fire that is a deliberately ignited flame or glow that escapes from its intended confines, such as sources of heating, cooking, or lighting units. A hostile fire, as contrasted with a friendly fire, is one that stays within its intended confines. A standard fire insurance policy only covers hostile fires. *See also* **Friendly Fire.**

Hot Flame Ignition—*See* **Ignition Temperature.**

Hot Spot—An area of intense heat, radiation, or activity.

Hot Surface Ignition—A surface that is sufficiently large and of high enough temperature (above the material's ignition temperature) that it may act as an ignition source and is considered to be hot surface ignition. *See also* **Ignition, Direct Electric; Ignition Source; Thermal Ignition.**

Hot Work—Any activity that uses a heat producing process (cutting, welding, grinding, brazing, soldering, and similar work) involving an open flame, heat application, that produces sparks, or constitutes a fire risk because it may act as an ignition source. Suitable fire safety precautions must be instituted where hot work occurs. Numerous major fires have occurred as a result of inadequate precautions undertaken during hot work activities. NFPA 51B, *Standard for Fire Prevention in Use of Cutting and Welding Processes,* provides guidance for safety precautions while undertaking hot work. See Figure H-8.

Figure H-8 Hot work with fire watch standby.

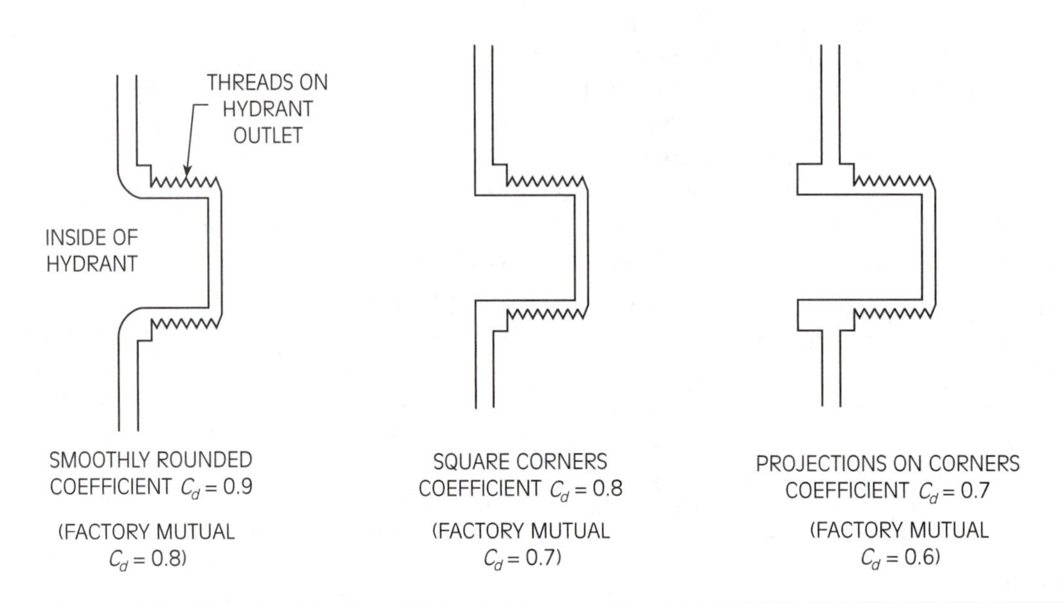

Figure H-9 Coefficients of hydrant outlets.

Hux Bar—A forcible entry tool used by firefighters. It is primarily used for prying. *See also* **Forcible Entry; Halligan Tool; Kelly Tool.**

Hydrant Coefficient—A mathematical factor used to account for the nature of a fire hydrant outlet (its smoothness) from which water flow measurements are taken through pitot tube measurements of the expelled water stream. See Figure H-9. *See also* **Coefficient of Discharge.**

Hydrant, Dry (or Suction) Fire—A permanent piping drafting source that provides a connection to static source water supplies below grade, such as lakes and ponds. In locations subject to ambient freezing conditions, the supply valve and piping are placed below the frost line. It may also be called a static source hydrant. See Figure H-10.

Hydrant, Fire—An upright pipe-shaped casting with an outlet or spout that is connected to a water supply main to enable the provision of water to fight a fire using hoses and mobile fire apparatus. Valves are provided to permit and control the flow of water. Depending on the potential for ambient freezing conditions, either dry barrel or wet barrel fire hydrants are provided. Wet barrel hydrants have a water control valve on the barrel of the hydrant for each outlet (AWWA *Standard C503*). Dry barrel hydrants (AWWA *Standard C502*), have a main valve for all outlets located below grade (below the frost line), which simultaneously controls the flow to all outlets on the hydrant. The first real fire hydrant in America was built by fireman George B. Smith of New York City in 1817. A *ball* hydrant, able to withstand higher pressures, was invented by English engineer Alfred Moore in 1848. NFPA specified its standard arrangement for hydrants in 1913–1914.

Fire hydrants are installed so accidental vehicle impacts allow the hydrant to break away at its connection point, limiting damage to both

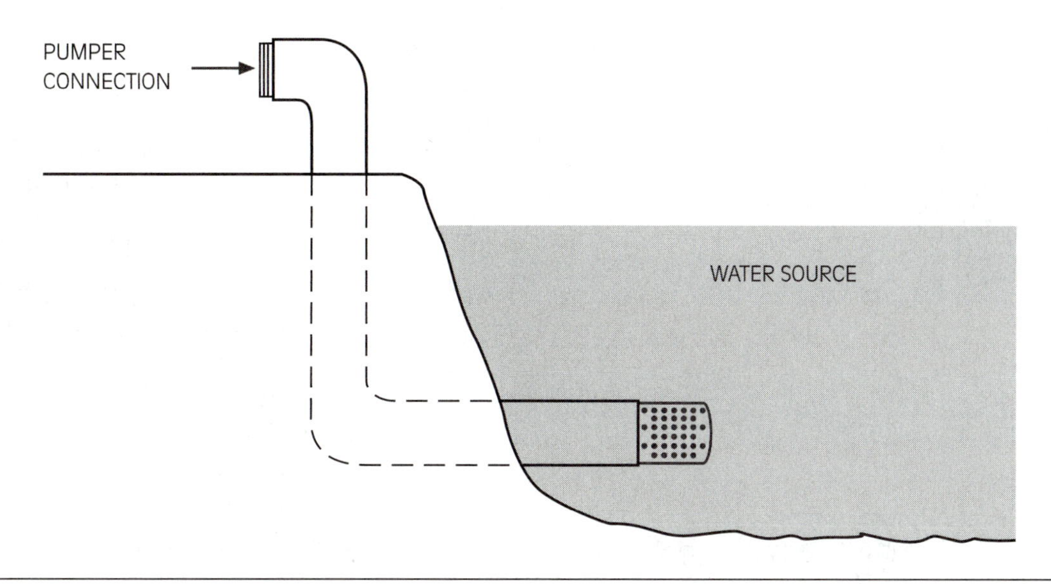

Figure H-10 Dry fire hydrant arrangement.

the vehicle and the water supply system. A gate valve is usually installed in the connecting pipe to the hydrant to shut off water to the hydrant in an emergency or for maintenance, and to maintain water supplies to the rest of the water distribution system. Local authorities may impose fines to individuals who use, damage, or obstruct fire hydrants for other than fire protection purposes. See Figures H-11 and H-12. *See also* **Fireplug.**

Hydrant, Flush Fire—A fire hydrant that is mounted below grade in a cast box. It is usually provided with a cover plate for aesthetic or normal vehicle access considerations. Flush hydrants were used for regular municipal service in cities in the United States, particularly New England, during the period of transition from the fire plug to the modern hydrant. Some of these consisted of a covered barrel that had a valve at the bottom. Fire vehicles carried a portable "chuck," which had a number of

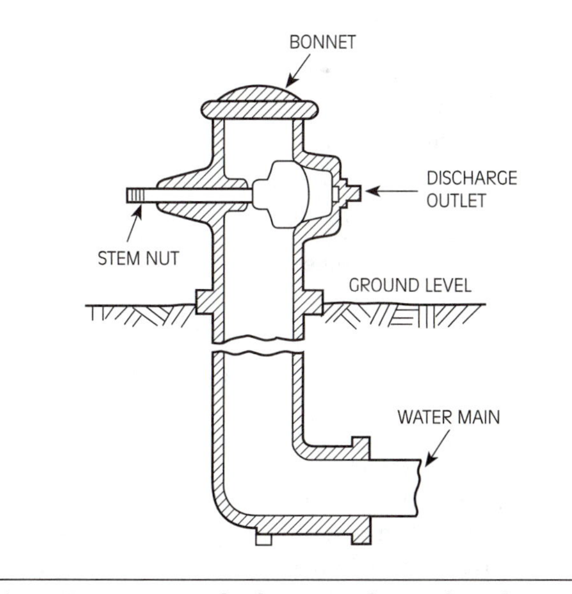

Figure H-11 Typical schematic of a wet barrel hydrant.

H

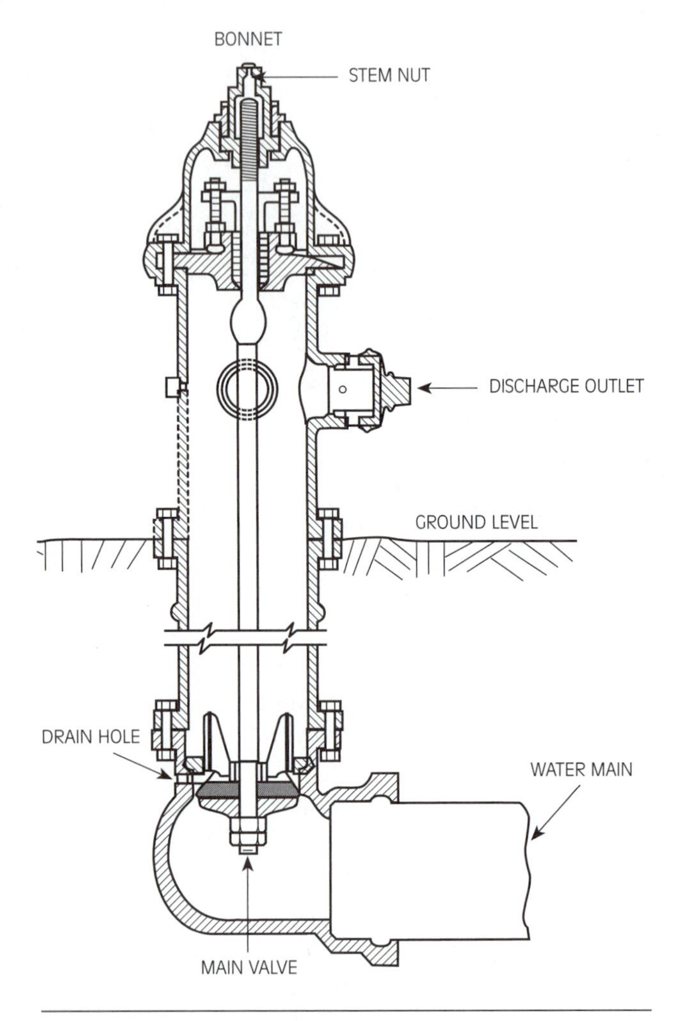

Figure H-12 Typical schematic of a dry barrel hydrant.

independently gated outlet nozzles that could be quickly connected to the top of the flush hydrant. Flush hydrants may be difficult to locate, due to other manhole or box covers similar to a flush hydrant cover. Vehicles may park over it, and winter conditions can obscure its location with snow.

Hydrant, Marking—Fire hydrants may be color coded to indicate the available flow available from the individual hydrant. NFPA 291, *Recommended Practice for Fire Flow Testing and Marking of Hydrants,* recommends that fire hydrants be classed in accordance with their rated capacities (at 20 psi [1.4 bar] residual pressure or other designated value) as follows:

Class AA—Rated capacity of 1,500 gpm or greater (5,680 l/min or greater)

Class A—Rated capacity of 1,000 to 1,499 gpm (3,785 to 5,675 l/min)

Class B—Rated capacity of 500 to 999 gpm (1,900 to 3,780 l/min)

Class C—Rated capacity of less than 500 gpm (1,900 l/min)

The barrels of public hydrants are recommended to be chrome yellow, except in cases where another color has already been adopted. The tops and nozzle caps should be painted with the following capacity-indicating color scheme to provide simplicity and consistency with colors used in signal work for safety, danger, and intermediate condition (the color scheme follows the general pattern of safety: green, intermediate condition; orange; and danger, red).

Class AA—Light blue

Class A—Green

Class B—Orange

Class C—Red

For rapid identification at night, it is recommended the capacity colors should be of a reflective-type paint.

Hydrant, Triple—A fire hydrant with three outlets, usually one 4.5 in. (11.5 cm) for connection to a mobile fire pumper, and two 2.5 in. (6.5 cm) outlets.

Hydrant, Wall Fire—A fire hydrant that is wall mounted. It is normally supplied by a fixed fire pump through a pipe extended through a wall. They are also commonly used to measure water flow from a fire pump (through a hose/nozzle pitot measurement) where a fixed flowmeter has not been provided in the pump discharge piping.

Hydrant, Yard—A fire hydrant provided at industrial facilities, typically with individual valves installed at each hose connection. A common yard hydrant has two 2.5 in. (6.5 cm) outlets and one 4.5 in. (11.5 cm) pumper connection. See Figure H-13.

Figure H-13 Yard hydrant at industrial facility.

Hydraulic Calculation—Mathematical computation of the amount and pressure of liquid in a contained distribution network, such as in a sprinkler system or a distribution network. Hydraulic calculations for fire protection purposes ensure water supplies will meet the water density requirements for fire control and extinguishment and allow economical water distribution designs to be obtained. *See also* **Hazen-William Formula; Hydraulically Designed System.**

Hydraulic Gradient—A graphical representation of residual pressure in a piping system that presents the flow characteristics of the system. It provides a visual representation of the pipeline elevation factors: pipe size and flow. *See also* **Hydraulic Profile.**

Hydraulic Profile—A graphical representation of water supply residual pressure condition, adjusted for a constant elevation factor. The hydraulic profile is used to calculate the friction factor that exists in individual sections of a water distribution system. The calculation of friction factors from the hydraulic profile is useful in determining the condition of the internal surfaces of the water mains. *See also* **Hydraulic Gradient.**

Hydraulic Ventilation—Employment of a ho
stream for the quick exclusion of smoke a
heat from an area or enclosure.

Hydraulically Balanced Sprinkler Syster
sprinkler system design to achieve the sar
ative flow from each sprinkler in a fire ar
system is designed through hydraulic
matical calculations that account for
losses and water quantities through each
of the pipe network.

Hydraulically Designed Sprinkler Sy
sprinkler system specifically designe
vide a prescribed water density when c
to an acceptable water supply. The

H

designed through hydraulic mathematical calculations that account for friction losses and water quantity changes through each section of the sprinkler system pipe network. The water supply is determined by calculation. Standards for hydraulically designed sprinkler systems first appeared in NFPA fire codes in 1972.

Hydraulically Designed System—A fire protection system that uses fluids for fire control and suppression and that engineered based on the prescribed application densities. Pipe sizes and to some extent, the overall arrangement is determined by engineering calculations to achieve adequate pressure within the pipe network to ensure a general uniformity of spray densities over the area of coverage. *See also* **Hydraulic Calculation.**

Hydraulically Most Demanding (or **Remote**) **Area**—The area of a sprinkler system that is shown by hydraulic design calculations to be the most demanding for water supply and pressure requirements to meet the required sprinkler density levels. The hydraulically most remote area for a gridded sprinkler system may not be the physically most remote area due to the hydraulic flows involved, whereas for most other systems it is normally the physically most remote area.

Hydraulics, Fire Service—*See* **Fire Hydraulics.**

Hydrocarbon Fire—A fire that is fueled by hydrocarbon compounds that achieve a high flame temperature almost instantaneously after ignition. A hydrocarbon fire will spread rapidly, burn fiercely, and produce a high heat flux. Liquid and gaseous hydrocarbon fires have highly luminous flames resulting from hot carbon particles. Combustion gases are usually fuel-rich because of limited mixing with oxygen, and produce nearly continuous radiation in the infrared region making the hot gas and soot particles behave as a gray body radiator. Petrochemical fires reach temperatures as high as 2,400°F (1,300°C), but average 1,850°F (1,000°C) due to various factors, such as radial cooling of the fire, wind, and fire geometry. Free-burning fires as a rule do not achieve the theoretical combustion temperatures for the fuel involved. Fires from a high-pressure source have convection as the major or primary mode of heat transfer. *See also* **Fire Barrier; Fire Curve; Time Temperature Curve, Cellulosic Fire.**

Hydrocarbon Fire Test—A fire test in a furnace using a time temperature curve that represents a hydrocarbon fire exposure and that results in an "H" rating for successful resistance to the fire exposure. Additionally, if the temperature rise on the unexposed surface is below 282°F (139°C) from ambient conditions, the time period is stipulated by the use of a numeric indication of minutes from 30 up to 240, in 30-minute increments. Hydrocarbon fire tests are conducted according to UL 1709, *Rapid Rise Fire Tests of Protection Materials for Structural Steel*. The furnace temperature reaches approximately 2,000°F (1,100°C) in five minutes and maintains this exposure during the entire test. The test limits the average steel temperature to 1,000°F (537°C). This temperature corresponds to the ability of steel to support its design load. Typically, the design load includes a design safety factor of approximately 50 percent. As the temperature of the steel increases, the ultimate load capacity of the steel decreases. At 1,000°F (537°C), the ultimate capacity and design load are approximately equal. *See also* **Fire Resistance Rating; Fire Test.**

Hydrostatic Pressure—The pressure (or force) exerted by the weight of water determined by its head (depth or height).

Hypergolic—Materials that ignite spontaneously when mixed together such as mono-methyl-hydrazine (MMH) and nitrogen tetra-oxide (N_2O_4). Hypergolic liquids are commonly used as spacecraft propellants to avoid the use of ignition systems. They have a high energy release to weight of the material ratio. These materials cannot be suppressed with ordinary extinguishing agents since the propellants contain their own oxidizer agent to independently support the combustion process. *See also* **Spontaneous Combustion.**

Hypergolic Ignition—*See* **Hypergolic.**

Hyperthermia—*See* **Heat Stress.**

Ignitability—The ease with which a material may ignite. Ignitability is a function of the material's properties, such as form and state. For example, small combustible particles are easier to ignite than a large object. For material fire propagation evaluations, ignitability is the time to sustain flaming at a specified heating flux.

Ignitable—A material or substance that is able to support a combustion process. *See also* **Incendive Arc.**

Ignitable Liquid—Any liquid or the liquid phase of any material that is capable of fueling a fire. It includes a flammable liquid, a combustible liquid, or any other material that can be liquefied and burn. Static charge is generated when liquids move in contact with other materials, such as flowing through pipes, mixing, pouring, pumping, spraying, filtering, or agitating. Under certain conditions, static may accumulate in the liquid. If the accumulation of charge is sufficient, a static arc may occur. If the arc occurs in the presence of a flammable vapor-air mixture, an ignition may result. Combustible liquids that have the potential to accumulate a static charge may be referred to as an ignitable liquid. *See also* **Arc; Spark; Static Ignition Hazard.**

Ignitable Mixture—A vapor-air, gas-air, dust-air mixture, or combinations of these mixtures that are within their flammable limits (capable of being ignited by an open flame, electric arc or spark, or device operating at or above the ignition temperature of the mixture). *See also* **Ignition Energy.**

Ignite—To initiate a combustion process from an ignition source.

Ignition—The process of starting a combustion process through the input of energy. Ignition occurs when the temperature of a substance is raised to the point at which its molecules react spontaneously with an oxidizer, and combustion occurs.

Ignition Delay—The time delay between the moment of exposure of a combustible substance to a high temperature or ignition source and visible combustion. It may also be used in the context of the time it takes for an (accidental) release of a flammable material to contact a source of ignition. *See also* **Ignition Time.**

Ignition, Direct Electric—Ignition of a material by an electric ignition source such as a high-voltage spark or a hot wire. *See also* **Hot Surface Ignition; Spark Ignition.**

Ignition Energy—The quantity of heat energy that must be absorbed by a substance to ignite and burn. Ignition energy for hydrocarbon vapors range from 0.00002 to 0.001 joules, whereas for coal and chemical dusts the range is from approximately 0.015 to 0.1 joules. When dust particle size is greater than 400 micrometers, even a high-energy source cannot ignite the dust cloud. *See also* **Dust Explosion; Frictional**

Ignition; Heat of Ignition; Ignitable Mixture; Ignition Source; Minimum Ignition Energy (MIE); Spark; Spark Ignition; Static Ignition Hazard.

Ignition Point—The minimum temperature at which a substance will continue to burn without additional application of external heat. It may also be called kindling point. *See also* **Kindling Temperature** or **Kindling Point.**

Ignition Sensitivity—A factor for the determination of the index of explosibility rating of a particular material. Ignition sensitivity is defined as the product of the material's ignition temperature, multiplied by minimum energy, multiplied by minimum concentration of Pittsburgh coal dust, divided by the product of ignition temperature, and multiplied by minimum energy, and multiplied by minimum concentration of the sample dust under consideration. *See also* **Explosibility Index; Explosion Severity; Index of Explosibility.**

I

Ignition Source—A source of energy such as static, open flame, electrical arcing, sparks, friction, hot surface, chemical reaction, lightning, and heat of compression that can or will act as an ignition point for a material if of sufficient ignition energy. Factors that influence ignition include temperature, exposure time, and energy. *See also* **Frictional Ignition; Hot Surface Ignition; Ignition Energy; Static Ignition Hazard.**

Ignition Temperature—The minimum temperature a material must attain in order to ignite under specific test conditions. Reported values are obtained under specific laboratory test conditions and may not reflect a measurement at the substance's surface. Ignition by application of a pilot flame above the heated surface is referred to as pilot ignition temperature. Ignition without a pilot energy source has been referred to as autoignition temperature, self-ignition temperature, or spontaneous ignition temperature. The ignition temperature determined in a standard test is normally lower than the ignition temperature in an actual fire scenario.

Advances in technology and further research into ignition temperatures have recognized that previous testing for ignition temperatures did not account for the detection of nonluminous or barely luminous reactions. As a result, further refinement in description of ignition temperature measurements of materials is being used, ASTM E659, *Standard Test Method for Autoignition Temperatures of Liquid Chemicals.*[5] These include the following:

Hot Flame Ignition—A rapid, self-sustaining, sometimes audible gas-phase reaction of the sample, or its decomposition products with an oxidant. A readily visible yellow or blue flame usually accompanies the reaction.

Cool Flame Ignition—A relatively slow, self-sustaining, barely luminous gas-phase reaction of the sample or its decomposition products with an oxidant. Cool flames are visible only in a darkened area.

Preflame Reaction—A slow, nonluminous gas-phase reaction of the sample or its decomposition products with an oxidant.

Catalytic Reaction—A relatively fast, self-sustaining, energetic, sometimes luminous, sometimes audible reaction that occurs as a result of the catalytic action of any substance on the sample or its decomposition products, in admixture with an oxidant.

---

[5] NFPA 325, *Guide to Fire Hazard Properties of Flammable Liquids, Gases, and Volatile Solids,* Definitions, National Fire Protection Association, Quincy, MA 1995.

Noncombustive Reaction—A reaction other than combustion or thermal degradation that is undergone by certain substances when they are exposed to heat. Thermal polymerization is an example of this type of reaction.

Reaction Threshold—The lowest temperature at which any reaction of the sample or its decomposition products occurs, for any sample oxidant ratio.

Autoignition Temperature (AIT)—The currently accepted term for the hot-flame ignition temperature.

Cool-Flame Reaction Threshold (CFT)—The lowest temperature at which cool-flame ignitions are observed for a particular system.

Preflame-Reaction Threshold (RTT)—The lowest temperature at which exothermic gas-phase reactions are observed for a particular system.

See also **Autoignition Temperature (AIT); Hot Flame Ignition; Kindling Temperature** or **Kindling Point; Self-Ignition Temperature.**

Ignition Time—The time between the application of an ignition source to a material and the onset of self-sustained combustion. *See also* **Ignition Delay.**

Impact Load—A load applied for a short time. A load delivered as a static load to a structure may not cause it to collapse, but applied as an impact load may cause collapse. Some firefighting activities (overturning heavy objects, firefighters jumping onto roofs, etc.) may apply impact loads that can cause a structure to collapse. *See also* **Dead Load; Live Load.**

Impingement Spray—The direct application of water droplets to a protected surface from a water spray nozzle for fire protection. Impingement spray is commonly provided for exposure protection against radiant heat or direct flame hazards (cooling of pressure vessels or storage tanks to prevent failure). Impingement sprays differ from area sprays, which are generally an overhead spray for general area coverage. See Figure I-1.

Incandescence—The emission of light by a substance due to its high temperature. *See also* **Glow.**

Incendiarist—An individual who starts disruptive and damaging (incendiary) fires in protest or means of rebellion against social or political establishments. *See also* **Arson; Pyromania.**

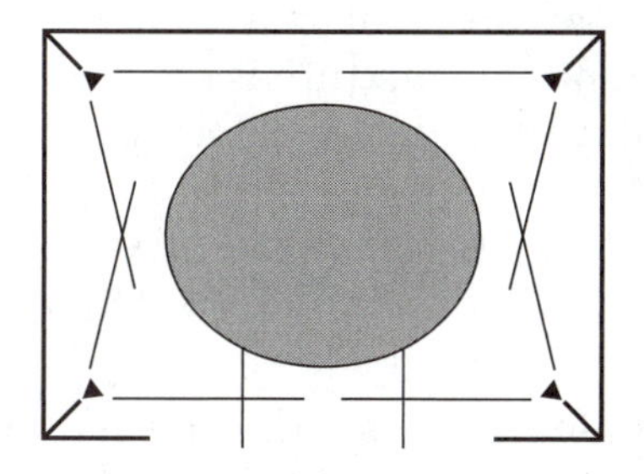

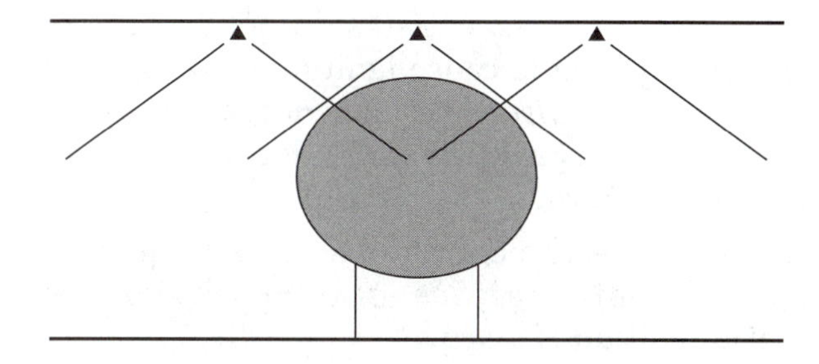

Figure I-1 Impingement spray versus area spray.

Incendiary—A substance causing or designed to cause fires.

Incendiary Energy—Hot particle energy that is sufficient to ignite a specific combustible mixture.

Incendiary Fire—A fire that has been deliberately ignited under circumstances in which the person knows the fire should not be ignited and is usually started for disruptive or deliberate damage.

Incendive—Something that is capable of starting a fire, generally through an arc or spark. *See also* **Non-Incendive.**

Incendive Arc—An arc is a luminous, high intensity electrical discharge in a gas or a vapor. An electrical arc that has enough energy to ignite an ignitable mixture is classified as incendive. A nonincendive arc does not possess the energy required to cause ignition, even if the arc occurs within an ignitable mixture. *See also* **Arc; Ignitable; Incendive Spark; Non-Incendive; Spark.**

Incendive Spark—A spark is a small, incandescent particle from an electrical failure or a moving ember from a fire source. A spark of sufficient temperature and energy to ignite an ignitable mixture is classified as incendive. A fire or explosion can result from an incendive spark. Nonincendive sparks do not possess the energy required to cause ignition of an ignitable mixture. *See also* **Arc; Incendive Arc; Non-Incendive; Spark; Spark Ignition; Static Ignition Hazard.**

Incendivity—The ability of a spark to ignite an ignitable mixture. *See also* **Spark Ignition; Static Ignition Hazard.**

Incident—An occurrence or event, either human-induced or by natural phenomena, that requires action by emergency response personnel to prevent or minimize loss of life or damage to property or natural resources.

Incident Action Plan (IAP)—A listing of tasks, prepared by an incident commander, for controlling, resolving, and adminstering and incident to its conclusion. Tasks are assigned to organized teams in the form of orders. Teams are also required to relay information about hazards or conditions that may be pertinent to the operation of the IAP.

Incident Command System (ICS)—The on-scene emergency management to allow participant to adopt an integrated organizational structure without being hindered by jurisdictional boundaries and complexities. It allows the combination of facilities, equipment, and personnel structure with responsibility for the management of resources to effectively accomplish states objectives required for an incident.

Incident Commander—The individual responsible and in charge of all the actions of individuals and equipment at an emergency incident. The term incident commander has generally replaced the term fireground commander in order to encompass all emergency incidents, not solely fire incidents. *See also* **Fireground Commander.**

Incinerate—To apply a combustion process to a material to destroy it and reduce it to ash.

Incineration—A controlled process by which solid, liquid, or gaseous waste material is burned and changed into waste gases, commonly leaving little or no residue.

Incinerator—A furnace used to burn waste materials.

Incipient Stage (Fire)—Ordinary combustible fires (cellulosic) generally start and progress through four stages: incipient (precombustion), visible smoke (smoldering), flaming fire, and intense heat. Hydrocarbon-based fires normally do not have several stages of fire growth, but almost instantaneously achieve high flame temperatures and heat flux after ignition. In the incipient stage, the combustion process produces some heat, but it has not spread to nearby materials. The incipient stage of smoldering ordinary combustible fires provides the widest window of opportunity for detection and control of the spread of fire, before it develops into devastating stages. An incipient stage fire is defined by OSHA as a fire that can be controlled or extinguished by a portable fire extinguisher; a Class II standpipe; that is, a 1.5 in. (3.8 cm) fire hose; or a small hose flowing up to 125 gpm (7.9 liters/second) without the need for protective clothing or breathing apparatus; or a fire brigade member being required to crawl on the ground or floor to stay below smoke and heat. *See also* **Fire Stages.**

Incombustible—Synonymous with noncombustible, but generally considered an obsolete and confusing term for fire protection due to the prefix "in." The term noncombustible is preferred over incombustible to avoid confusion. *See also* **Noncombustible (Material).**

Incomplete Combustion—*See* **Combustion, Incomplete.**

Index of Explosibility—The index of explosibility is the product of the ignition sensitivity and the explosion severity. The indices are dimensionless quantities and have a numerical value of one for a dust equivalent to the standard Pittsburgh coal. An explosibility index greater than one indicates a hazard greater than that for coal dust. *See also* **Deflagration Index; Explosion Severity; Explosibility Index; Ignition Sensitivity.**

Indirect Application or **Indirect Attack**—A firefighting technique for enclosed areas whereby water is applied to produce steam and distribute unvaporized water droplets to cool and suppress fires out of direct reach of fire streams. An indirect attack may subsequently allow the use of direct firefighting attack methods. Water fog is applied into the upper atmospheric level of the fire within a confined space to generate steam and distribute the unvaporized water droplets. *See also* **Direct Attack.**

Indraft—Air drawn to a combustion process due to the reaction process and thermal currents that arise. Firestorms are characterized by destructively violent surface indrafts around a continuous area of intense fire. *See also* **Plume; Updraft.**

Industrial Risk Insurers (IRI)—A US-based association, maintained by a group of leading insurance companies to provide a market for fire and allied peril insurance on high valued, protected risks in the industrial, petroleum, chemical, and related industries and service organizations. It provides specialized underwriting and advisory engineering services. It was established in 1890 in response to a need to provide insurance for large, highly protected risk (HPR) properties. It provides property insurance for about 40 percent of the *Fortune* 1000 corporations. *See also* **Factory Insurance Association (FIA); Oil Insurance Association (OIA).**

Inergen—A fire suppression gas developed to replace Halon, technically known as IG-541. It consists of 52 percent nitrogen, 40 percent argon, and 8 percent carbon dioxide (CO_2). It has a zero ozone depletion potential (ODP) rating thus also does not have any global warming potential. It requires a design concentration of 34 to 50 percent for fire suppression (as compared with 5 percent for Halon applications). It is considered safe for human exposure and leaves no residue. *See also* **Argonite™; Clean Agent; Clean Agent Fire Suppression System (CAFSS); FM-200; Halon.**

Inert Gas—Any gas that is nonflammable, nonreactive, and noncontaminating for its intended use, and oxygen deficient to the extent required.

Inerting—The process of removing an oxidizer (usually air or oxygen) to prevent a combustion process (explosion or fire) from occurring or continuing. Inerting is normally accomplished by purging with a gas that will not support combustion, such as nitrogen or carbon dioxide (CO_2). Normal oxygen level in the atmosphere is 21 percent (approximately 20.9 percent oxygen, 78.1 percent nitrogen, 1 percent argon, carbon dioxide, and other gases). Combustion of stable hydrocarbon gases and vapors will usually not continue when the ambient oxygen level is below 15 percent. Acetylene is an unstable gas and requires an oxygen level below 4 percent before extinguishment will occur. For ordinary combustibles (wood, paper, cotton, etc.), the oxygen concentration level must be lowered to 4 or 5 percent for total fire extinguishment. Inerting for fire extinguishment results in an asphyxia safety hazard for personnel. Inerting, in effect, reduces the oxygen content of the air in the enclosed space below the lowest point at which combustion can occur by replacing the oxygen in the enclosure with an inert gas. The inert gas chosen should not react with the medium contained within the enclosure. *See also* **Blanketing (or Padding), Gas; Gas Blanketing; Purging.**

Inferno—A location that has intense heat or fire, such as the roaring heat of a blast furnace.

Inflame—To start on fire or burst into flames.

Inflammable—An identical meaning as flammable, however because the prefix "in" most often indicates a negative in many words and causes confusion (for example, incomplete), the use of flammable is preferred over inflammable. *See also* **Flammable; Nonflammable.**

Infrared (IR)—Descriptive of invisible heat rays located beyond the long wavelength end of the visible spectrum (beyond 7,600 angstroms) and possessing high penetrating power. They are wavelengths of the electromagnetic spectrum that are no longer than those in visible light and are shorter than radio waves (10^4 to 10^1 centimeters). IR radiation is emitted by the combustion process, and fire detection devices are available to detect these emissions. See Figure I-2. *See also* **IR Fire Detector.**

Infrared Fire Detector—*See* **IR Fire Detector.**

Infrared Scanner—A portable device used to detect hidden fire hot spots by searching an area for infrared radiation. *See also* **IR Fire Detector.**

Inherent Flame Resistance—Terminology used to describe textiles that exhibit flame resistance due to the essential characteristic of the fiber or polymer from which the textile is made. The flame

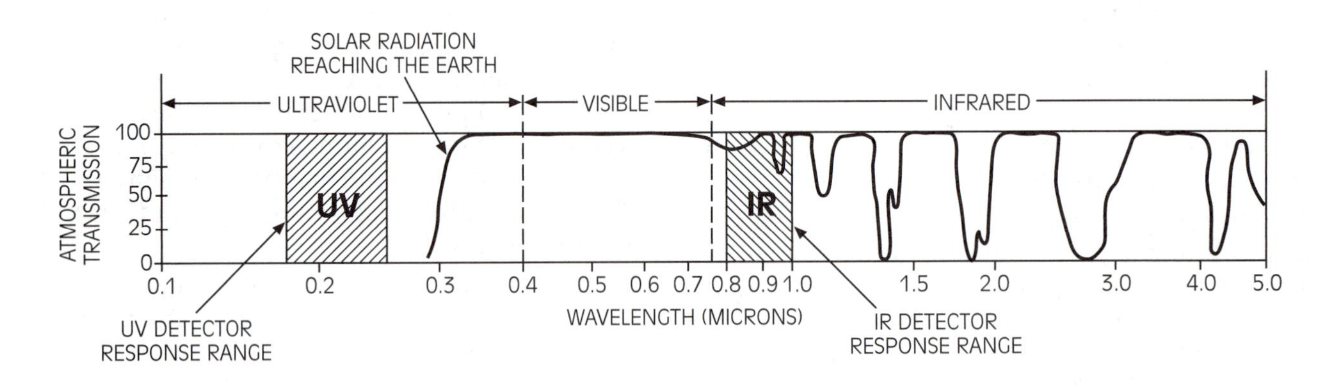

Figure I-2 Radiation spectrum.

resistance for textiles used as clothing is determined by *Federal Test Method Standard 191A, Textile Test Methods, Method 5903.1, Flame Resistance of Cloth: Vertical.*

Inherently Safe—An essential characteristic of a process, system, or equipment that makes it without risk or very low in hazard or risk. For example, the use of noncombustible liquids instead of combustible liquids for an operation (transformer insulating oil) makes a process inherently safer by removing the hazard of fire should the liquid contact an ignition source.

Inhibitor—A substance that reduces explosive overpressures or prevents burning by chemically interfering with the combustion process.

Initial Attack—The first activity of manual fire suppression activities with fire hose lines. The primary objective of the initial attack is to safeguard life and prevent the spread of the fire. Additional fire hoses and equipment may be provided and arranged while the initial attack on the fire is occurring. *See also* **Attack, Fire.**

Initial Attack Apparatus—An automotive fire service vehicle that is equipped with a permanently mounted fire pump, a water tank, and hose. Its primary purpose is to initiate a fire suppression attack on structural, vehicular, or vegetation fires, and to support associated fire department operations.

Initiating Device—A device, either automatic or manual, that originates a signal upon detection of a fire incident, and sends it to the fire alarm control panel or other fire protection monitoring equipment.

Initiating Device Circuit (IDC)—A fire alarm circuit that connects a fire alarm control panel to initiating devices, either automatic or manual. The circuit specifies the alarm condition but does not identify the specific device that initiated the alarm.

In-Line Balanced Proportioner (ILBP)—A foam concentrate proportioner system used with positive displacement foam concentrate pumps and atmospheric foam concentrate storage tanks. It provides for correct foam concentrate proportioning over a wide range of water flow and pressure conditions. The system operates by maintaining equal pressure between the foam concentrate and water within the proportioner. This automatic balancing of the water and concentrate allows the proportioner to be used over a wide range of flows. The system operates by passing the foam concentrate from a positive displacement foam pump through a diaphragm balancing valve and into the proportioner. The diaphragm valve senses the water and foam concentrate pressures, balances these pressures, and meters the correct amount of foam concentrate through the inlet orifice to the proportioner. The orifice is sized according to the type of foam concentrate being used and the flow rate required. See Figure I-3. *See also* **Balanced Pressure Pump Proportioning; Proportioner.**

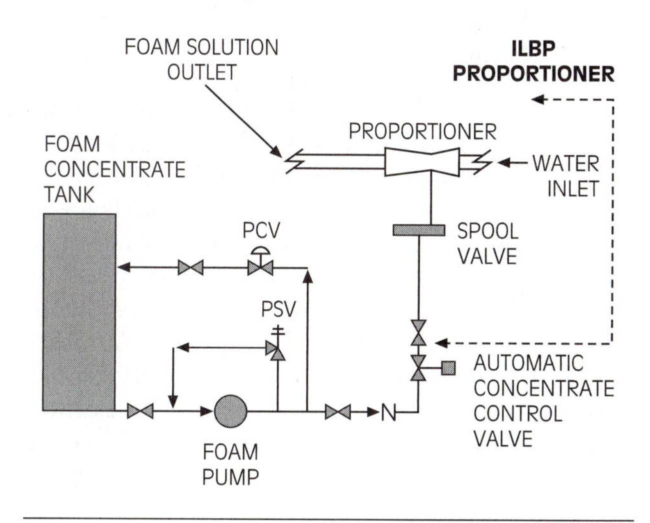

Figure I-3 In-line balanced proportioner.

In-Line Eductor—An eductor device (venturi) that is placed in series with the flow in a firewater line. The eductor is pre-calibrated and adjustable types are available.

Inspection Hole—*See* **Smoke Indicator Hole.**

Inspector's Test (Valve) Connection—A section of piping fitted with an isolation (test) valve provided on a water-based, piped fire protection system, such as a sprinkler system, to periodically test the system. Inspector's test connections are to be located at the hydraulically most demanding point on the system to verify water flow through the system, to test alarms, and also to flush the system. The outlet of the valve is sized to simulate the discharge of one sprinkler and to ensure an alarm will be activated from such a flow. Every water flow switch provided on a water-based fire protection system should be provided with a inspector's test connection. *See also* **Sight Flow Connection** or **Drain.**

Insurance Services Office (ISO), Inc.—A nonprofit association of insurance companies that provides services for participating companies. It provides statistical, actuarial, and underwriting information for numerous affiliated insurance companies and more than a dozen lines of insurance. It maintains one of the largest private databases in the world for insurance premiums and losses paid. Included in these lines are liability, automobile, boiler and machinery, homeowner's, farm, and commercial fire insurance. ISO is a voluntary, nonprofit, unincorporated association of insurers. Before 1971, the functions performed by ISO were undertaken by various insurance organizations in different states. ISO gathers data that are used to establish rates for fire protection policies for residential and commercial properties. ISO developed the municipal grading schedule that is commonly used to establish a basis of insurance rates for municipalities. It is an evaluation of the fire protection features of cities and towns based on seven factors: climatic conditions, water supply, fire department, fire service communications, fire safety control, building codes, and a survey report. *See also* **Municipal Grading Schedule; Rating Bureau.**

Interior Finish—The exposed interior surfaces of buildings. Interior finish materials are grouped into Class A or Class B for the purpose of fire protection. Class A materials have a higher degree of fire safety than Class B materials. Class A materials have a flame spread rating of 0 to 25 and a smoke developed rating of 0 to 50 when tested in accordance with NFPA 255, *Standard Method of Test of Surface Burning Characteristics of Building Materials.* They include any material classified at 25 or less on the flame spread test scale and 450 or less on the smoke test scale when any element thereof, when tested, does not continue to propagate fire. Class B materials have a flame spread rating of 26 to 75 and a smoke developed rating of 0 to 450 when tested in accordance with NFPA 255, *Standard Method of Test of Surface Burning Characteristics of Building Materials.* They include any material classified at more than 25, but not more than 75, on the flame spread test scale and 450 or less on the smoke test scale.

International Association of Arson Investigators (IAAI)—The International Association of Arson Investigators was founded in 1949 by a group of public and private officials to address fire and arson issues. The purpose of the association is to strive to control arson and other related crimes through education and training, in addition to providing basic and advanced fire investigator training. IAAI chapters are located throughout the world.

International Association of Black Professional Fire Fighters (IABPFF)—An organization for the assistance of black firefighters. It provides counseling in matters of working conditions, advancement, and interracial programs. The organization was founded in 1970.

International Association of Fire Chiefs (IAFC)—The association was established in Baltimore, Maryland in 1873 as the National Association of Fire Engineers. In 1926, at the 54th annual conference in New Orleans, the name of the organization was changed to the International Association of Fire Chiefs. It was the first national fire service organization in the United States and one of the oldest professional management organizations. It represents fire chief interests and aims to professionally advance the fire service. It conducts an annual conference to discuss matters of concern to fire chiefs.

International Association of Fire Fighters (IAFF)—A union of firefighters that was established in 1918. It is concerned with fire prevention, the protection and safety of firefighters and emergency medical and rescue workers, and it represents them politically on governmental legislation. It is also concerned with the education, training, health, and welfare of its members. The IAFF is affiliated with the American Federation of Labor–Congress of Industrial Organizations (AFL–CIO) and the Canadian Labor Council (CLC). The national office provides support to over 2,500 local chapters in North America. It also serves as a liaison with the local, national, and international media. The IAFF produces several publications internally that support the interests of its membership.

International Association of Fire Safety Science (IAFSS)—The International Association of Fire Safety Science was established at the first international symposium on fire safety science held at the National Institute of Standards and Technology (NIST) at Gaithersburg, Maryland, in 1985. It was founded to encourage research into the science of preventing and mitigating the adverse effects of fires and also to provide a forum to present the results of such research. IAFSS perceives its role to lie in the scientific bases for achieving progress in unsolved fire problems. It seeks cooperation with other organizations concerned with applications or with the sciences that are fundamental to fire research. The IAFSS promotes high standards, encourages and stimulates scientists to address fire problems, and provides the necessary scientific foundations and means to facilitate applications aimed at reducing life and property losses. It organizes and supports symposia and publishes their proceedings and other fire safety educational materials. It maintains an international communication forum and information exchange (e-mail) for those interested in fundamental, strategic, and applied research on fire safety and fire prevention.

International Fire Service Accreditation Congress (IFSAC)—A self-governed system that provides accreditation for the certification of fire service programs. It was organized in 1991 and is based at Oklahoma State University. *See also* **National Professional Qualifications Board (NPQB).**

International Fire Service Training Association (IFSTA)—A nonprofit educational association, founded in 1934, whose objective is the upgrading of firefighting techniques and safety through training and educational materials. It publishes training manuals used worldwide for the fire service through Fire Protection Publications, located in Stillwell, Oklahoma, at Oklahoma State University.

International Shore Connection—Standardized firewater distribution piping connection available for a fireboat to pump into a system or for the connection of a vessel's firewater system to a land-based firewater system. The connection is required on all vessels over 500 gross tons (454 metric tons) subject to Safety of Life at Sea

(SOLAS) regulations and on US inspected vessels over 1,000 gross tons (907 metric tons). It was required by the SOLAS regulations for vessels to eliminate the problem of hose threads of the vessels not matching those of a port facility to which it docks. ASTM F 1121, *Standard Specification for International Shore Connections for Marine Fire Applications,* provides guidance on its design and construction. It may also be called a ship-to-shore connection.

International Society of Fire Service Instructors (ISFSI)—An organization formed in 1960 for the exchange of ideas, information, and training techniques for the fire service, fire safety organizations, educational institutions, and other interested parties.

Intrinsically Safe (IS)—A circuit or device in which any spark or thermal effect in normal or abnormal conditions is incapable of causing ignition of a mixture of flammable or combustible material in air under prescribed hazardous atmosphere conditions (combustible vapors) due to insufficient ignition energy. Intrinsically safe devices or equipment are used where it would be impractical to install or use explosionproof equipment due to its cost. The intrinsically safe principle is applied to devices and wiring in which the output or consumption of energy is small; the limited amount of supply voltage, resistance, capacitance, and inductance; the manner in which a circuit is broken, and; its material, shape of contacts, and type of gas or vapor to be encountered. Intrinsically safe devices in the United States are commonly tested to UL 913, *Standard for Safety, Intrinsically Safe Apparatus and Associated Apparatus for Use in Class I, II, and III, Division 1, Hazardous (Classified) Locations* and installed per ANSI/ISA-RP 12.6, *Recommended Practice for the Installation of Intrinsically Safe Systems for Hazardous Classified Locations. See also* **Classified Area; Non-Incendive; Non-Incendive Circuit.**

Intumescence—A swelling or bubbling of a material caused by heat. Intumescent materials are used as fireproofing due to the insulative layer from the swelling or bubbling process that is produced when exposed to heat. The quality and thickness of the intumescent layer determines its effectiveness as a fireproofing material. *See also* **Fireproofing; Intumescent.**

Intumescent—A fireproofing material such as an epoxy coating, sealing compound, or paint. It foams or swells several times its volume when exposed to heat (thermal loading) from a fire and simultaneously forms an outer char covering that together form an insulating thermo layer against a high temperature fire. Intumescent fireproofing materials may be composed of different configurations to suit the penetration to be sealed. An intumescent sealant may contain a catalyst that, when heated (usually at approximately 212°F [100°C]), decomposes to produce phosphoric acid. This in turn reacts with carbonific components to give a phosphate. The decomposition of the phosphate produces carbon dioxide (CO_2) and water (H_2O). This causes a carbonaceous residue to bubble and foam to make a thermoset resin. As a result, a thick, insulative fire and smoke seal are formed. Intumescent materials are also used in honeycomb fire dampers and in laminated fire-resisting glass. The glass turns opaque when heated, providing additional insulation.

Proprietary products such as fire pillows, plugs, bricks, or fire prevention cushions are designed to seal large cable, pipe, and ducts in fire-rated barriers that are altered on a frequent basis or to seal temporarily an opening needed during construction. These products swell when exposed to heat to completely seal penetrations to prevent fire, smoke, and gases from passing through. Some manufacturers' pillows or cushions do not meet

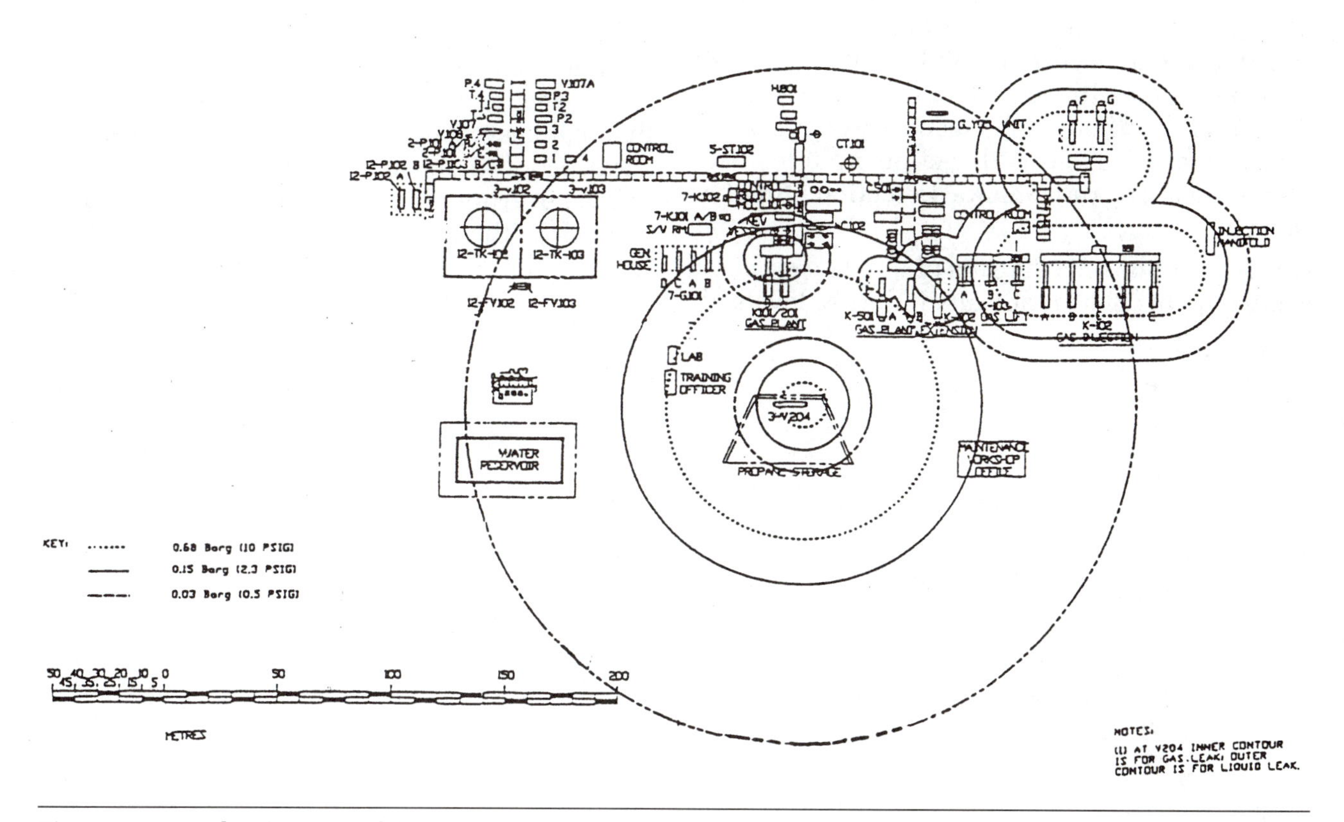

Figure I-4 Isoplat. (an example)

requirements for the provision of a smoke barrier with their product and are only rated as a fire barrier. *See also* **Fireproofing; Intumescence.**

Ionization Smoke Detector—*See* **Smoke Detection.**

IR Fire Detector—An optical flame-sensing device used to detect a fire by the presence of infrared (IR) radiation with wavelengths of 8,500 to 12,000 angstroms. IR detectors utilize a photocell to sense IR radiation from a fire source. Lighting, arc welding, x-rays, and gamma rays have no effect on IR detectors. High intensity lighting, sunlight, and high-temperature objects may affect IR detectors. They are most effective for approximately 50 ft. (15 m) from the source (they are subject to the inverse square law of optics). The lens of the device must also be kept clear for optimum sensing ability. *See also* **Infrared (IR); Infrared Scanner; Ultraviolet-Infrared (UV/IR) Fire Detection.**

I

IR Smoke Detection—A method of smoke detection that utilizes an infrared beam to detect the occurrence of smoke. The beam is projected over an area from an emitter to a receiver unit. Smoke particles cause the beam to be disrupted and the smoke occurrence will be detected. IR smoke provides linear smoke detection within the limits specified by the manufacturer. It may also be called a beam detection system. *See also* **Linear (Beam) Smoke Detection; Smoke Detector.**

Isochar—A line on a diagram of fire damage connecting points that graphically indicates levels of equal char depth. The graphical depiction of char levels (isochars) can help determine the duration of fire at various locations. They are used by fire investigators to determine fire origin and growth. *See also* **Char.**

Isoplat—The graphical representation on a plot plan, elevation, or three-dimensional drawing that connects points of equal blast overpressures. Isoplats show the theoretical calculation of blast overpressures from the center point of an explosion (the ignition point), which may differ in an actual blast occurrence due to the variable conditions in effect at the site. See Figure I-4.

Jaws-of-Life—Hydraulically assisted mechanical mechanism used to spread heavy steel construction apart (car frames or parts). They are used by the rescue services particularly for assisting in the extraction of individuals from vehicles involved in automotive crashes. The mechanism is shaped in the form of a huge plier, which is opened to pry structural members apart. This allows the victim to escape, giving the mechanism the nickname of jaws-of-life. *See also* **Porta-Power.**

Jet Fire—Combustion occurring at the release of liquid, vapor, or gas under pressure from a leakage point (orifice), the momentum of which causes entrainment of the surrounding atmosphere. The jet fire has a high heat flux, turbulent flame, and a capability of eroding the material it impacts. The stability, magnitude, and distance of a jet fire flame are a function of the release pressure, wind direction, leakage (orifice) size, orientation, and other site-specific influences. A jet fire has higher combustion efficiency than pool fires because the jet entrains air during its discharge. Flame radiation from a jet fire comes from hot gases and incandescent soot particles. The relative contributions depend strongly on the fuel and its release conditions. There is no significant convective heat transfer to objects that are outside of the flame. A jet fire is very localized but very destructive to anything close to it. A standard test for a jet fire exposure has not been adopted by any standards organization, however several individual company jet fire exposures have been proposed. It may also be called a spray fire.

Jet Siphon (Valve)—A pumping device used to transfer water. A small high-pressure (jet) water stream is inserted into a large stream of water to push it along by momentum force. May also be called a jet dump. See Figure J-1.

Jockey Pump—A fixed water pump that maintains constant pressure on firewater system piping to compensate for small leakages and to supply incidental first aid usage. It is sometimes referred to as a pressure maintenance pump. It is mainly provided to handle water

Figure J-1 Jet siphon.

uses without having the main firewater pumps unnecessarily cycling on and off and possibly producing a water hammer effect. Jockey pumps are normally set to activate after some small amount of pressure drop has occurred in the fire main. This is normally about 5 to 15 psi (34 to 135 kPa) above the startup pressure of the main fire pump(s), with 10 psi (7 kPa) generally selected. Jockey pumps are generally electrically driven and have small capacities compared to the main water pumps. Jockey pumps do not meet the same levels of approval, integrity, and reliability as required for main firewater pumps.

Joint Council of National Fire Service Organizations (JCNFSO)—An organization formed to provide joint determination of fire service organization goals and objectives for national fire service organizations. It was composed of various national fire service organizations and established the National Professional Qualifications Board for firefighters. The organization was formed in 1970 but disbanded in 1989. *See also* **National Professional Qualifications Board (NPQB).**

Kelly Tool—A forcible entry tool used by the fire service to pry or force open doors, windows, or other openings in a building to gain access for rescue or firefighting operations. It has a forked blade or chisel at one end and an adz blade at the other end. *See also* **Claw Tool; Forcible Entry; Halligan Tool; Hux Bar.**

K-Factor—(1) Coefficient specified for individual sprinklers based on their orifice design and used for hydraulic calculations of the sprinkler system. K-factors are determined by the design and manufacturer of the sprinkler head. *See also* **Conduction; Thermal Conductivity.**

K-Tool—A prying/extraction hand tool used by the fire services to perform through-the-lock forcible entry on locked doors. It is designed to pull out the lock cylinders and expose the mechanism in order to open the locked door.

Kindle—The process of starting a fire from small quantities of combustible materials in order to set larger quantities of combustible materials on fire.

Kindling—A common term for small pieces of wood used to start larger materials on fire. The kindling is ignited, which in turn develops a small fire that is sufficient to ignite large combustible materials.

Kindling Temperature or **Kindling Point**—The temperature at which a substance must be heated to burst into flames, similar to ignition temperature. Every burnable substance has its own kindling temperature. The lower the kindling temperature, the more easily a substance catches fire. Paper has a kindling temperature of approximately 450°F (232°C), cellophane 468°F (242°C), wood 375°F to 510°F (190°C to 266°C), and cotton 511°F (266°C). It has been demonstrated that wood and wood-based products have several other features that influence when they may ignite. These include the specific gravity or density of the material, its physical characteristics, moisture content, rate and period of heating, nature of the heat source, and the air supply and its velocity. *See also* **Autoignition Temperature (AIT); Ignition Point; Ignition Temperature; Self-Ignition Temperature.**

Kink—A method or occurrence in a firewater hose that stops the flow of water. It generally consists of arranging a fold or bend back onto the hose, similar to an "S" shape, and applying pressure to collapse the hose to stop the flow of water. Purposely applied kinks are used when an isolation valve is not immediately available or the disruption of water flow may hinder other firefighting activities, for example, at a burst hose section. Unwanted kinks occur due to improper arrangement or layout of hoses. *See also* **Hose Rack.**

Kink (Pressure) Test—A pressure test for fire hose performed in conjunction with the application of a sharp kink to the hose. The hose is pressurized to 10 psi (69 kPa) and then sharply kinked 18 in. (45.7 cm) from the free end by tying the hose back against itself as close to the fittings as practicable. The pressure is then raised to the proof test pressure and immediately released. *See also* **Burst Test Pressure; Proof Test Pressure.**

Knockdown—During firefighting activities, the reduction of flame and heat to a point (flames have been "knocked down") where further extension of a fire has been overcome and the overhaul stage can begin.

Knockdown Speed—During the application of firefighting foam to a spilled liquid fuel, it is the time the foam spreads across it's surface.

Knots—A method of securing a rope or line. The fire service uses various knots for specific purposes. The principle knots include a clove hitch (two half hitches used to attach a rope to an object) and bowline knot, used to form a loop in a natural fiber rope.

Kortick Tool—A forest service firefighting tool used primarily for the construction of a fireline. It is a combination of a hoe and rack and is similar to the McLeod Tool.

L

Labeled—Equipment or materials to which is attached a specified label, symbol, or other identifying mark of an organization acceptable to the authority having jurisdiction and concerned with product evaluation, which maintains periodic inspection of production of labeled equipment or materials. The provision of a label by the manufacturer on a fire protection product indicates compliance with appropriate standards or performance in a specified manner for fire safety. *See also* **Approved; Classified; Factory Mutual (FM); Listed; Underwriter's Laboratories (UL).**

Ladder Pipe—A heavy or master stream nozzle attached at the top of an aerial ladder and fed from the ground for elevated water application.

Ladder Truck—A fire truck that is provided with aerial ladders or elevating platforms. An aerial ladder truck has a metal extension ladder mounted on a turntable. Most trucks can raise the ladder to 100 ft. (30 m), or eight stories. The elevating platform truck has a cage-like platform that can hold several people. The platform is attached to a lifting device, either an articulating boom or a telescoping boom, which is mounted on a turntable. A built-in hose is provided for the length of the boom and is used to direct water onto a fire. Several variations are available which are referred to by various names, e. g., snorkel, snozzle, straight stage aegal platform novo quintec, etc, *See also* **Snorkel** or **Snorkel Truck.**

Laminar Flame—A smooth-surfaced flame with low burning velocity and without turbulent conditions in the flame area. A candle flame is an example of a laminar flame. *See also* **Diffusion Flame.**

Large Diameter Hose (LDH)—Firefighting hose whose diameter is larger than that normally utilized in firefighting operations. It is normally used to relay water in large quantities from one location to another through the use of fire pumpers. It is commonly provided in sizes of 3.5 to 6.0 in. (9.0 to 15.0 cm). *See also* **Medium Diameter Hose (MDH).**

Large Drop Sprinkler—A type of sprinkler head that produces large droplets and is used for fire control of high challenge fire hazards, particularly high-pile storage facilities. It normally has a K-factor between 11.0 and 11.5 and has prescribed water penetration, cooling, and distribution criteria. Large drop sprinklers generate large droplets of a size and velocity to enable effective penetration of a fire plume with high velocity. Large drop sprinklers were developed by the Factory Mutual Research Corporation and practically demonstrated in 1971.

Large Orifice (LO) Sprinkler—A standard spray sprinkler head that generally discharges 140 percent more water at the same water pressure. It is normally used where high density is required or low residual water pressure may be present.

Latent Heat—The characteristic amount of energy absorbed or released by a substance during a change in its physical state that occurs without changing its temperature. The latent heat associated with melting a solid or freezing a liquid is called the heat of fusion; that associated with vaporizing a liquid or a solid or condensing a vapor is called the heat of vaporization. The structure of a crystalline solid is maintained by forces of attraction between the individual molecules or ions, which oscillate slightly around their mean positions in the array. When heat is absorbed, these motions increase until at the melting point the attractive forces can no longer preserve the orderly arrangement, and the solid changes into a liquid, in which the individual particles move about independently, attracted to each other only by forces much weaker and less specifically directed in space. When a substance is heated sufficiently, even the weak forces that hold the particles together in the liquid state are overcome, and at the boiling point, the liquid transforms into vapor. Latent heat is associated with processes other than changes between solid, liquid, and vapor phases of a single substance. Many solids exist in different crystalline modifications, and the transitions between these are generally attended by absorption or evolution of latent heat. The process of dissolving one substance in another sometimes involves heat; if the solution process is a strictly physical change, the heat is a latent heat. However, the process sometimes is accompanied by a chemical change, and part of the heat is associated with the chemical reaction. The latent heat of the fusion of water at 32°F (0°C) is 143.4 Btu/lb. (79.7 calories/gram). The latent heat of the vaporization of water at 212°F (100°C) is 970.3 Btu/lb. (53.6 calories/gram). *See also* **Boilover.**

L

Life Safety—The preservation of life from a fire event or its associated hazards. *See also* **Life Safety Code (LSC).**

Life Safety Code (LSC)—A fire code developed by the National Fire Protection Association (NFPA) for the preservation of life from a fire event or its associated hazards. It is primarily concerned with exit facilities and arrangements and protection against fire events. As a result of the large loss of life in the Triangle Shirtwaist Fire in New York in 1911, the NFPA was challenged to provide life safety measures for factories, loft buildings, control of smoking in hazardous areas, improved exits, and provision of fire drills. NFPA 101, *Life Safety Code*, is the standard that is used to delineate the appropriate fire exit requirements. *See also* **Life Safety.**

Life Safety Systems—The embodiment of measures taken or prescribed to prevent the endangerment of individuals threatened by a fire event or its associated hazards. In buildings, life safety systems primarily consist of exit facilities, emergency lighting, fire sprinklers, and standpipe systems. In high-rise buildings, stairways serve as vertical emergency exits; all elevators are automatically shut down to prevent the possibility of people becoming trapped in them. Emergency generator systems permit the operation of one elevator at a time to rescue people trapped in them by a power failure. Generators also serve other vital building functions, such as emergency lighting and fire pumps. Fire suppression systems often include sprinklers, but if none are required by building codes, a separate piping system is provided with electric pumps to maintain pressure and to bring water to fire hose cabinets throughout the building. There are also exterior connections at street level for portable fire truck pumps. The fire hoses are so placed that every room is accessible; the hoses are intended primarily for professional firefighters but may also be used by the building occupants.

Lightning—The visible discharge of atmospheric electricity that occurs when a region of the atmosphere acquires an electrical charge, or potential difference, sufficient to overcome the resistance of the air. A typical lightning flash involves a potential difference between cloud and ground of several hundred million volts, with peak currents on the order of 20,000 amperes. Temperatures in the lightning arc are on the order of 50,000°F (30,000°C).[6] Lightning strikes may cause an ignition of combustible materials because of their high-voltage discharges. The tendency of strikes to occur at high points enables lightning rods of conductive metal to dissipate the strikes harmlessly into the ground. *See also* **Lightning Rod.**

Lightning Protection System—A system designed to protect structures and buildings from destructive lightning strikes by draining lightning surge-induced charges harmlessly to ground. A lightning protection system usually consists of strike termination devices, electrical conductors, ground terminals, and surge suppression devices.

Lightning Rod—A metallic rod (usually copper) that protects a structure from lightning damage by attracting the lighting strike and transmitting its current to the ground. Lightning tends to strike the highest object in the vicinity. Because of static charge buildup and attraction, the rod is placed at the top of the structure; from there it is connected to the ground by low-resistance cables. In the case of a building, the soil is used as the ground; on a ship, the water is used. A lightning rod affords protection because it diverts the current from the nonconducting parts of the structure, allowing it to follow the path of least resistance and pass harmlessly through the rod and its cables. It is the high resistance of the nonconducting material that causes it to be heated by the passage of electric current, leading to fire and other damage. A lightning rod provides a cone of protection whose ground radius approximately equals its height above the ground. NFPA 780, *Standard for the Installation of Lightning Protection Systems*, provides guidance in the need and installation of lightning rods for fire protection of buildings. *See also* **Lightning.**

Light Water—*See* **Foam, Aqueous Film Forming (AFFF).**

Limited-Combustible Material—A building material that does not meet the definition of a noncombustible material. It has a potential heat value not exceeding 3,500 Btu/lb. (8,141 kJ/kg), when tested in accordance with NFPA 259, *Standard Test Method for Potential Heat of Building Materials*, and complies with either of the following:

1. A material that has a structural base of a noncombustible material, with a surfacing not exceeding a thickness of 0.125 in. (0.32 cm) that has a flame spread index of 50 or less.
2. A material that has a flame spread index of less than 25 and no evidence of continued progressive combustion. Its composition is such that surfaces that would be exposed by cutting through the material on any plane would have a flame spread index of less than 25 and no evidence of continued progressive combustion.

Linear Beam Fire Detection—A detection device that utilizes an infrared (IR) beam to detect the presence of smoke or heat. The beams

[6] *Encyclopedia Britannica* CD-97, CD-ROM, Encyclopedia Britannica, Inc., London, UK, 1997.

L

have a minimum and maximum length and have to be oriented in a straight line. Linear beam fire detection is useful for protecting hazards that are arranged in row, such as electrical panels, hazardous pumps, etc., and for perimeter coverage requirements or improved room coverage over point detectors. It may also be called a beam smoke detection (BSD) system. *See also* **Linear (Beam) Smoke Detection (LSD) System; Pneumatic Fire Detection System.**

Linear (Beam) Smoke Detection (LSD) System—Smoke detection over the entire length of a straight path, affected by the use of infrared (IR) beam detection devices. Linear smoke detection systems are usually more sensitive and cover a larger area than point smoke detection systems. See Figure L-1. *See also* **IR Smoke Detection; Linear Beam Fire Detection; Smoke Detection.**

Linear or **Line-Type Heat Detection (LHD)**—A fixed temperature or rate of temperature rise detection device that is able to detect heat along a line (curved or straight). A linear or line-type heat detector may be pneumatic or electrical in nature (that is, using tubing or wire). It is utilized where long distances or numerous locations inside a process area require fire detection.

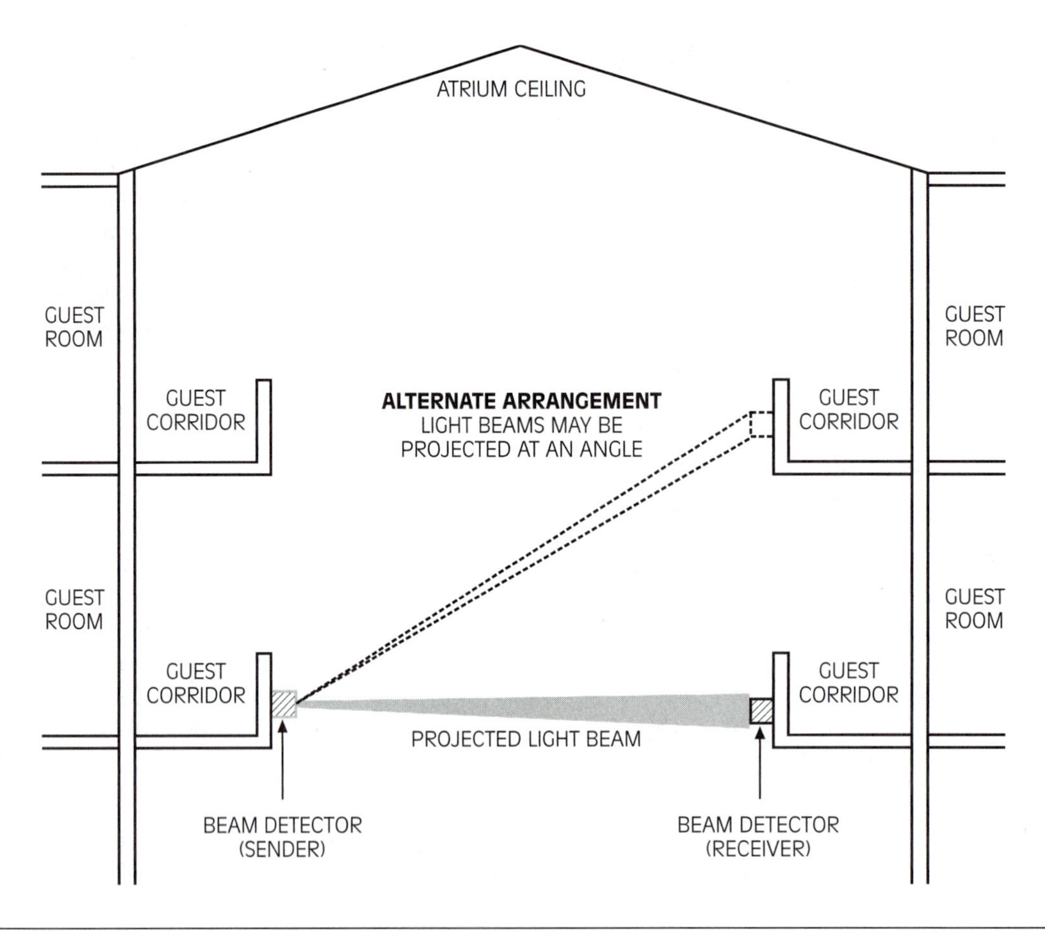

Figure L-1 Linear (beam) smoke detector.

The electrical system uses a pair of wires that sense the generation of heat and cause a signal to be sent to a monitoring device. Based on the same principle as a point sensing thermocouple, heat will cause an electromagnetic force to occur in a pair of wires composed of two pieces of different metals along any point in its length. The electromotive force appears at the free end of the wire and is translated into a fire alarm. The wires do not have to be run in a straight fashion and can be routed in any direction as required to protect the hazard.

The pneumatic system uses flexible tubing that is pressurized with compressed air or an inert gas (such as nitrogen). A pressure sensor is mounted on the line. A fire occurrence causes the tubing to melt and release the system pressure causing the pressure sensor to alarm. See Figure L-2 and L-3. *See also* **Bimetallic Wire.**

Line Fire—An elongated fire on a horizontal fuel surface. *See also* **Pool Fire.**

Linen Hose—Firefighting hose that does not have a rubber inner liner and is made of flax or linen fabric. It is normally provided for standpipe systems and forest firefighting operations. It may also be called an unlined hose. *See also* **Hose, Fire.**

Listed—Term used by fire codes specifying equipment or materials tested for fire safety concerns that is included in a list published by an organization acceptable to the "authority having jurisdiction." A listing recognition is concerned with product evaluation in that it maintains a periodic inspection of the production of listed equipment or materials and whose listing states either that the equipment or material meets

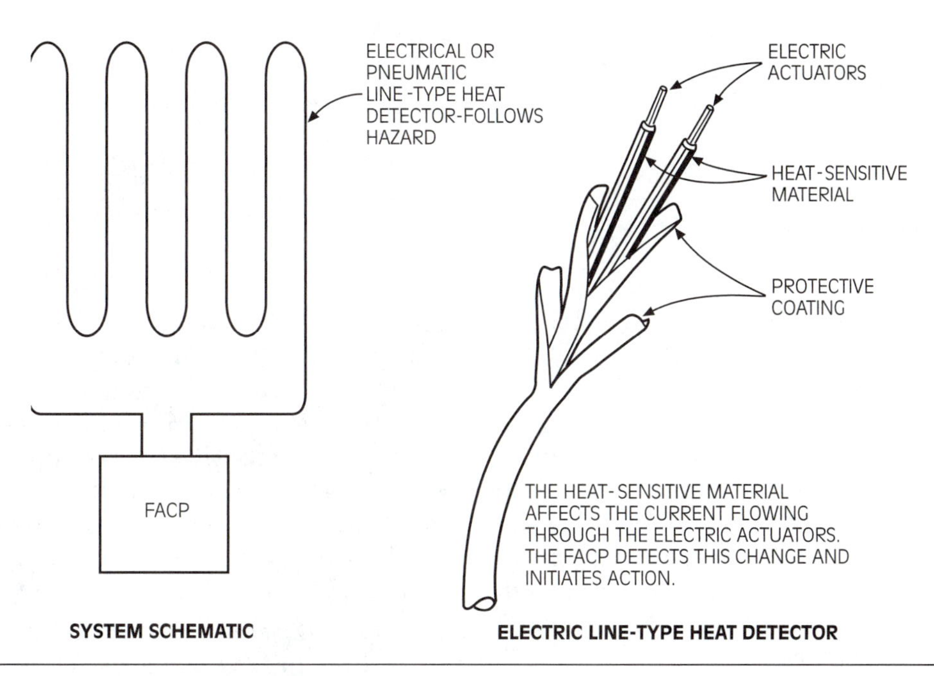

Figure L-2 Line-type heat-detection devices.

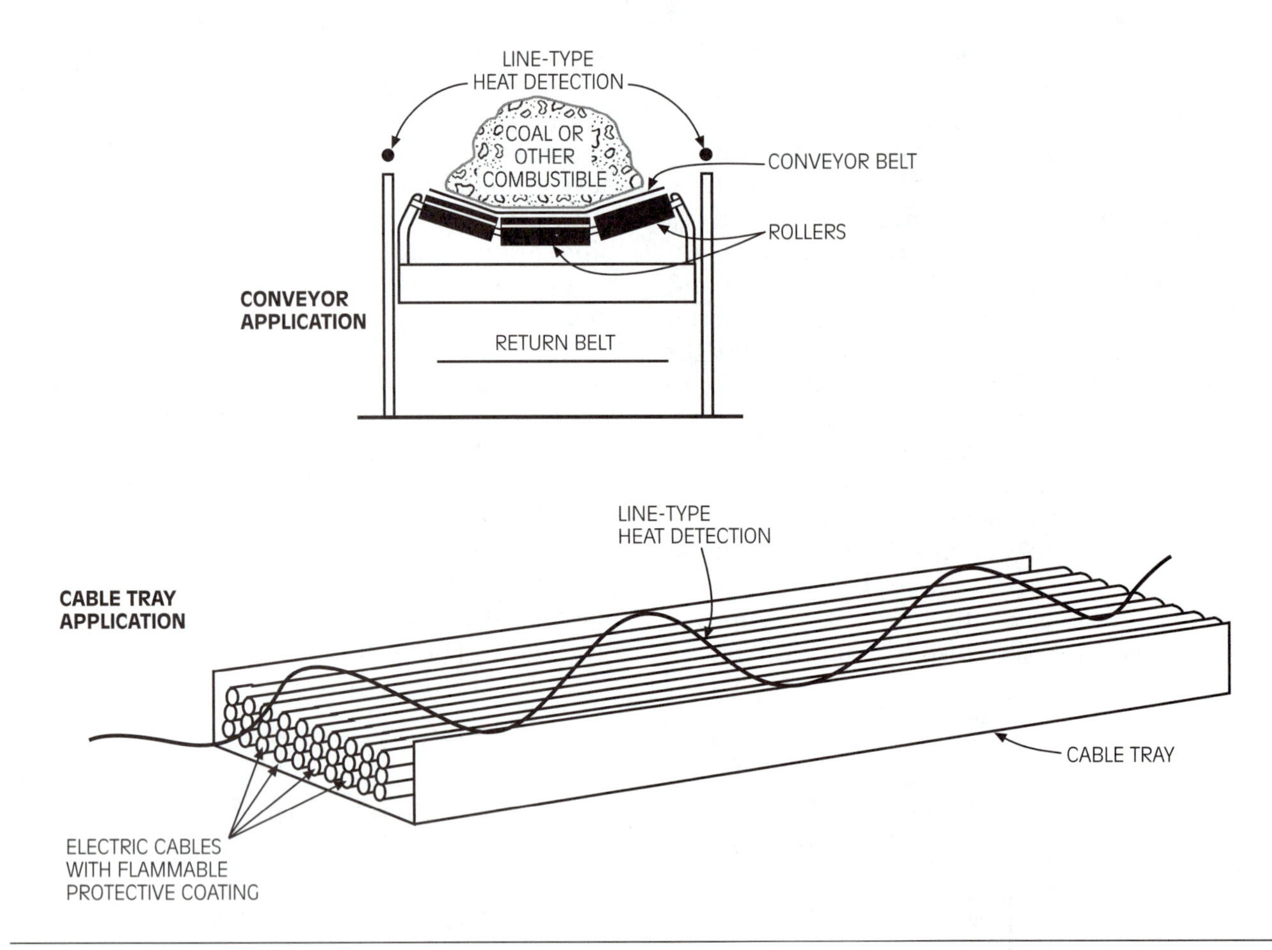

Figure L-3 Applications of line-type heat-detection devices.

applicable standards or has been tested and found suitable for use in a specified manner. The means for identifying listed equipment may vary for each organization concerned with product evaluation, since some of them do not recognize equipment as listed unless it is also labeled. The authority having jurisdiction should utilize the system employed by the listing organization to identify a listed product. *See also* **Approved; Classified; Factory Mutual Global (FM); Labeled; Underwriter's Laboratories (UL).**

Live Load—The weight of the non-integral and nonstructural parts and contents of a building or structure. Live loads generally consist of furnishings, movable equipment, occupants, etc. Firefighting activities (water application, equipment, personnel, etc.) may increase live loads imposed on a structure or building, which may inadvertently and unknowingly lead to its collapse. *See also* **Dead Load; Impact Load.**

Lloyd's of London—A unique insurance enterprise that has evolved from the underwriting transactions carried on in the 1680s between ship owners, merchants, and insurers in the Coffee House of Edward Lloyd (1648–1713), which has adopted the name "Lloyd's." The risk and profit of the insurance underwriting is shared among a large number of members, who form groups termed "syndicates." Syndicates may be composed of a few members or up to a hundred. The underwriting members of a syndicate are known as "names," whose entire personal wealth is theoretically available to support claims against the policies. Members are only liable for the amount they individually underwrite and are never liable for the losses of other members or the syndicate. In the 1990s, large claims were made as a result of several catastrophic accidents, which resulted in huge financial losses for Lloyd's and much acrimony. In other periods, members have been accustomed to receive a large annual bonus, while their capital is invested elsewhere.

Members originally gathered at Lloyd's Coffee House because Edward Lloyd published a list, known as *Lloyd's List*, stating the sailing dates and known whereabouts of merchant ships that were important for underwriters, ship owners, and merchants. (The original Lloyd's Coffee House was destroyed during a London fire in 1838.) Lloyd's now underwrites all kinds of insurance policies, including fire risks, and is considered a world class underwriting organization, although its primary line is marine insurance. Officially, Lloyd's is acknowledged to date from 1771. Merchants and brokers who were using the Coffee House entered into their first formal agreement and the following year a committee was formed to govern the affairs of the subscribers in their capacity as underwriters. It operated as a private concern until 1871, at which time Parliament passed the Lloyd's Act, incorporating members of the association into a single corporate body.

Local Application—The provision of a fire suppression system to discharge directly onto or near a hazard that is not enclosed and is isolated from other hazards or exposures so that a fire will not spread. Application is limited to the fire hazard area itself. It is primarily intended to apply fire suppression agent directly to the burning surface. An example of local application is the provision of overhead nozzles for the protection of deep fat fryers from a fixed wet chemical system.

Loss Control—A program whose objective is to minimize accident-based losses. Total loss control is based on studies of near misses or non-injury or damage incidents, and on an analysis of both direct and indirect accident causes. Both injuries and property damage are included in the analysis. Loss control activities include fire prevention, education, inspection, overhaul, and salvage. Examples include fire training, housekeeping, education, fire brigade organization, and disaster control and recovery. *See also* **Fire Prevention; Overhaul; Salvage.**

Loss Prevention—A program to identify and correct potential accident problems before they result in an injury or financial impact. Loss prevention also encompasses the activities of fire prevention.

Low-Energy Circuit—Electrical circuits that do not contain sufficient energy to produce incendiary sparks such as signal and communication circuits. *See also* **Non-Incendive Circuit.**

Lower Explosive Limit (LEL)—The minimum concentration of combustible gas or vapor in air, below which propagation of flame does not occur on contact with an ignition source. The lower limits of flammability of a gas or vapor at ordinary ambient temperature expressed in percent of the gas or vapor in air by volume. *See also* **Explosive Limits; Lower Flammable Limit (LFL); Upper Explosive Limit (UEL).**

Lower Flammable Limit (LFL)—Synonymous with lower explosive limit (LEL). *See also* **Flammable Limit** or **Flammability Limits; Lower Explosive Limit; Upper Flammable Limit (UFL); Vapor.**

Low Expansion Foam—Low expansion foams protect two basic types of fuels: hydrocarbon fuels containing molecules of hydrogen and carbon, and polar solvent/water miscible fuels that contain molecules of oxygen, nitrogen, sulfur, chlorine, phosphorus, and/or any combination of these. These polar solvent/water miscible fuels readily mix with water. Examples are alcohol, methanol, ethanol, etc. *See also* **Foam Concentrate; Water Additive.**

Low (Fire) Hazard—A condition or state where the amount of combustibles present are lower than would be present for an ordinary fire hazard. *See also* **Extra (Fire) Hazard.**

Low Order Explosion—A slow rate of pressure rise or low force explosion characterized by a pushing or dislodging effect upon the confining structure or container and short missile distances.

Low Smoke-Producing Cable—Commonly, electrically insulated cable that produces smoke to a maximum peak optical density of 0.5 and a maximum average optical density of 0.15. Low smoke-producing cables are tested in accordance with NFPA 262, *Standard Method of Test for Fire and Smoke Characteristics of Wires and Cables. See also* **Fire-Resistant Cable.**

M

Maltese Cross or **Cross Pattee-Nowy**—Insignia of the fire services that was adopted from the Knights Hospitallers (Knights of St. John of Jerusalem) because of their humanitarian service and heroic fire extinguishment feats. The Knights existed during the 11th and 12th centuries. In addition to their humanitarian service, they provided military assistance to the Knights of the Crusades (in their efforts on the island of Malta) when their enemies threw glass bombs containing naphtha onto their armor and ships. Many knights performed heroic deeds extinguishing these fires. In acknowledgement, the cross worn by these knights was decorated and inscribed. The Maltese Cross symbolizes the ideals of saving lives and extinguishing fires. See Figure M-1.

Figure M-1 Fire service Maltese cross.

Management Oversight and Risk Tree (MORT)—A system safety concept that applies analytical procedures to all phases of a safety program. It uses a MORT "tree" to identify significant elements of a safety program or safety management system.

Manual Activation Callpoint (MAC)—Synonymous with manual pull station. It is frequently used in the British industry. *See also* **Break Glass Unit.**

Manual Pull Box—Synonymous with manual pull station. *See also* **Fire Alarm Signal; Manual Pull Station (MPS).**

Manual Pull Station (MPS)—A switch provided on a fire alarm system that is manually activated to indicate a fire event. The switch is configured to conspicuously identify it as a fire alarm device and is usually fitted with a tamper device (break glass, rod, or cover) to discourage or prevent false activation. It sends a signal to a central monitoring station for notification of location and activation of alarms. See Figure M-2. *See also* **Automatic Fire Alarm System; Break Glass Unit; Manual Pull Box; Pull Station.**

Marine Fire Protection—The Safety of Life at Sea (SOLAS) is the primary international treaty concerning vessel safety measures including fire protection measures. The US Coast Guard also enforces the Code of Federal Regulations (CFR) concerning commercial vessels in US waters. NFPA 1405, *Guide for Land-Based Fire*

Figure M-2 Manual pull station.

Fighters Who Respond to Marine Vessel Fires, provides a comprehensive marine firefighting response program that includes vessel familiarization, training considerations, pre-fire planning, and special hazards for land-based firefighters. Fixed structures (such as oil platforms) in territorial waters generally must adhere to national fire safety regulations that cover these designated areas.

Mass Burning Rate—The mass burning rate for a fire is the mass of fuel supplied to the flame per unit time, per unit area of the fire. The measurement calculation units are usually kg m^2/second.

Mass Optical Density—The optical property of smoke related to the measured yield of solid and liquid particulates of smoke generation. The mass optical density is used to determine smoke visibility (ability to see through smoke). *See also* **Smoke Visibility.**

Master Stream—A large capacity portable or fixed firefighting water delivery appliance classification formed by either multiple fire hose lines or fixed piping. Master streams are delivered by monitor nozzles, hydrant-mounted monitor nozzles, or portable deluge sets. It generally has a capability of flowing in excess of approximately 250 to 350 gpm (946 to 1,140 l/min) of water or water-based extinguishing agent with nozzles pressures of 80 to 100 psi (551 to 689 kPa) into large streams of extended reach for firefighting applications. Master streams are beyond the capability to control through manual manipulation by hose nozzles because of the reaction forces involved. It may also be called a heavy stream. See Figure M-3. *See also* **Deck Gun (Deluge Set); Heavy Stream; Hose Stream.**

Figure M-3 Terminator nozzle, capable of flowing approximately 2,000 gpm (7,500 l/m).

Material Safety Data Sheet (MSDS)—An information sheet on properties of a material that meet certain combustible, toxic, or other hazardous threshold criteria that are specified by the Environmental Protection Agency regulation Title III of the Superfund Amendments and Reauthorization Act (SARA). MSDS are required to be provided and maintained by organizations that have hazardous materials. They are also required to provide copies to the local fire department for the purpose of firefighter protection and pre-planning. *See also* **Hazardous Material; Hazardous Materials Identification System (HMIS).**

Maximum Occupancy Allowances—The maximum allowable number of individuals that may occupy a structure. Maximum occupancy postings are primarily provided at structures to ensure the available exits for the facility are adequate and panic will not ensue during an emergency evacuation. The maximum occupancy postings are based on measurable standards for exit door provisions. These provisions are most commonly cited in the Life Safety Code (NFPA 101) from the National Fire Protection Association (NFPA).

Maximum Possible Loss (MPL)—The total amount of coverage an underwriter is willing to write on an insured peril. It is an estimate of the largest loss to be expected under adverse circumstances. It excludes catastrophic conditions, but considers protection equipment not functioning, with damage limited only by spacing of the structures, by firewalls, or by a lack of continuity of combustibles. Primarily used in the insurance underwriting industry for valuing insurance policies.

McLeod Tool—A forest service firefighting tool consisting of a combination of a hoe and rake. It is used primarily for the construction of a control line. It is similar to a Kortick tool. See Figure P-10.

Means of Egress—A continuous and unobstructed way of exit travel from any point in a building or structure to a public way for the purposes of evacuation. *See also* **Evacuation; Exit; Means of Escape.**

Means of Escape—A way out of a building or structure. It may not comply with the definition of means of egress but does provide an alternative avenue to evacuate a facility. *See also* **Evacuation; Exit; Means of Egress.**

Mean Time Between Failure (MTBF)—The average time between successive failures, estimated by the total operating time of a population of items divided by the total number of failures within the population during the measured time period. The MTBF of a repairable item is estimated as the ratio of the total operating time to the total number of failures. The MTBF is useful for fire protection systems to determine the need for backup or standby equipment, such as a standby fire pump.

Mechanical Foam—Firefighting foam that is produced by mixing water, foam concentrate, and air to produce a foam consisting of bubbles of air instead of carbon dioxide (previously provided by "chemical" foams). *See also* **Chemical Foam; Foam Concentrate; Nozzle, Aspirating.**

Mechanically Pumped Extinguisher—A portable fire extinguisher in which the firefighting agent is mechanically pumped from its storage container and the container itself does not become pressurized. A hand water pump is an example of a mechanically pumped extinguisher. *See also* **Fire Extinguisher, Portable.**

Medium Diameter Hose (MDH)—A fire hose of average size that is intended for firefighting attack and relaying purposes. Normally 2.5 or 3.0 in. (6.3 or 7.7 cm) in diameter. *See also* **Large Diameter Hose (LDH).**

Medium Expansion Foam—*See* **High Expansion Foam.**

Melt—To change from solid to liquid, or to become consumed by the action of heat input.

Metal Fire—Combustible metal fires, such as sodium, titanium, uranium, zirconium, lithium, magnesium, and sodium-potassium alloys. Metal fires are difficult to extinguish. Dry powder fire extinguishers are normally provided to extinguish metal fires. *See also* **Dry Powder.**

Minimum Explosible Concentration (MEC)—A concentration of combustible dust (suspended in air) that is the minimum that will support a deflagration.

Minimum Ignition Current (MIC)—The minimum current in a specified spark test apparatus and under specified conditions that is capable of igniting the most easily ignitable mixture.

Minimum Ignition Energy (MIE)—The minimum energy required from a capacitive spark discharge to ignite the most easily ignitable mixture of a gas or vapor. *See also* **Ignition Energy.**

Misfire—The failure of an explosive or propulsive charge to ignite at the proper time.

Mobile Water Supply Apparatus—A fire service support vehicle, such as a tanker or tender, designed primarily for transporting a specified amount of water to a fire emergency scene. The water is transferred to other fire service vehicles or pumping equipment for use at the fire emergency.

Moeller Tube—A firefighting foam delivery device provided at the interior of a combustible liquid storage tank. It consists of a flexible porous tube kept coiled up and stored at the interior top of the tank wall. When required, it is designed to unroll and fall to the storage tank liquid level from the system pressure. Foam is inserted into the tube from the foam-water delivery pipe for the system. The tube is porous so the foam can travel from it and spread out onto the surface of the combustible liquid. It is classed as a Type I outlet by NFPA 11, *Standard for Low Expansion Foam. See also* **Foam Chamber; Type I (Foam) Discharge Outlet.**

Monitor—A water delivery device used to project very large quantities of water or foam-water through a nozzle for the purpose of providing fire protection (cooling, control, or fire suppression). They can be portable or fixed as well as remotely operated. It may also be called a "cannon" due to the large water streams it produces. See Figure M-4.

Monitor Nozzle—A classification of a large capacity firewater nozzle controlled by wheel-operated gears. It can be a fixed installation or a portable device. Monitors discharge large amounts of water

Figure M-4 Monitor.

M

and have a good straight stream range. A 1.0 in. (2.5 cm) diameter nozzle at 100 psi (689 kPa) has a flow of about 300 gpm (18.9 liters/second) with a range of 140 to 150 ft. (42.7 to 45.7 m) when the wind is less than 5 mph (8 kph). Beyond this range the water stream loses its continuity, but some of the water is thrown somewhat further in the form of rain droplets, which may be carried by the wind. In adverse winds 10 mph (16 kph) or greater, the range of monitor may be shorted by as much as 40 percent. The effective range of a spray pattern nozzle is about 40 ft. (12.2 m) at 500 gpm (31.5 liters/second) and 100 psi (689 kPa) nozzle pressure to about 125 ft. (38.1 m) at 100 psi (689 kPa) nozzle pressure with a straight stream. See Figure M-5.

Monitor Wagon—Term for a fire truck provided with heavy stream appliances and large diameter hose. *See also* **Heavy Stream.**

Monitor, (Water) Oscillating (WOM)—A firewater application nozzle that is permanently fixed but is able to automatically swing back and forth (oscillate) to provide a wide area coverage. Automatic oscillation is provided by a Pelton wheel, which uses a small stream of water to impulse a cog wheel that slowly directs movement of the monitor.

Figure M-5 Typical monitor nozzle.

Moody Diagram—A diagram to determine the piping friction loss factor when using the Darcy-Wieback friction loss formula in water distribution networks. The Moody diagram relates the piping Reynolds number, pipe diameter, and roughness to determine a friction loss factor. It is named for American engineer Lewis F. Moody (1880–1953), who investigated friction losses in pipes in the 1940s. *See also* **Hazen-Williams Formula.**

Multipurpose Dry Chemical—A dry chemical fire suppression agent that is effective on Class A, B, and C fires. It is based on a mixture of ammonium phosphate or ammonium phosphates and sulfates. Multipurpose dry chemical agents suppress a fire by interfering with the chemical chain reaction of the fire and also by coating Class A combustible surfaces with a glassy coating during the reaction of the chemical during application, which excludes oxygen from the combustion process. A Class A fire extinguished by a multipurpose dry chemical application may re-ignite since these fires may be deep-seated in the material and the chemical agent may not reach these areas. Multipurpose dry chemical extinguishers are the most widely used due to their versatility. *See also* **Dry Chemical; Potassium Bicarbonate; Sodium Bicarbonate.**

Multipurpose Piping System—A water distribution piping network provided for residential buildings for the purpose of supplying both domestic and fire protection applications.

Municipal Fire Alarm Box (Street Box)—An enclosure that houses a fire alarm transmitter and is located on a public thoroughfare. It is used to send an alarm to the fire services.

Most fire alarm boxes have been removed due to high incidents of false alarms and the improvements made in other avenues of communications available to report a fire (telephones, hardwired systems, etc.). *See also* **Fire Alarm Box; Fire Alarm System.**

Municipal Fire Alarm System—A fire alarm system provided for cities and towns that uses street boxes placed throughout the municipality to receive fire alarms at the fire service headquarters monitoring station. Each box sends a specially coded signal that identifies its location. Municipal fire alarm systems are gradually becoming obsolete due to other advanced avenues of communication available to report a fire and increased incidents of vandalism and false alarms reported from the street boxes. See Figure M-6.

Municipal Grading Schedule—An evaluation of the fire protection features of cities and towns by the Insurance Services Office (it was transferred from the American Insurance Association [AIA] in 1971), based on seven factors: climatic conditions, water supply, fire department, fire service communications, fire safety control, building codes, and a survey report. It is used to establish a basis of insurance rates.

Climatic Conditions—Susceptibility to fire from special environmental effects such as dry brush or earthquakes.

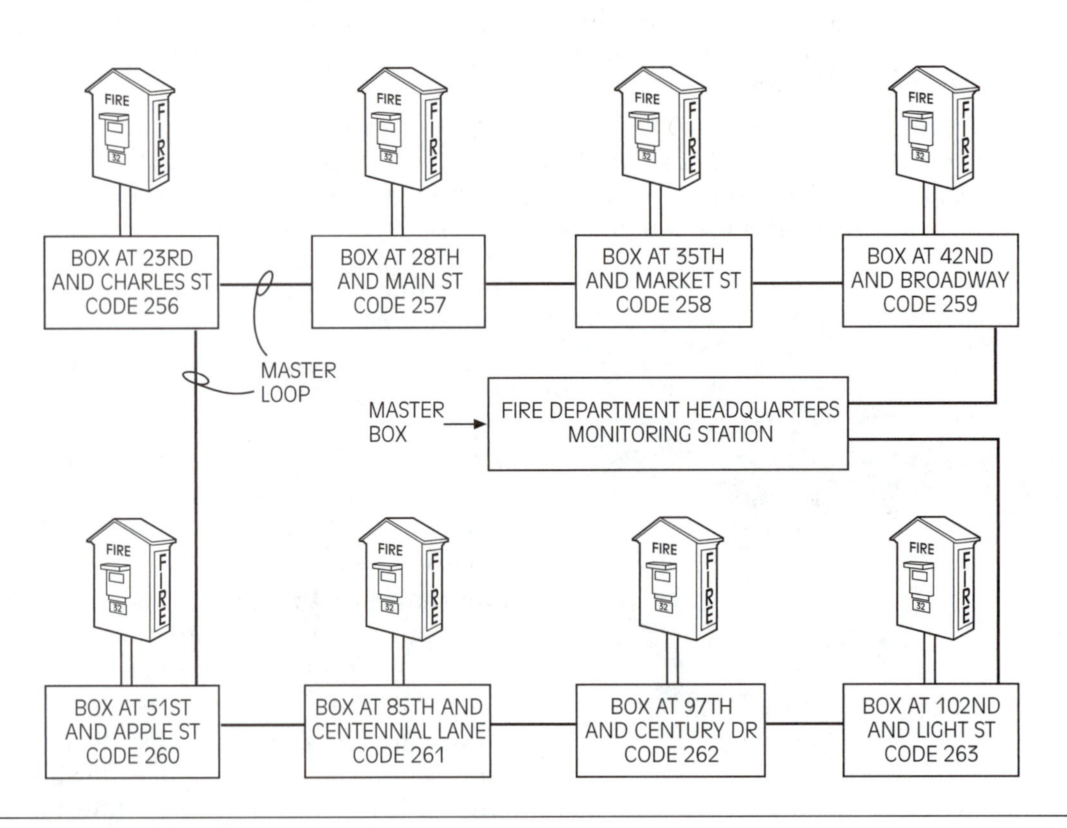

Figure M-6 Municipal fire alarm system.

Water Supply—The structure of the facility, distribution capability, and amount available.

Fire Department—Type, quantity, and distribution of equipment, pumper capacity, training programs and facilities, pre-fire planning programs, hose supplies and hose drying facilities, and personnel availability and use.

Fire Service Communications—Supervision and maintenance of the system, location, and construction of alarm headquarters, technical design of the system, reliable power sources, number and distribution of alarm boxes, number and type of telephone trunk lines, and number and training of alarm operators.

Fire Safety Control—Number of personnel involved, frequency and extent of inspections, permit requirements set by municipal ordinances, existence of fire code and prevention provisions in plumbing and electrical codes.

Building Codes—Use of fire resistive materials in areas of higher fire risk.

Survey Report—The conduction of an on-site inspection of the city or town to determine violations. The evaluations were originally started in 1917 by the National Board of Fire Underwriters (NBFU), now the AIA, as an aid in evaluating and underwriting municipalities. The impetus to establish the grading schedule was as a result of several major city conflagrations in the United States in the early 1900s, primarily the Baltimore fire of 1904. *See also* **Insurance Services Office (ISO), Inc.**

Mutual Aid—Mutual aid and regional mobilization plans that are prearranged commitments used among adjacent (but separate governmental jurisdiction) fire departments to assist each other in fighting fires (or other emergencies) that exceed the capacity to control and extinguish by the equipment and manpower available by the individual fire department. Each party agrees to assist each other when called upon to support firefighting efforts for major incidents (or other designated emergencies). It may also be called assistance agreements, mutual assistance, outside aid, a memorandum of understanding, or letter of agreement. *See also* **Automatic Aid.**

Mutual Aid Box Alarm System (MABAS)—A county, regional, or state-wide mutual aid system. Mutual aid is predetermined to five or more alarm levels based on occupancy, type of alarm, etc.

Naked Light—General terminology used in British industry and applied to all exposed ignition sources, such as flames, hot incandescent material, and mechanically produced sparks.

National Arson Prevention Initiative (NAPI)—A program to coordinate governmental resources to prevent arson fires. It was established by President Clinton (1946–) in 1996 in response to a series of arson fires at places of worship. The Federal Emergency Management Agency has led the effort with the Department of Justice, Department of Treasury, and the Department of Housing and Urban Development in coordination with state and local agencies.

National Association of Fire Equipment Distributors (NAFED)—A trade association of fire equipment dealers, affiliates, and associated members founded in 1962 by a few fire equipment distributors. It is active on National Fire Protection Association (NFPA) committees and industry problems. It promotes state licensing of fire equipment companies.

National Association of Fire Investigators (NAFI)—The objectives of the association are to increase the knowledge and improve the skills of persons engaged in the investigation and analysis of fires, explosions, or in the litigation that ensues from such investigations. It certifies fire and explosion investigators and fire investigation instructors. The association was organized in 1961.

National Association of State Fire Marshals (NASFM)—An organization that provides information on conferences, committee activities, legislative actions, etc., of interest to state fire marshals within the United States.

National Board of Fire Underwriters (NBFU)—A US-based organization established to protect the interests of fire-insurance companies, establish safety standards in building construction, and repress incendiarism and arson. It also provided engineering, statistical, and educational services. The NBFU was established in 1866 and was maintained by stock fire insurance companies. As a result of their requirements for municipalities to obtain low insurance rates, fire safety measures for building construction and water supplies began to improve in the United States in the late 19th century and the occurrence of large municipal conflagrations began to decrease. In 1964, it became part of the American Insurance Association (AIA). *See also* **American Insurance Association (AIA).**

National Building Code—A set of building regulations developed through the consensus process of technical committees for adoption by governmental bodies.

National Burglar and Fire Alarm Association (NBFAA)—An organization that provides consumer information, tips on security, false alarm prevention, and general information on burglar and fire alarm devices and systems.

National Electrical Code (NEC)—A code for the design and installation of electrical wiring and devices to safeguard life and property from fires. The original national electrical code was developed in 1897. It is currently issued as NFPA 70, *National Electrical Code (NEC).*

National Fire Academy (NFA)—An educational and research institute of the United States Fire Administration (USFA) located in Emittsburg, Maryland. It was established by the US Federal Fire Prevention and Control Act of 1974 (Public Law 93-498). It provides professional development for firefighters and related professionals. It offers courses in executive development, fire prevention, leadership, incident management, public education, fire service education, arson, infection control, and hazardous materials. *See also* **Federal Emergency Management Agency (FEMA); Fire Academy.**

National Fire Alarm Code (NFAC)—A fire code of the National Fire Protection Association (NFPA) for the provision of fire alarm systems, NFPA 72, *National Fire Alarm Code. See also* **Fire Alarm Control Panel (FACP).**

N

National Fire Codes (NFC)—Standards, codes, and guidelines maintained and published by the National Fire Protection Association (NFPA) concerning fire safety that are available for adoption by governmental bodies. The codes themselves have no enforcement authority. Only when they are adopted by an organization that has jurisdictional authority do they become enforceable or applicable by law. The codes are consensus documents. Their content is arrived at through committee action and is publicly voted upon by representatives of various industries, manufacturers, governmental bodies, and private individuals.

National Fire Danger Rating System (NFDRS)—A fire hazard rating system used by wildland fire agencies in the United States that identifies a set of wildland fuel complexes from which fire behavior and spread estimates can be calculated. Each fuel model has a defined fuel depth, density, and load that represents fuel conditions found in various parts of the United States.

National Fire Data Center (NFDC)—An office of the United States Fire Academy (USFA). It acts to gather, analyze, publish, and disseminate data related to the prevention, occurrence, control, and results of fires. The analysis of the fire data identifies major problem areas, assists in setting priorities, determines solutions to problems, and monitors the progress of programs to reduce fire losses. The center uses information provided by the National Fire Incident Reporting System (NFIRS).

National Fire Incident Reporting System (NFIRS)—A fire data collection system maintained by the United States Fire Administration (USFA), National Fire Data Center (NFDC). The NFIRS is the largest fire data collection and database system in the world. The fire incident data are voluntarily provided through participating fire departments within the United States. Data are recorded into the system based on National Fire Protection Association coding and reporting standards. On average, one million fires are reported into the system each year. A standard NFIRS package is available that includes fire incident casualty forms, a hazardous material form, a coding structure for data processing purposes, manuals, computer software and procedures, documentation, and a National Fire Academy training course for utilizing the system. *See also* **Fire Injury; United States Fire Administration (USFA).**

National Fire Information Council (NFIC)—A US-based organization whose goal is to establish the United States as the number one nation in fire safety.

National Fire Protection Association (NFPA)—A nonprofit organization based in the United States (Quincy, Massachusetts), organized in 1896, and the leading worldwide authority on fire safety. It is dedicated to protecting lives and property from the hazards of fire. Its stated objective is "to promote the science and improve the method of fire protection, to obtain and circulate information on this subject, and to secure cooperation in matters of common interest." NFPA is responsible for the development and distribution of the National Fire Codes® (through consensus committees), related handbooks, and numerous fire service training and public education materials. NFPA was organized to set up uniform sprinkler rules as a result of concern by several insurance agencies about the numerous types of automatic sprinklers and installation practices. NFPA published its first fire code in 1896 concerning automatic sprinklers. In 1900, the National Board of Fire Underwriters joined the NFPA and undertook the expense of publishing the NFPA fire codes. This continued until 1964. Over 68,000 fire protection professionals compose the international membership of the NFPA with representation from more than 100 organizations from over 70 countries. *See also* **United States Fire Administration (USFA).**

National Fire Protection Research Foundation—An organization of the National Fire Protection Association (NFPA) that serves as a focus and catalyst for obtaining usable solutions to fire problems. Its primary activities are in research for fire risk assessments, new technologies, and strategies. It was established by the NFPA in 1982.

National Fire Sprinkler Association (NFSA)—A trade association composed of manufacturers and installing contractors of automatic fire sprinkler systems and related equipment. Its mission is to "create a market for the widespread acceptance of competently installed automatic fire sprinkler systems in both new and existing construction, from homes to high-rise." It acts as a spokesperson on sprinkler issues globally. It was organized in 1905 as the National Automatic Sprinkler Contractors Association. In 1914, it changed its name to the National Automatic Sprinkler Association. To signify its work for non-water fixed systems, it changed its name again in 1944 to National Automatic Sprinkler and Fire Control Association. In 1983, the organization again changed its name to what it is now: the National Fire Sprinkler Association. The NFSA has departments in labor relations, engineering, training and education, field operations, public fire protection, and membership and communications. *See also* **American Fire Sprinkler Association (AFSA).**

National Hose Thread (NH)—A standard fire hose thread that has dimensions for inside and outside screw threads. The thread is defined in NFPA 1963, *Standard for Fire Hose Connections.* It may also be called national standard thread (NST).

National Incident Management System (NIMS)—An organization approach from FEMA to assist and help coordinate in a major incident help from other jurisdictions, the state and the federal government. NIMS was developed so responders from different jurisdictions and disciplines can work together to better respond to natural disasters and emergencies, including acts of terrorism. NIMS hopes to achieve a unified approach to incident management, standard command and management structures, and emphasis on preparedness, mutual aid and resource management.

National Institute of Standards and Technology (NIST)—An agency of the US government whose mission is to advance measurement science and develop standards to support industry, commerce, scientific institutions, and the US federal government. NIST conducts research into fire events, fire modeling, and protective devices to prevent and reduce fire losses through its Center for Fire Research Division. The National Institute of Standards and Technology (formerly the National Bureau of Standards) was founded in 1901.

National Interagency Fire Center (NIFC)—America's primary logistical support center for wildland fire suppression. Located in Boise, Idaho, it is the headquarters of the federal government's wildland fire experts in fire ecology, fire behavior, technology, aviation, and weather. NIFC's role is to provide a national response to wildland fires and emergencies. It also serves as a clearinghouse on wildland fire information and technology. The center was established in 1965 in cooperation with the Bureau of Land Management (BLM), the Forest Service, and the National Weather Service. Later, the Department of the Interior, the Office of Aircraft Services, the National Park Service, the Bureau of Indian Affairs, and the US Fish and Wildlife Service became members. *See also* **Fire Ecology.**

National Professional Qualifications Board (NPQB)—A board, originally established by the Joint Council of Fire Service Organizations, that accredits programs for firefighter training organizations. It is composed of various national fire service organizations in the United States. *See also* **International Fire Service Accreditation Congress (IFSAC); Joint Council of National Fire Service Organizations (JCNFSO).**

National Standard Thread (NST)—Threads used for hose and appliance screwed connections as specified in NFPA 1963, *Standard for Screw Threads and Gaskets for Fire Hose Connections.* May also be called standard thread. *See also* **American National Fire Hose Connection Screw Thread.**

National Volunteer Fire Council (NVFC)—An organization representing the interests of volunteer fire, EMS, and rescue services concerned primarily with activities, training, and legislation. Founded in 1976, it represents volunteer firefighters in matters of self-interest.

National Wildfire Coordinating Group (NWCG)—An organization of the United States whose purpose is to coordinate the programs of the participating wildfire management agencies to avoid duplication of effort and to provide a means of constructively working together. The NWCG is composed of the Department of Agriculture Forest Service (FS); four Department of the Interior agencies: the Bureau of Land Management (BLM), the National Park Service (NPS), the Bureau of Indian Affairs (BIA), and the Fish and Wildlife Service (FWS); the United States Fire Administration (USFA), and state forestry agencies through the National Association of State Foresters (NASF). Because of a number of wildland fires in the early 1970s and subsequent concerns, the National Wildfire Coordinating Group (NWCG) was created. It was renamed in 1994 to the National Wildland Fire Coordinating Group.

Net Positive Suction Head (NPSH)—The suction head absolutely available to a pump less the sum of the suction system and component friction losses. The static height the fluid must be lifted below the centerline of the pump impeller (on static lift operations), and the lessening absolute pressure when the suction vessels is under a vacuum. *See also* **Cavitation; Centrifugal Pump; Net Positive Suction Heat Available (NPSHa).**

Net Positive Suction Head Available (NPSHa)—A hydraulic term similar to net positive suction head (NPSH), except that the vapor pressure of the fluid at pumping temperature has been deducted. *See also* **Net Positive Suction Head (NPSH).**

Net Positive Suction Head Required (NPSHr)—A characteristic of a pump design configuration to avoid cavitation effects and ensure the pump performs efficiently. It is normally shown on the pump's performance curve. *See also* **Cavitation.**

Neutral Plane—The height in a compartment above which smoke will or can flow out during a fire event. A neutral plane may change from one-half to one-third of the compartment height as the fire becomes fully involved in flames. However, the smoke interface can extend very close to the floor of the compartment. See Figure N-1. *See also* **Smoke Interface.**

Newton-Raphson Method—A mathematical calculation method that uses an iterative scheme that starts with an estimate of the solution and repeatedly computes better estimates. The Newton-Raphson method uses quadratic convergence as opposed to linear convergence and therefore fewer iterations are needed to achieve a solution. Newton-Raphson computational techniques are used to mathematically analyze the flow in pipe networks, particularly in the fire protection profession for water distribution systems. *See also* **Hardy Cross Method.**

Nomex™—A Du Pont Company registered trademark for an inherently heat- and flame-resistant fiber. It provides protection against heat, flame, flash, electric arc, and static discharge. It is used as personal attire by workers at specialized locations or who are involved in specialized activities that have a high potential for fire exposure, such as firefighters, aircraft servicers, test pilots, race car drivers, oil and chemical plant workers, etc. *See also* **Fire-Retardant Coveralls/Clothing (FRC).**

Non-Aspirating Discharge Device—A foam-water delivery device that provides a specific water discharge pattern. When used with AFFF or FFFP foams, it provides a discharge pattern similar to the water discharge pattern.

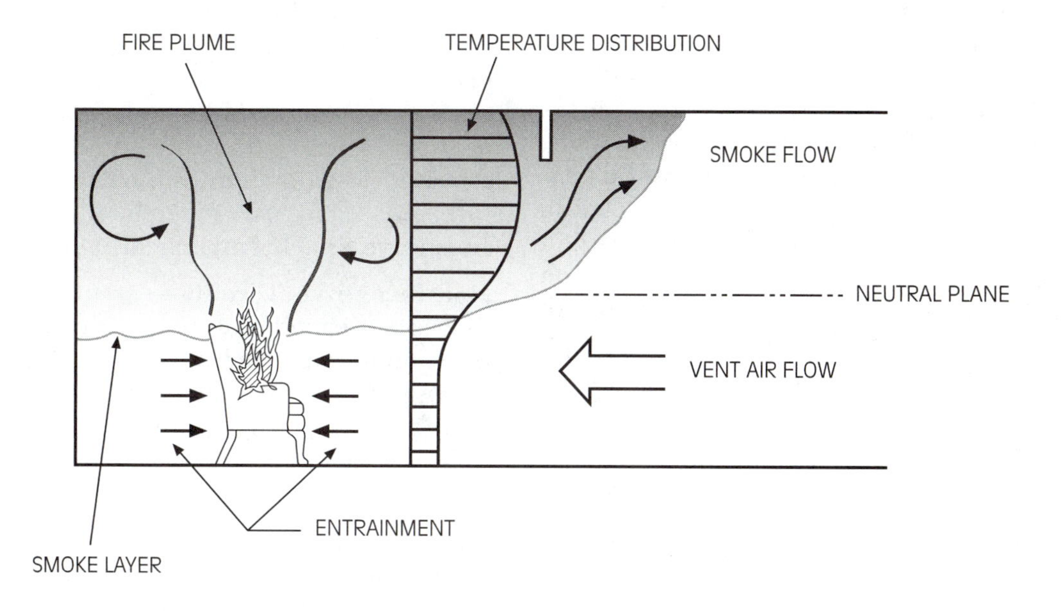

Figure N-1 Neutral plane location from induced flows, room fire.

Non-Aspirating Nozzle—*See* **Nozzle, Non-Aspirating.**

Nonburning—A testing term used to indicate that a material will not ignite and combust when exposed to a flame. It is also used to describe a material that is not on fire at the time of exposure, such as a nonburning fuel.

Noncombustible (Material)—A material that does not ignite, burn, support combustion, or release flammable vapors when subjected to fire or heat. Noncombustible materials also do not aid combustion or add appreciable heat to an ambient fire. Materials are tested for noncombustibility in accordance with ASTM E 136, *Standard Test Method for Behavior of Materials in a Vertical Tube Furnace at 750°C.* This test places the sample material in a furnace chamber for three trials. The chamber is a tube or refractory material with a diameter of 2.95 in. (7.5 cm) and a height of 5.9 in. (17 cm). The tube is heated by electrical heating coils. Before testing the material, the temperature in the tube is stabilized uniformly at 1,382°F (750°C) for a period of at least ten minutes (except for the ends). A material sample of 3.6 by 1.9 in. (4 by 5 cm) is placed in the center of the tube and two thermocouple readings are taken: one from the center of the sample and one from a specific location in the furnace. Materials are classified noncombustible if, during the three separate tests, the test does not (a) cause the temperature readings of the furnace thermocouple to rise by 122°F (50°C) or more above the initial furnace temperature, (b) cause the temperature readings of the specimen thermocouple to rise 122°F (50°C) or more above the initial furnace temperature, and (c) flame for 10 seconds or more. Remember that all materials may burn if exposed to the appropriate conditions (high temperatures, pressures, etc.). In practical terms, noncombustible materials do not contribute to the incipient stage of a fire nor in general to its temperature or duration. *See also* **Incombustible.**

Nonflammable—A term that describes a substance that does not burn when exposed to a flame or is not readily capable of burning with a flame under normal circumstances. Many substances classified as nonflammable in air become flammable if the oxygen content of the gaseous medium is increased above 0.235 absolute atmospheres. *See also* **Inflammable.**

Nonflammable Gas—Gases that are known to be nonflammable at any temperature (nitrogen, helium, etc.).

Nonflammable Liquid—A fluid that does not have a flash point and is not flammable in air.

Non-Incendive—Something that is incapable of causing ignition due to the low energy availability contained within it. It is usually used to describe devices that may spark but will not be a hazard as an ignition source. A spark that has enough energy to ignite an ignitable mixture is classified as incendive. A fire or explosion can result from an incendive spark. Non-incendive arcs or sparks do not possess the energy required to cause ignition of an ignitable mixture. *See also* **Arc; Incendive; Incendive Arc; Incendive Spark; Intrinsically Safe (IS); Spark.**

Non-Incendive Circuit—An electrical circuit that is not capable of producing sufficient ignition energy, such as from arcs or thermal effects, during its intended operating conditions including opening, grounding, or shorting of field wiring circuits. *See also* **Intrinsically Safe (IS); Low-Energy Circuit.**

Non-Incendive Equipment—Equipment, which in its normal operating condition will not ignite a specific hazardous atmospheric mixture in its most

N

easily ignited concentration. This equipment may contain circuits with sliding or open-closed contacts that release insufficient energy to cause ignition.

Nonrechargeable Fire Extinguisher—A portable fire extinguisher that is designed and designated for one-time use and is not expected to undergo extensive maintenance (hydrotesting, refilling, etc.). Nonrechargeable fire extinguishers are marked to indicate they are not to be reused.

Normal Loss Expectancy (NLE)—An estimate of the largest loss to be expected under normal circumstances, excluding a catastrophic condition, with all available means of protection functioning as intended. Primarily used in the insurance underwriting industry to estimate property losses for the purpose of insurance policy values. *See also* **Probable Maximum Loss (PML).**

Nozzle—A device for directing a pressurized water stream in a desired pattern, density, and direction. Various nozzles are capable of projecting solid, heavy streams of water, curtains of spray, or fog. Nozzles have three main functions: they control flow, and they provide shape and reach for firewater application. The flow of water is controlled by the size of the orifice of the nozzle. The nozzle itself creates a restriction at the end of the waterway that changes the water pressure into velocity, which allows the water to travel distances. The design features of the nozzle create the shape of the water spray that is produced from straight stream to fog patterns. Different situations require different methods of applying water or foam. There are four basic types of firefighting nozzles: the smooth bore, the single gallonage (sometimes called variable pressure/variable flow), the adjustable gallonage, and the automatic or constant pressure. The latter three nozzles make up a group of nozzles called combination nozzles because they have the capability to produce both a straight stream and a fog stream.

Nozzles are selected according to the amount of heat that must be absorbed. Manual firefighting nozzles can apply water in the form of streams, spray, or fog at rates of flow between 15 gallons (57 liters) to more than 100 gallons (380 liters) per minute. Straight streams of water have greater reach and penetration, but fog absorbs heat more quickly because the water droplets present a greater surface area and distribute the water more widely. Fog nozzles may be used to disperse vapors from flammable liquids, although foam is generally used to extinguish fires in flammable liquids and seal the liquid surface to prevent vapor release.

The first firewater nozzles on fire engines were provided directly to the outlet pipe of the pump. They were usually 6 or 7 ft. (1.8 or 2.1 m) long and made of copper or brass. As fire hose became available to provide water directly to a fire source, the length of nozzles decreased. Early nozzles had a diameter of 0.875 to 1.5 in. (2.2 to 3.2 cm). Fog or spray nozzles were not developed until the 1930s and were not widely adopted by fire departments until the early 1950s. See Figure N-2. *See also* **Nozzle, Combination; Nozzle, Fog.**

Nozzle, Aspirating—A nozzle designed to draw air into the fluid (foam-water) stream when discharging. The purpose of aspirating nozzles is to provide air into a foam-water solution for fire suppression. *See also* **Nozzle, Non-Aspirating; Mechanical Foam.**

Nozzle, Cellar—*See* **Cellar (Pipe) Nozzle.**

Nozzle, Combination—A manually adjustable firefighting nozzle capable of delivering water from a straight stream to wide fog pattern. They are

N

Figure N-2 Assortment of handheld nozzles.

the most widely used nozzles with the fire services. See Figure N-3. *See also* **Nozzle.**

Nozzle, Foam-Water Spray—A firefighting foam-water spray nozzle designed to air aspirate the foam and to distribute foam or water in the pattern specific to the spray nozzle (in contrast to the pattern produced by a foam-water sprinkler).

Nozzle, Fog—A firefighting water nozzle capable of producing a spray of water that contains water droplets with a mass medium diameter smaller than 0.03 in. (0.075 cm) in diameter. (A spray of water that contains water droplets with a mass medium diameter larger than 0.03 in. (0.075 cm) in diameter is technically considered a "spray" nozzle.) Water fog is most effective for heat absorption when the droplet size is small and therefore highly beneficial for fire suppression. Water fog and spray are produced by the impingement of converging water jets or by forcing the water through specially designed teeth, which break it up into fine particles. Chief Glen Griswold of the Los Angeles County Fire Department was responsible for the early testing and development of the fog nozzle for combating oil fires. He conducted various experiments

Figure N-3 Example of a combination nozzle.

on fog nozzles in the 1930s and developed a fan pattern, impinging spray fog nozzle, which he patented in 1931. This was the first practical fog nozzle made commercially in the United States. Spray nozzles for firefighting have been mentioned in historical literature on firefighting as early as 1863. Fog nozzles, also commonly referred to as spray nozzles, were tested and adopted for widespread firefighting applications use in the 1940s and early 1950s. They were tested for marine firefighting applications during and just after WWII by both the US Navy and Coast Guard. These applications would eventually lead to improvements in the automatic sprinkler, culminating in the standard spray sprinkler design in 1950 that produced a smaller droplet size than the earlier manufactured sprinklers. The use of water sprays (fog) for firefighting was considered one of the major advances in firefighting technology, after the application of power-driven drives for pumping equipment. *See also* **Nozzle; Nozzle, Straight Stream; Water Fog; Water Spray.**

Nozzle, Non-Aspirating—A nozzle designed so that it does not draw air into the fluid (foam-water) stream when discharging a foam solution for aeration. It discharges foam in the pattern

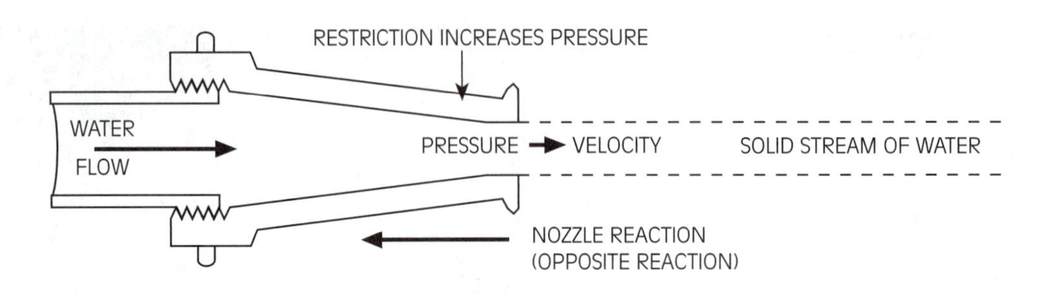

Figure N-4 Straight stream nozzle, smooth bore.

designated by the discharge device. *See also* **Nozzle, Aspirating.**

Nozzle, Partition—A firefighting nozzle designed to discharge water between partition studs or joists in concealed building construction. They have a piercing point or applicator head for penetrating through wall construction surfaces.

N

Nozzle, Penetrating—A firefighting nozzle designed to penetrate the sheet metal covering of an aircraft exterior and inject a fire extinguishing agent into the interior of the craft. Most penetrating nozzles require human manpower to force the nozzle tip through the metal covering. The latest penetrating nozzles use pneumatic force to provide entry through the aircraft exterior.

Nozzle Pressure—The level of pressure that occurs at a nozzle from the velocity of water during flow from its outlet. *See also* **Pitot Pressure.**

Nozzle Reaction—The amount of force produced by the velocity of water flowing through a nozzle. It acts in the opposite direction of the direction of the water leaving a nozzle. The reaction force is dependent on the water pressure and nozzle outlet diameter. For straight stream nozzles,[7]

Nozzle reaction in pounds =
$1.57 \times d^2 \times$ (base nozzle pressure)

For fog nozzles,

Nozzle reaction in pounds =
gpm $\times \sqrt{}$(base nozzle pressure) $\times$ 0.0505

Nozzles, Revolving—*See* **Cellar Pipe.**

Nozzle, Smooth Bore—*See* **Nozzle, Straight Stream.**

Nozzle, Spray—*See* **Nozzle, Fog.**

Nozzle, Straight Stream—A firefighting water nozzle that produces a straight stream of water for transversing large distances or penetrating a material to reach deep-seated fires. Straight stream nozzles were generally the only nozzle

[7] NFPA, *Fire Protection Handbook,* 18th Edition, Chapter 5, Fire Streams, National Fire Protection Association, Quincy, MA, 1997.

available to firefighters from the invention of the fire pump to about the 1930s and 1940s. A solid or smooth bore is used to produce the straight stream. It may also be called smooth bore nozzle or solid stream nozzle. See Figure N-4. *See also* **Nozzle, Fog.**

Nozzle, Turret—A large capacity firefighting water or foam nozzle provided on the roof or bumper of a fire service vehicle and remotely controlled from inside the vehicle. Turret nozzles are commonly provided on airport crash rescue vehicles.

Occupancy—The purpose for which a building, or part thereof, is used or intended to be used.

Occupancy Classification—A method of determining the expected level of fire severity in a building for the purpose of defining the required fire safety features. The occupancy fire severity is based on historical fuel loadings for similar facilities. Occupancy classifications are defined in various building or fire codes that have been adopted by the individual governmental authority, such as the Uniform Building Code, Uniform Fire Code, and Life Safety Code, etc. See Figure O-1.

Occupancy Load—The measure of the number of personnel hours per unit time spent in a building. The occupancy load is primarily calculated to determine the number and size of exits required for a facility. It may also be applied for determining other life safety features for a particular location, such as fire alarm, explosion resistance (if exposed to such occurrences), etc.

Offensive Firefighting or **Offensive Attack**—Manual fire suppression activities that are concentrated on controlling a fire to reduce its size in order to accomplish extinguishment. It usually refers to interior firefighting. *See also* **Advanced Exterior Firefighting; Defensive Firefighting.**

Oil Insurance Association (OIA)—A former association of insurance underwriters for the petroleum industry. The OIA merged with the Factory Insurance Association (FIA) to become Industrial Risk Insurers (IRI) in 1975. *See also* **Industrial Risk Insurers (IRI).**

Old Style Sprinkler—A sprinkler that discharges up to 60 percent of its water upward, in a radius of 2 to 3 ft. (1.5 to 3 m). The remaining 40 percent is discharged downward with a radius of up to 5 to 6 ft. (1.5 to 2 m), possibly up to 8 ft. (2.5 m). The distribution pattern of the "old style" sprinkler was generally less uniform and droplet size less favorable. Present day spray sprinklers discharge all of the water downward. Old style sprinklers were generally manufactured before 1953. Although old style sprinkler heads may still be in existence, sprinkler heads older than 50 years are normally recommended for replacement. Sprinkler heads manufactured during WWI also contain suspect alloys. *See also* **Spray Sprinkler; Sprinkler.**

Open Burning—A combustion process that is uncontrolled. Open burning is commonly associated with the burning of waste materials outside. *See also* **Free-Burning.**

Open Head System—A sprinkler system provided with sprinklers that are open (without fusible elements). Water is released through the open heads from the opening of an isolation valve to the system. The isolation valve is controlled manually or automatically through the use of a fire detection system. Also commonly called a deluge sprinkler system. *See also* **Closed Head Foam-Water Sprinkler System; Sprinkler System.**

COMPARISON OF:
- FIRE LOAD DENSITY
- COMBUSTIBILITY OF CONTENTS
- RATE OF HEAT RELEASE

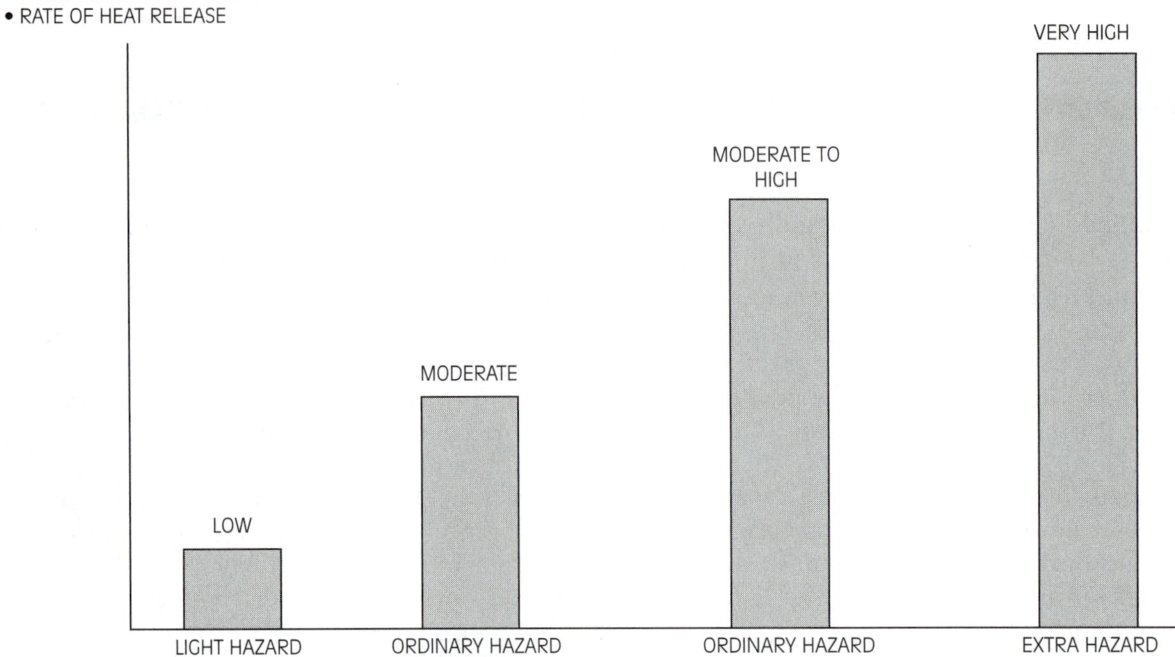

LIGHT HAZARD — EXAMPLES:
CHURCHES
CLUBS
EDUCATIONAL
HOSPITALS
PRISONS
LIBRARIES
MUSEUMS
NURSING HOMES
OFFICES
RESIDENTIAL
RESTAURANT SEATING
THEATERS
ATTICS

ORDINARY HAZARD GROUP I — EXAMPLES:
PARKING GARAGES
CAR DEALERS
BAKERIES
BEVERAGE MANUFACTURING
CANNERIES
DAIRIES
ELECTRONIC PLANTS
GLASS MANUFACTURING
LAUNDRIES
RESTAURANT SERVICE AREAS

ORDINARY HAZARD GROUP II — EXAMPLES:
CEREAL MILLS
CHEMICAL PLANTS
CONFECTIONARY
DISTILLERIES
DRY CLEANERS
FEED MILLS
HORSE STABLES
LEATHER GOODS
LIBRARIES-LARGE STACKS
MACHINE SHOPS
METAL WORKING
RETAIL AREAS
PAPER MILLS
PIERS & WHARVES
POST OFFICES
PRINTING & PUBLISHING
REPAIR GARAGES
STAGES
TEXTILE MANUFACTURE
TOBACCO PRODUCTS
WOOD SHOPS

EXTRA HAZARD — GROUP I:
FLAMMABLE LIQUIDS
FLAMMABLE METALS
WOOD DUSTS
PRINTING INK (LOW FLASH POINTS)
PLASTIC FORMS

GROUP II
ASPHALT SATURATING
FLAMMABLE LIQUIDS SPRAYING
SOLVENT CLEANING
VARNISH & PAINT DIPPING

Figure O-1 Occupancy classifications. (Source: NFPA 13)

Optical Flame Detector—A fire detector that sets off its alarm at the presence of light from the flame of a fire, usually in the ultraviolet or infrared range of the electromagnetic spectrum. The detectors are normally set to detect a flicker of a flame. They may be equipped with a time delay feature to avoid the occurrence of false alarms from transient light sources not associated with fire events. There are six types of optical flame detectors currently available: ultraviolet (UV), single frequency infrared (IR), dual frequency infrared (IR/IR), ultraviolet/ infrared simple voting (UV/IR), ultraviolet/ infrared ratio measurement (UV/IR), and multiband.

Ordinary Building Construction—*See* **Building Construction Types.**

Orifice—An opening for the passage of a fluid or gas. An orifice is often used as a metering device or for restriction of flow (orifice flow meter or orifice plate). Most firewater protection application devices use an orifice to control the amount of water applied to meet prescribed application rates or densities. An orifice will also have a coefficient of discharge depending on its design and arrangement that influences its performance. *See also* **Coefficient of Discharge.**

OS & Y Valve—An isolation control valve that has an outside valve screw and yoke (OS & Y). The position of the screw indicates whether it is open or closed. If it is extended, the valve is open; inserted into the valve body, it is closed.

Overhand Safety Knot—A type of knot used by the fire service. It is an additional knot placed on another knot of any type to prevent the rope end from slipping back thought the original knot. See Figure O-2.

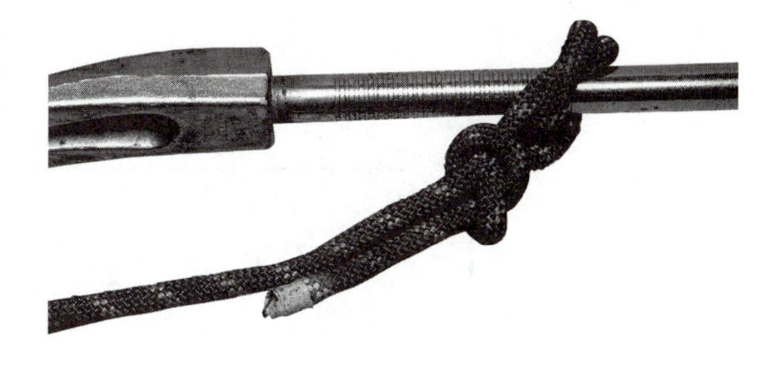

Figure O-2 Overhand safety knot.

Overhaul—Actions taken during firefighting efforts to ensure that all embers are extinguished after a fire is controlled. They are considered the final extinguishment actions. It generally consists of searching for hidden and remaining fires or embers after the main fire has been extinguished. Inadequate or incomplete overhauling may cause a fire to rekindle. *See also* **Loss Control.**

Overhead Foam Injection—A method of foam application for the fire protection of atmospheric or low pressure storage tanks that contain hazardous materials that may ignite. Two types of designs are commonly used. For cone roof tanks or internal floating roof tanks with other than pontoon decks, multiple foam makers are mounted on the upper edge of the tank shell. These systems deliver and protect the entire surface area of the liquid in the tank. For open and covered floating roof tanks with pontoon decks, the foam system is designed to protect only the seal area. The overhead foam system consists of a foam solution pipe extended from the proportioning and supply source, which is in a safe location. A foam aspirating mechanism is provided just upstream of the foam chamber or application piping to the seal area and its dispersing nozzle. Foam chambers installed on the tank are provided on the shell just below the roof joint. A foam chamber

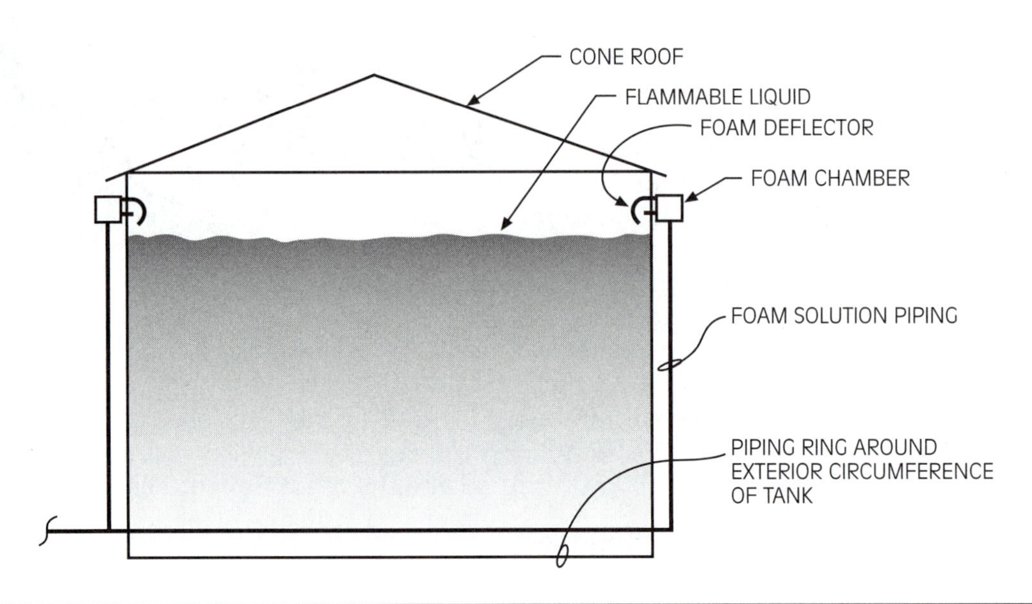

Figure O-3 Overhead foam application onto the surface of liquid in a cone-roof storage tank.

is usually provided with a deflector positioned on the inside tank wall at the foam chamber. It is used to deflect the foam against the tank wall and allow it to run down onto the surface of the tank and shell seal area. A foam dam is provided on the pontoon to contain the foam to the seal area. See Figure O-3. *See also* **Foam; Foam Dam; Subsurface Foam Injection.**

Overheat—The destruction of a material by heat without self-sustained combustion. Removal of the heat source will stop the destruction from continuing or leading to combustion. Overheating is the stage before ignition of a fire.

Overload—Operation of equipment in excess of its normal, full-load rating. If it persists for a sufficient length of time, it would cause damage or dangerous overheating. A fault, such as a short circuit or ground fault, is not an overload. For firewater pumping applications, it describes a fire pump that is discharging more water than it is designed to discharge. *See also* **Overload, Electrical; Overload Point.**

Overload, Electrical—Overcurrents that are large enough and persist long enough to cause damage or create a danger of fire. These currents in excess of rated ampacity produce effects in proportion to the degree and duration of the overcurrent. Electrical overloads cause internal heating of the electrical conductor. Heating occurs in the entire length and cross-section of the conductor from the power source to the load. If the overload is greater than five times the ampacity rating, the conductor may become hot enough to ignite fuels in contact with it as the insulation melts off. *See also* **Circuit Breaker; Overload.**

Overload Point—A data point on a firewater pump curve where the pump supplies 150 percent of its rated capacity at 65 percent of its rated pressure. *See also* **Fire Pump Curve; Overload.**

Overpressure—Any pressure relative to ambient pressure caused by an explosive blast either positive or negative. *See also* **Blast.**

Oxidant—A chemical material that supports the combustion reaction process to combine with a fuel, such as oxygen, nitrous oxide, nitric oxide, chlorates, and chlorine.

Oxidizer—Any material that combines with a fuel source to support the combustion process. Free-burning fires need an oxygen level of 16 to 21 percent (12 percent for a smothering fire) to support combustion.

Oxygen Index (OI)—The minimum concentration of oxygen, expressed as percent by volume, in a mixture of oxygen and nitrogen that will just support combustion of a material under conditions of test. ASTM D2863, *Method for Measuring the Minimum Oxygen Concentration to Support Candle-Like Combustion of Plastics (Oxygen Index),* is commonly used to define the Oxygen Index. In the ASTM procedure, a small, vertically oriented sample is burned downward in an oxygen-nitrogen mixture. The oxygen level of the gas mixture is increased to determine the minimum percent of oxygen that will begin to support combustion of the sample. The minimum oxygen concentration that supports combustion is called the Oxygen Index (OI). Burning in an upward direction may take place more readily and will produce a lower OI The OI test, as determined by ASTM D2863, is limited to nonmetals (plastics) at ambient pressure. The OI concept is now being utilized at elevated pressures (and temperatures) and for metals as well as nonmetals.

Panic—The word panic is often applied to a strictly individual, maladaptive reaction of flight, immobility, or disorganization stemming from intense fear. Individual panic frequently occurs as a unique individual response without triggering a similar reaction in others. Panic as collective behavior, however, is shared behavior. The US sociologists Kurt Lang and Gladys E. Lang view panic as the end point in a process of demoralization in which behavior becomes privatized and there is a general retreat from the pursuit of group goals.[8] Current research by prominent fire experts indicates that panic behavior (i.e., physical competition between participants) has typically not occurred in fire incident evacuations based upon interviews and questionnaires from survivors of major fire incidents where such actions might have been expected. *See also* **Evacuation.**

Panic Bar—A bar that extends at least half the width of an exit door leaf and will open the door if subjected to adequate pressure. They are called panic bars because it alleviates the need to know precisely how to open a door in an emergency, when personnel may be under stress, disoriented, or consist of an assemblage of individuals that may unknowingly block access to common door opening devices. The pressure to open the door is not to exceed 15 lbs. (6.75 kg). Guidance in the requirements and application of panic bars are provided in NFPA 101, *Life Safety Code*.

[8] *Encyclopedia Britannica* CD-97, CD-ROM Edition, Encyclopedia Britannica, Inc., London, UK, 1997.

Panic Hardware—A locking bolt with a hinged bar fitted on the inside of an exit door so that it is always operable from the inside in an emergency. It may also be called an exit device.

Parapet Wall—An extension of a fire wall above the roof line to prevent fire from spreading to an adjacent building. The higher the wall the greater the protection; however, aesthetics, cost, and practicality generally limit the height. Parapet walls of 1.5 ft. (45.7 cm) are considered standard for most areas. A fire may also extend around the end of the firewall at the building eave. To provide adequate protection, the firewall is extended outward from the building to the height of the parapet wall or a wider section is added to the firewall around the eave portion of the building.

Partitions, Fire—*See* **Fire Barrier; Firewalls.**

Passive Fire Protection (PFP)—Protection measures that prolong the fire resistance of an installation before an eventual fire occurrence from the effects of smoke, flames, and combustion gases. These can consist of insulation (fireproofing) of a structure, choice of noncombustible materials of construction, use of fire-resistant partitions, and compartmentation to resist the passage of fire. It includes coatings, claddings, or free-standing systems that provide thermal protection in the event of fire and that require no manual, mechanical, or other means of initiation, replenishment, or sustainment for their performance during a fire incident. Passive systems also embrace the basic requirements for area separation and classification. *See also* **Active Fire Protection (AFP); Compartmentation; Fireproofing.**

Passive Smoke Detection System—A fire detection system where smoke is transported to and into a sensing chamber by outside forces, that is, fire plume strength or environmental airflows. A passive smoke detection system may have difficulty detecting smoke from smoldering types of fires because this smoke may not be hot enough to rise to the smoke detector location. *See also* **Active Smoke Detection System; Cold Smoke.**

Penetration Seal—A purposely made seal (or seals) formed in situ to ensure that penetrations or "poke throughs" to fire barriers do not impair its fire resistance. Wiring, cable or piping openings, ducting through floors, ceilings, walls, and building joints must be provided with fire-rated penetration seals to prevent the spread of fire or its effects. The penetration sealing material is to be made of limited-combustible or noncombustible material that meets the requirements of ASTM E-814, *Fire Tests of Through-Penetration Fire Stops,* or UL 1479, *Standard for Safety Fire Tests of Through-Penetration Firestops. See also* **Firestop; Firestopping; Unprotected Opening.**

Pensky-Martens Tester—A test apparatus to measure the flash point of a liquid. Ref. ASTM D 93, *Standard Test Methods for Flash Point by the Pensky-Martens Closed Tester.*

Perfect Combustion—A combustion process in which the exact amount of air and fuel are present in quantities to react completely in a combustion process. *See also* **Combustion, Incomplete; Stoichiometric Combustion; Stoichiometric Mixture.**

Perforated Pipe—Pipe positioned at the ceiling that has been drilled with holes to allow water to spray out to provide for the control and extinguishment of fire. It was used before the provision of sprinklers to provide for fire suppression in buildings. Sir William Congreve invented and patented perforated pipes for fire protection in London in 1812. They were first used for fire protection purposes in America in 1852 and continued in use until 1885. *See also* **Sprinkler System.**

Performance-Based Fire Protection Design—An engineering approach to fire protection design based on (1) agreed upon fire safety goals, objectives, and criteria; (2) deterministic and/or probabilistic evaluation of fire scenarios for the occupancy; (3) the physical and chemical properties of fire and fire effluents; and (4) a quantitative assessment of design alternatives against the fire safety goals and objectives. One primary difference between a prescriptive and a performance-based design is that a fire safety goal—life safety, property protection, mission continuity, and environmental impact—is explicitly stated. Prescriptive requirements may inhibit fire safety components from effectively meeting the fire safety goals as an integrated system. Additional guidance on performance-based design for fire safety considerations is found in *ICC Performance Code for Buildings and Facilities*™ or the *SFPE Code Official's Guide to Performance-Based Design Review. See also* **Fire Code; Fire Performance; Performance-Based Regulation.**

Performance-Based Regulation—A code or standard that expresses requirements for a building, structure, or system in terms of functional objectives and performance requirements. This is in contrast to prescriptive requirements, which require specific features and are mandatory. *See also* **Fire Code; Performance-Based Fire Protection Design.**

Personal Alert Safety Systems (PASS)—An individual protective device that emits an audible alarm to notify others and assist in locating a firefighter in danger. The personal alert safety system (PASS) device includes a motion detector that senses movement and automatically sounds an alarm signal if no movement is sensed for 30 seconds, in case a firefighter is incapacitated and cannot activate the alarm. Requirements for PASS devices are specified in NFPA 1982, *Standard on Personal Alert Safety Systems (PASS) for Fire Fighters*.

Phlogiston Theory—A hypothetical substance, representing flammability, postulated in the late 17th century by the German chemists Johann Becher (1635–1682) and Georg Stahl (1660–1734) to explain the phenomenon of combustion. According to the phlogiston theory, every substance capable of undergoing combustion contains phlogiston, and the process of combustion is essentially the process of losing phlogiston (it escaped into the air). Because it was known that a substance such as mercury becomes heavier during combustion, it was assumed that phlogiston had negative weight; that is, the substance became heavier when it lost phlogiston. It had also been observed that a substance would not burn long in a closed container. Becher and Stahl thought that combustion stopped because the air in the closed container had become so saturated with phlogiston that it could not absorb any more of it. Substances such as coal and sulfur were believed to be composed almost entirely of phlogiston. In experiments with the gas now known as oxygen, the English chemist Joseph Priestley (1733–1804) discovered its properties of supporting combustion. When oxygen was discovered, it was found that combustible substances burned much better in it than they did in air. It was then mistakenly assumed that oxygen had to be a gas completely devoid of phlogiston so that it could absorb whatever was released from the burning substance. The newly discovered oxygen was thus described as air that was dephlogisticated. The phlogiston theory was disproved by the French chemist Antoine Lavoisier (1743–1794), who demonstrated through his quantitative experiments that combustion is a process in which oxygen combines with another substance. By 1800, virtually all chemists had recognized the validity of Lavoisier's work, and the phlogiston theory was discredited. *See also* **Fire.**

Photoelectric (Light Scattering) Smoke Detector—A method of smoke detection that uses the scattering of a light beam from the presence of smoke particles onto a photosensitive detector to sense a fire condition and send a signal for alarm. *See also* **Projected Beam Smoke Detector; Smoke Detector.**

Pickup Tube—A flexible hose or tube that draws foam concentrate or a wetting agent liquid from a storage container into a piped water stream (usually by a vacuum caused by an eductor venturi fitting) to allow a foam or wetting agent solution to be produced and proportioned in the proper percentage for fire protection applications. Pickup tubes are mainly used with manual application devices (nozzles) for ease of use and mobility, otherwise a fixed piping system is provided. *See also* **Eductor; Eductor, Foam Nozzle; Proportioning.**

Piercing Nozzle—*See* **Puncture Nozzle.**

Pike Pole or **Ceiling Hook**—A long pole with a prong and hook attached, primarily designed for fire service use to pierce a ceiling to allow water to escape and prevent collapse of the ceiling. It is also used to turn over debris, to probe combustibles for hot spots, and to pull and drag materials. The prong or hook is shaped in the fashion of an old battle pike, hence its name. See Figure P-1.

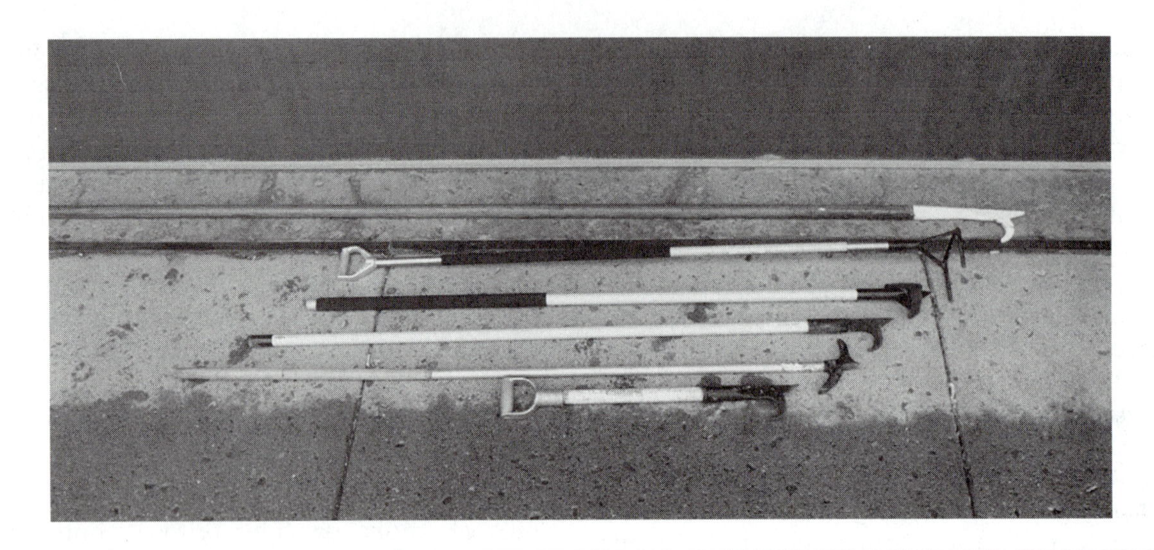

Figure P-1 Pike pole.

Pill Test—A fire test for floor covering in which a small pill is ignited and the resultant fire conditions and burn pattern are noted. The pill test simulates small ignition sources such as a cigarettes and is used to determine if sustained burning of the floor covering will occur. Since 1971, the US government requires that all carpets manufactured in the United States meet the *Federal Flammability Test, FF-1-70, Pill Test.* The test uses a 9 sq. in. (58 cm^2) sample of floor covering. A methanamine tablet is ignited in the middle of the sample. If the flame advances to within 1 in. (2.54 cm) of the edge of the sample, it does not meet the acceptance criteria for the test. Floor covering burning is also influenced by the presence of underlying padding.

Piloted Ignition—A deliberately provided ignition source, such as a flame front generator, pilot light, etc.

Pilot Head Detection System—A fire detection system that uses fusible heads on a pneumatic charged system placed over the area of protection or hazard. Activation of the fusible head releases the system pressure, which normally is linked by a pressure switch to a water suppression trip valve to activate water flow to a deluge water spray system. The system provides for automatic fire detection and activation of protective devices or alarms. *See also* **Pneumatic Fire Detection System; Sprinkler, Pilot.**

Pipeman—A firefighter who is designated to direct and control a firewater stream from a nozzle.

Pipe Schedule—A table provided in NFPA 13, *Standard for the Installation of Sprinkler Systems,* that specifies the sizes of the pipe distribution network based on the number of sprinklers that can be installed on the pipe. Pipe schedule designed sprinkler systems were specified before hydraulically designed sprinkler systems were devised. Pipe schedule systems are generally less economical because smaller pipes can be specified when the system is hydraulically designed.

Pitot Pressure—The pressure in a stream of water at the point of its discharge from the orifice of a hose nozzle or hydrant butt into the open air. A pitot pressure is obtained by the insertion of a pitot tube in the center of the water stream from an outlet. With the stream of water open to the atmosphere, there is no pressure head, so the indicated reading on the pitot tube is velocity head alone. From the velocity of the water stream, a corresponding flow rate can be calculated directly based on the outlet size. *See also* **Nozzle Pressure; Pitot Tube.**

Pitot Tube—A device that measures the velocity head of flowing liquid. Pitot tubes are manually inserted, or provided as part of a fixed installation, into a flowing stream of water from an orifice to determine the amount of water flow during testing of firewater protection devices (firewater pumps, condition of distribution network, etc.). They are based on the flow formula Q = AV, whereby if the area (A) of an orifice is known and the velocity (V) measured, the quantity (Q) of water flow can be determined.

$$V = \sqrt{(2gh)}$$

Where:

V = velocity in feet/second

g = gravity constant 32.2 ft/sec^2

h = head in feet producing velocity

Therefore:

$$V = 12.14 \sqrt{P}$$

P = velocity pressure

See Figure P-2. *See also* **Pitot Pressure.**

Placard UN/NA/POT—Identification board (10¾″ × 10¾″) used for uniform marking of vehicles for transport of hazardous materials. *See also* **Hazardous Materials Identification System.**

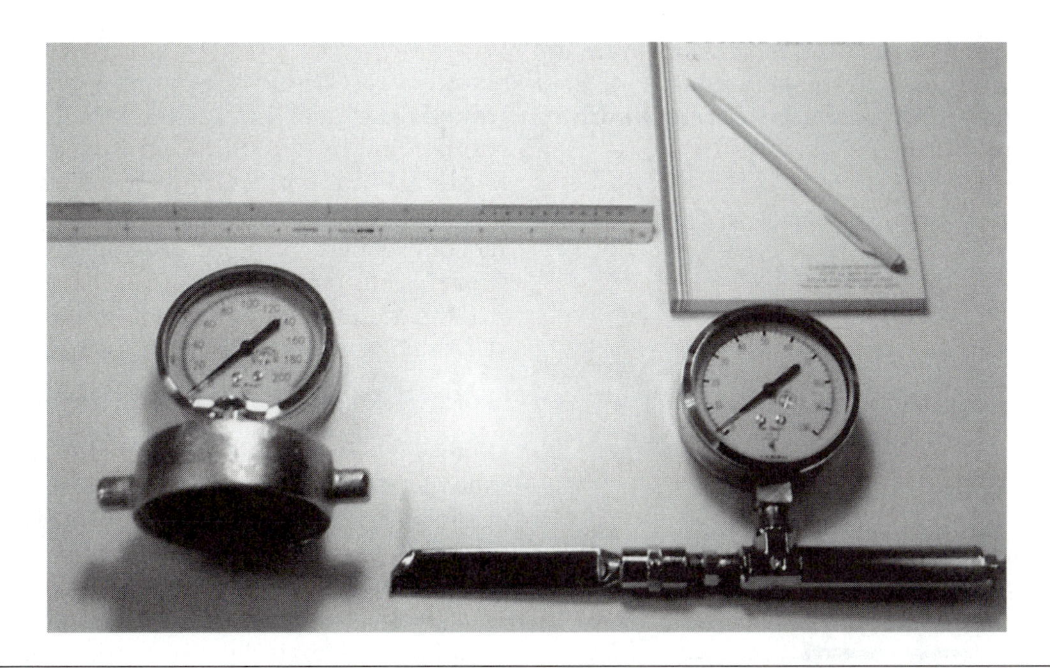

Figure P-2 Water flow testing equipment: pitot tube with Bourdon gauge on right and hydrant cap with Bourdon gauge on left.

P

Playpipe—*See* **Underwriters Playpipe.**

Plosophoric Material—Two or more unmixed materials that are not classified as explosives, but when mixed or combined together form an explosive compound. *See also* **Explosive.**

Plug—*See* **Hydrant.**

Plume—The column of hot gases, flames, and smoke rising above a fire. In a confined area a fire plume rises almost vertically. In an outside, unconfined area the configuration of a fire plume is affected by ambient conditions (wind, temperature, etc.). A fire plume consists of a flame plume, a thermal column of combustion gases, and smoke particles. A fire plume's temperature decreases rapidly after the combustion process due to the entrainment of air. Therefore, the ignition hazard from a fire plume is primarily dependent on the flame height of the plume. Objects located above a flame are not likely to ignite unless large amounts of radiated heat are present or flame contact is made. It may also be called a convection column, thermal updraft, or thermal column. See Figures P-3 and P-4. *See also* **Convective Column; Fire Plume; Indraft; Thermal Convection.**

P

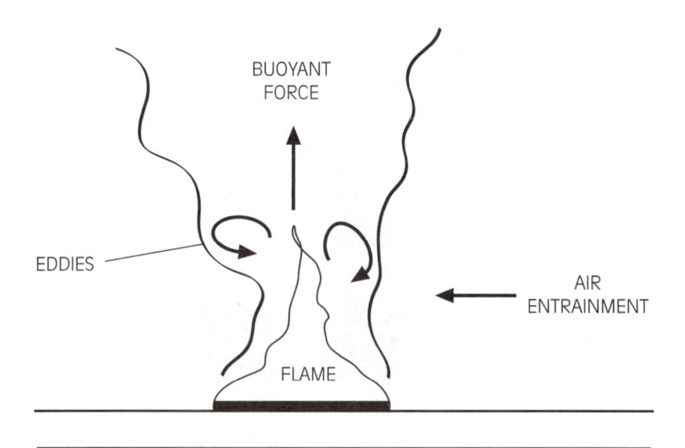

Figure P-3 Fire plume and buoyancy forces.

Pneumatic Fire Detection System—A fire detection system that detects fire from heat, which either melts fusible elements (spot-type detection) in the system or a low melting point pneumatic (plastic) tubing (linear detection). Loss of pressure in the system activates a pressure switch that sends a signal for an alarm and fixed fire suppression system activation. *See also* **Heat Detection; Linear Beam Fire Detection; Pilot Head Detection System.**

Point of Demand-Supply (PDS)—The physical location in a water-based fire protection system at which the volume and pressure for both the water supply and demand have been hydraulically calculated.

Point of Origin—The exact physical location where an ignition source and a fuel come in contact with each other and combustion begins. Point of origin locations are useful in arson investigations and the prediction of fire events to determine fire growth and damage.

Point Protection—The provision of concentrated, high-speed water spray onto a likely ignition point by an ultra-high-speed water spray system. Point protection is primarily provided to a likely ignition source location. It is commonly accomplished by the provision of two or more nozzles at positions physically as possible close to the anticipated point of ignition. *See also* **Ultra High Speed Water Spray System.**

Poke Through—*See* **Firestop; Penetration Seal.**

Pool Fire—A turbulent diffusion fire burning above an upward facing horizontal pool of vaporizing liquid fuel under conditions where the fuel vapor or gas has zero or very low initial momentum. Pool fires are of primary concern in the process industries where large volumes of combustible materials are processed, handled, or stored. A water spray is not very effective in extinguishing pool fires because the liquid fuel

Figure P-4 Fire and smoke plume.

P

normally floats on the water and continues to burn. Foam fire suppressants are the preferred extinguishing medium because they float on the surface of the liquid fuels due to their lower density. Foams are then able to starve the pool fire of oxygen and cause its extinguishment, in addition to cooling the fuel. *See also* **Line Fire; Running Fire; Skin Fire.**

Portable Fire Extinguisher—*See* **Fire Extinguisher.**

Portable Pump—Generally a small electrical or internal combustion-driven pump capable of supplying a small water hose at a rate of about 50 gpm (3.2 liters/second) at 90 psi (620 kPa). It is used for small fires or in the very early incipient stages of a fire.

Porta-Power—A manually operated hydraulic tool used by the fire service for rescue operations. It can be fitted with a variety of tool accessories for various applications. *See also* **Jaws-of-Life.**

Positive Pressure Ventilation (PPV)—The application of positive air ventilation to an enclosed fire event to influence the degree of ventilation, aid in firefighting activities, and influence burning activity. Mechanical ventilators (fans) are used to blow fresh air into an enclosure in sufficient amounts to create a pressure differential within the enclosure that forces the existing air or products of combustion through an exit opening in the enclosure. Positive pressure ventilation has been used to assist in firefighting operations.

Post Incident Trauma—Psychological stress that may occur to individuals after they return from a response to a stressful emergency incident. *See also* **Critical Incident Stress Debriefing (CISD); Fire Trauma.**

Post Indicator Valve (PIV)—A post extension from an underground nonrising stem gate valve on a water distribution network. The post indicates the open or closed position of the underground valve through the viewing of an "Open" or "Shut" label observable in the windows in the post. It also provides a means of locking or sealing the underground valve in a particular position to prevent unauthorized use. PIVs are provided in firewater distribution networks to readily determine the operating condition of the system water supply network and aid in emergency functions of the system. They may also be located in a wall as a wall post indicator valve (WPIV). See Figure P-5.

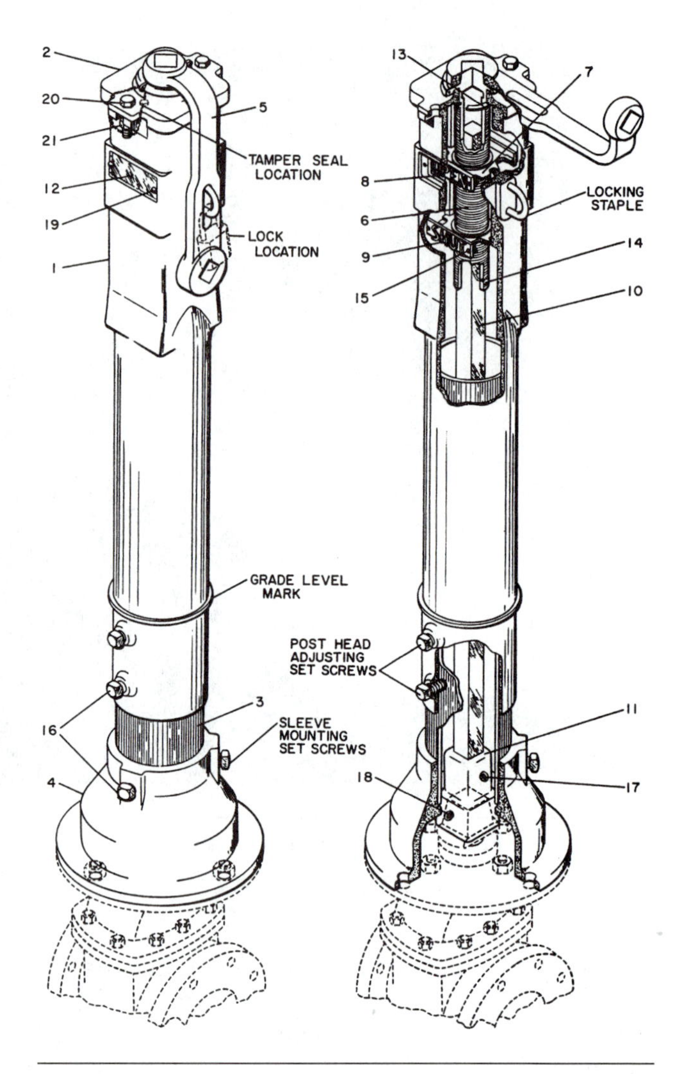

Figure P-5 Post indicator valve (PIV). *(Courtesy Grinnel Fire Protection Systems Company, Inc.)*

P

Post-Traumatic Incident Debriefing—*See* **Critical Incident Stress Debriefing.**

Potassium Bicarbonate—A dry chemical agent used for rapid fire suppression. It was developed at the US Naval Research Laboratories (circa 1959) to apply simultaneously with aqueous film forming foams (AFFF) for aircraft crash fires. It is commonly referred to as Purple-K, a trademarked name by Wormald, U.S., Inc. It is generally about twice as effective as sodium bicarbonate. It is compatible with aqueous film forming foam and is sometimes combined in a twin nozzle arrangement to simultaneously apply foam and dry chemical for rapid fire knockdown, and vapor suppression for liquid fuel fires for protection of aircraft operations. Purple-K powder is dyed purple to differentiate it from other dry chemical agents; the K is the chemical symbol for potassium. It was originally named Purple K for the purple flame spectrum it exhibits when it encounters flame temperatures. *See also* **Dry Chemical; Multipurpose Dry Chemical; Sodium Bicarbonate; Twin Agent Unit** or **Twinned Agent Systems.**

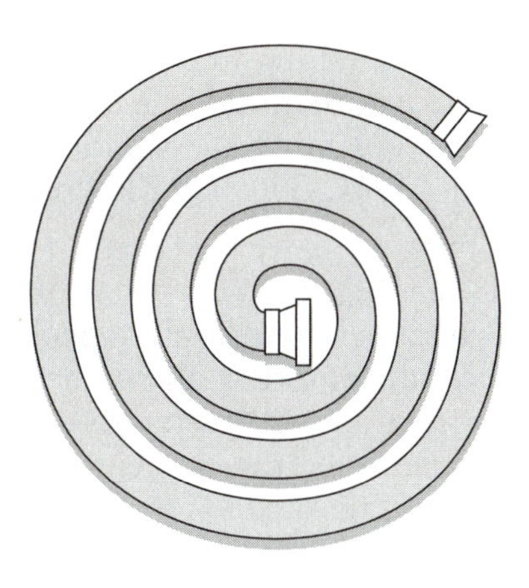

Figure P-6 Potato hose roll (simplified for clarity).

Potato Roll—A method of storing fire hose wherein it is rolled up similar to ball of string. See Figure P-6. *See also* **Donut Roll.**

Practical Critical Fire Area (PCA)—Terminology use in aircraft firefighting that is representative of actual aircraft accident conditions. The purpose of the critical area concept is not to define fire attack procedures but to allow a basis for calculating the quantities of extinguishing agents necessary to achieve protection within an acceptable period of time. The PCA is an area that is two-thirds of the Theoretical Critical Fire Area (TCA) that is adjacent to an aircraft in which fire must be controlled for the purpose of ensuring temporary fuselage integrity and providing an escape area for its occupants. *See also* **Theoretical Critical Fire Area (ITA).**

Pre-Action Valve—*See* **Sprinkler System, Pre-Action System.**

Preconnected—A description of fire protection hoses that are generally permanently attached to a source of supply prior to the occurrence of a fire. Use of preconnected hoses reduces the amount of time required to provide equipment available for fire protection activities. It is commonly used to describe fire hoses that are routinely carried and normally connected to the discharge outlets of fire trucks (pumpers). These hoses are then immediately available for use once the pump is operational and a water supply is available to the pump. *See also* **Attack Line.**

Pre-Engineered System—A fixed fire protection system (normally dry or wet chemical) that has predetermined flow rates, nozzle pressures, and quantities of extinguishing agent. The system has a specific pipe size, maximum and minimum pipe lengths, flexible hose specifications, number of fittings, and numbers and types of nozzles. The hazards protected by these systems are specifically limited by a testing laboratory based on actual fire tests. Application limitations are identified in the specifications of the system. Pre-engineered systems are normally stored pressure or cartridge operated. Typically these can be found at restaurant exhaust hoods and ducts, deep-fat fryers, vehicle fueling systems, and mobile self-propelled equipment. They offer advantages of cost savings by providing standard designs and components. *See also* **Wet Chemical Fire Suppression System.**

Pre-Fire Plan—A description of the potential hazard and recommended actions to be taken by firefighting personnel during a fire or explosion event for a specific hazard, location, or facility based on previous inspections and surveys of identified hazards. Pre-fire plans note the structural features, physical layout, special hazards, installed protection systems, fire hydrant locations, water supplies, and similar features pertinent to firefighting operations. Pre-fire plans should be routinely updated or revised as changes occur in a facility or location. *See also* **Deployment Plan.**

Pre-Incident Planning—A description of the potential hazard and recommended actions to be taken by public emergency response agencies or private industry for response to emergency incidents at a specific facility or location.

Premixed Flame—A flame for which the fuel and oxidizer are intimately mixed prior to ignition or a combustion process, such as in a process heater gas burner or home gas cooking range. Propagation of the flame is governed by the interaction between flow rate, transport processes, and chemical reaction. *See also* **Diffusion Flame.**

Premixed Foam Solution—A stored solution of a measured amount of foam concentrate and water to provide the proper concentrate percentage as specified by the manufacturer for foam fire protection applications. It is prepared in advance of a fire incident to avoid the need of proportioning at the time of the incident and is provided in a storage tank. *See also* **Proportioning.**

Preprimed System—Water or water-based fire protection system using open heads or nozzles in which the distribution piping is prefilled with water up to the nozzles/sprinklers or a rupture disc provided within the system (within a union fitting). There is an immediate discharge of agent when the sprinkler operates, improving response time. Nozzles or sprinklers are fitted with blow-off caps or a rupture disc is fitted to retain the priming water in the system. When the system is activated, the available water pressure bursts the rupture disc or forces off the blow-off caps or plugs. Foam solutions react with steel pipe over a period of time, causing them to lose their capability to produce a fire-resistant foam. Therefore, in a preprimed system charged with a foam-water agent, there could be a delay in the discharge of effective foam until all the preprimed solution has been flushed out and fresh foam solution reaches the opened sprinklers, unless corrective steps are taken (periodic flushing and recharging of the system). The foam manufacture should be consulted before prepriming is considered to determine its limitations. *See also* **Blow-Off Caps** or **Blowout Plugs.**

Prescribed Burning or **Fire**—Controlled application of fire to wildland under specified environmental conditions that allow the fire to be confined to a predetermined area and at the same time to produce the intensity of heat and rate of spread required to attain planned resource management objectives by the appropriate authority for the wildland. These objectives may be the removal or modification of woodland fuels, provision of a firebreak or path, and killing unwanted plant growth. *See also* **Broadcast Burning.**

Pressure Gauge—An instrument or device used to measure fluid pressure. *See also* **Compound (Pressure) Gauge.**

Pressure Governor—A pressure control device that controls the speed of an engine driving a pump to prevent excessive discharge pressures from occurring and maintain a steady pressure. When discharge pressures increase, the governor reduces engine speed, and when the pressure decreases, the governor increases engine speed to maintain the desired pressure. *See also* **Relief Valve.**

Pressure Loss—A decrease in water pressure due to friction in conduits, fluid viscosity, lift, or other factors.

Pressure Proportioning Tank—A proportioning method for the introduction of foam concentrate or a wetting agent into a water delivery system for fire protection applications. It consists of displacing foam concentrate from a closed tank (with or without a diaphragm separator) by applying the delivery system water pressure and using water flow through a venturi orifice for metering the foam concentrate in the appropriate percentage. *See also* **Bladder Tank; Eductor; Proportioning.**

Pressure Sandwich—A method of smoke control in a building. The smoke control system exhausts smoke from the floor of the fire and shuts down its supply air (fresh makeup air). The arrangement creates a positive pressure on the floors above and below the fire floor, preventing the spread of smoke. *See also* **Smoke Management System (SMS).**

Pressure Treated Wood—*See* **Fire-Retardant Impregnated Wood.**

Pressurized Stairway—*See* **Smoke Management System.**

Probable Maximum Loss (PML)—The amount an underwriter considers the largest loss likely to occur. It is similar to the insurance term Normal Loss Expectancy (NLE). *See also* **Normal Loss Expectancy (NLE).**

Process Safety Management (PSM)—A comprehensive set of plans, policies, procedures, and practices, and administrative, engineering, and operating controls designed to ensure that barriers to major incidents are in place, in use, and effective.

Products of Combustion—*See* **Combustion Products.**

Professional Engineer (PE), Fire Protection—An individual who has successfully passed prescribed engineering tests (Engineer-in-Training, EIT, and Professional Engineer, PE) and attains a level of experience recognized by society as an engineer in the discipline of fire protection engineering. A record of registration as a Professional Engineer with the appropriate state or states in which he or she has a license to practice engineering is required as part of the licensing statutes. Examination tests are approved and administered by the ABET/NCEE. Registration as a Professional Engineer demonstrates competency and is a requirement of firms offering engineering services to the public within the United States. Various states offer registration as a Professional Engineer in the field of Fire Protection after meeting specific examination and experience requirements. *See also* **Fire Protection Engineer (FPE).**

Progressive Hose Lay—The provision and arrangement of fire hoses with double-gated Ys provided at intervals in the main hose line to permit the continuous application of water to a fire incident, while lateral lines are provided to the fire edges and the main line is being extended.

Projected Beam Smoke Detector—A device that senses the presence of smoke by the transmission and reception of a light beam. The amount of light transmitted between a light source and a photosensitive sensor is monitored. When smoke particles are introduced into the light path, some of the light is scattered and some absorbed, thereby reducing the amount of light reaching the receiver, causing the detector to respond. *See also* **Photoelectric (Light Scattering) Smoke Detector; Smoke Detector; Very Early Smoke Detection and Alarm (VESDA) System.**

Proof Test Pressure—A test pressure for fire hose. Proof test pressures are specified in NFPA 1961, *Standard on Fire Hose,* and are not to be less than two times the specified service test pressure. Proof test pressures are held for at least 15 seconds and not more than 1 minute. *See also* **Burst Test Pressure; Kink (Pressure) Test.**

Proportioner—A device used to proportion foam concentrate or a wetting agent and water at the proper percentage. May also be called an eductor. *See also* **Foam Proportioning; In-Line Balanced Proportioner (ILBP); Ratio Flow Controller; Wetting Agent.**

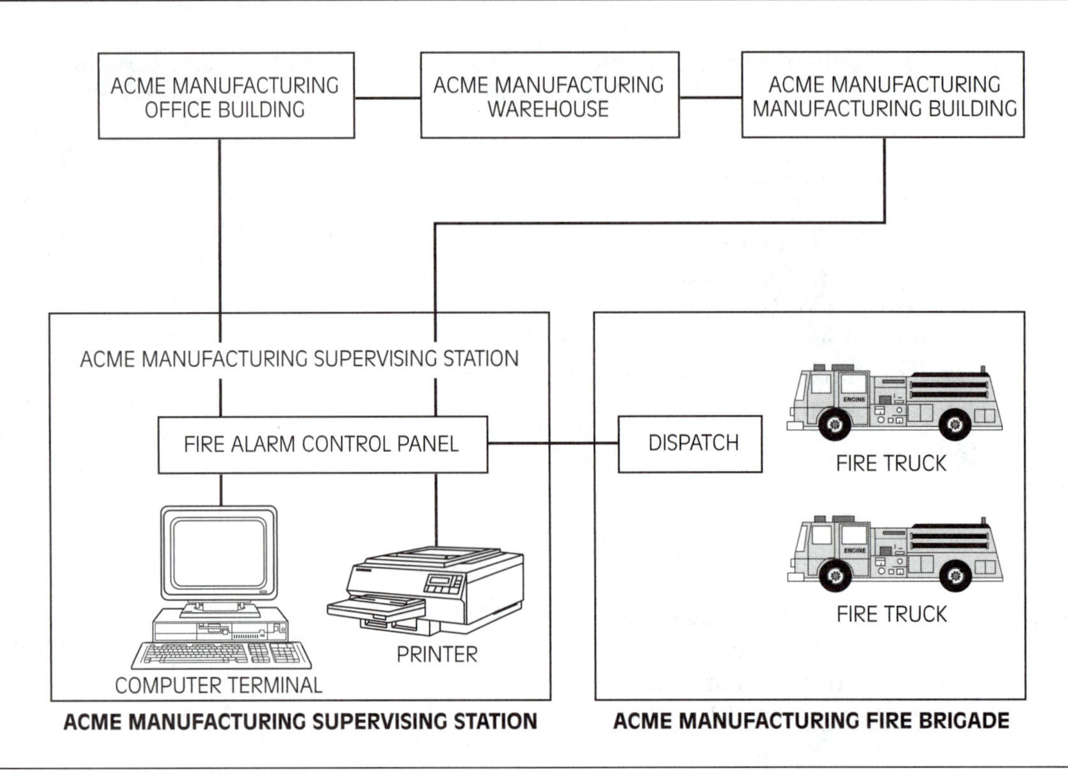

Figure P-7 Proprietary supervising station fire alarm system.

P

Proportioning—Methods and devices used to provide the prescribed amount of foam solution or wetting agent to the water delivery system (piping or nozzle), in order to create a foam-water or wetting agent mixture for firefighting activities. Proportioning methods include a coupled water-motor pump, a nozzle eductor, an in-line eductor, metered proportioning, a pressure proportioning tank, and a pump proportioner (around-the-pump proportioner). See also **Around-the-Pump Proportioning; Balanced Pressure Pump Proportioning; Pickup Tube; Conductivity Method; Coupled Water-Motor Pump Proportioning; Eductor; Eductor, Foam Nozzle; Fixed Foam Application Systems; Premixed Foam Solution; Pressure Proportioning Tank.**

Proportioning, Metered—The use of orifices or venturies to inject the correct amount of foam concentrate or wetting agent into a water delivery system (piping or nozzle) for firefighting activities. Pumps or pressurized tanks are used to force the agent through the metering device into the water system.

Proprietary Supervising Station Fire Alarm System—A private fire alarm system that is not connected to a public or central station fire alarm system, but reports to its own central monitoring station. See Figure P-7. *See also* **Central Station System** (or **Central Station Fire Alarm System); Fire Alarm System.**

Protect in Place—*See* **Shelter in Place.**

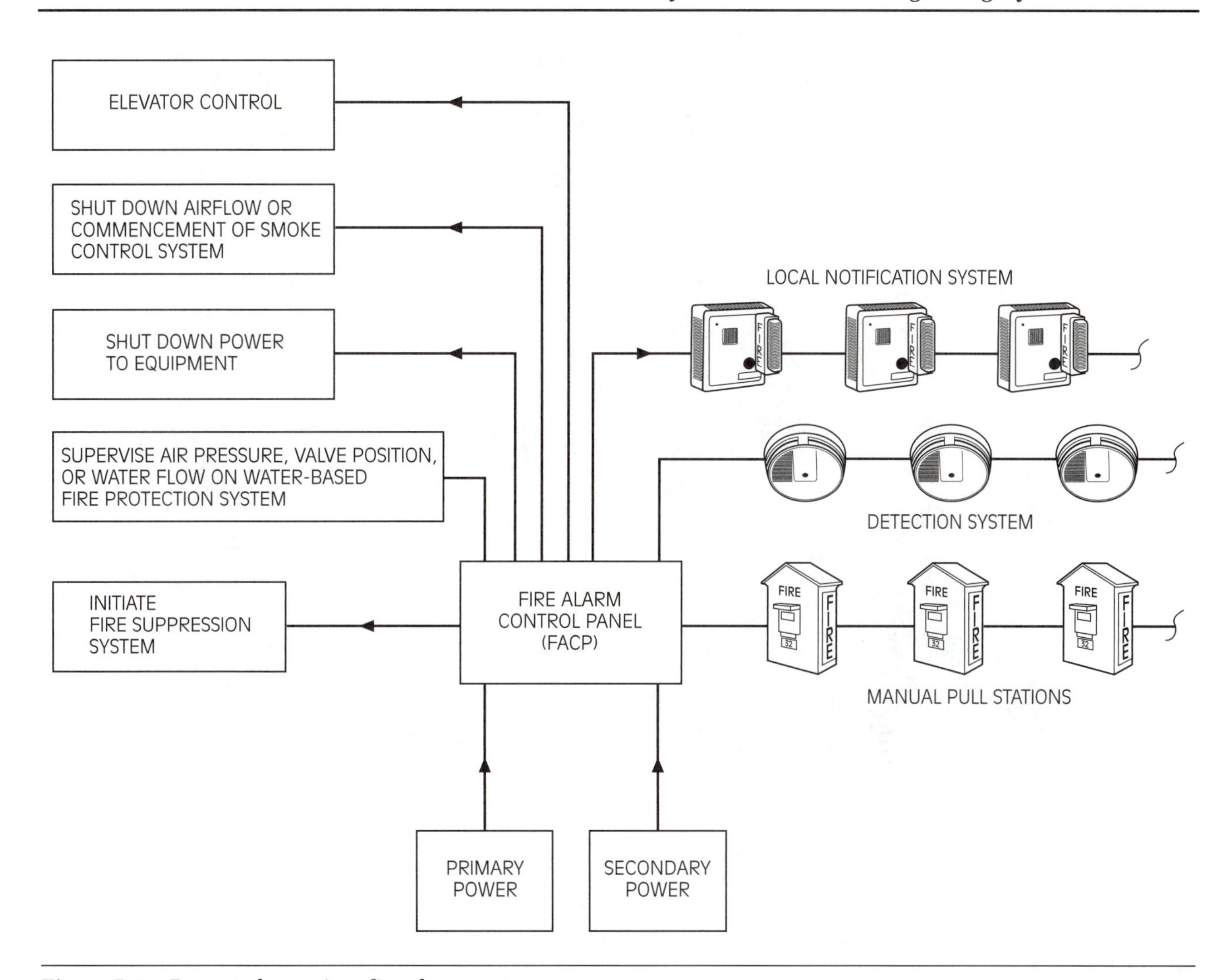

Figure P-8 Protected premises fire alarm system.

Protected Premises Fire Alarm System—A local building or enclosure fire alarm system. It provides alarm initiation, notification, and control within or directly outside the building or enclosure. See Figure P-8.

Protective Fired Vessel—A fired vessel that is provided with equipment (such as flame arrestors, stack temperature shutdowns, forced draft burners with safety controls, and spark arrestors) designed to eliminate the air supply and exhaust as sources of ignition.

Protective Signaling System—Electrically operated circuits, instruments, and devices together with the necessary electrical energy designed to transmit alarms and supervisory trouble signals necessary for the protection of life, property, continued business operation, and the environment.

P

Figure P-9 Proximity suit.

P

Protein Foam—The first type of mechanical firefighting foam developed (circa 1937) and generally marketed in the United States in the early 1940s. It consists of a protein source (hoof, horn, blood meal) that has undergone alkaline hydrolysis. The protein hydrolzate is then stabilized with metal salts and has corrosion and bacterial inhibitors added. If required, freezing point depressants are added. It is normally available in 3 or 6 percent proportioning concentrations. The foam generated from these concentrates is very thick and viscous and exhibits poor fuel tolerance. It spreads very slowly over a fuel surface. It needs to be applied very slowly over a fuel surface to prevent breakdown. Use of water with industrial or petroleum waste in foam concentrates will seriously affect the quality of the foam produced. Protein foams have been superceded by more modern types of foams (AFFF, FFFP, etc.). It may also be called regular foam. *See also* **Foam.**

Proximity (Ensemble) Suit—Firefighting clothing that has its outer surfaces provided with enhanced heat reflection properties rather than traditional firefighter bunker clothing, and therefore allows the wearer to fight a fire at a closer distance than normally would be expected. See Figure P-9.

Puff Test—A test performed on fixed fire suppression system piping to determine if the piping is blocked or mostly obstructed. Compressed air,

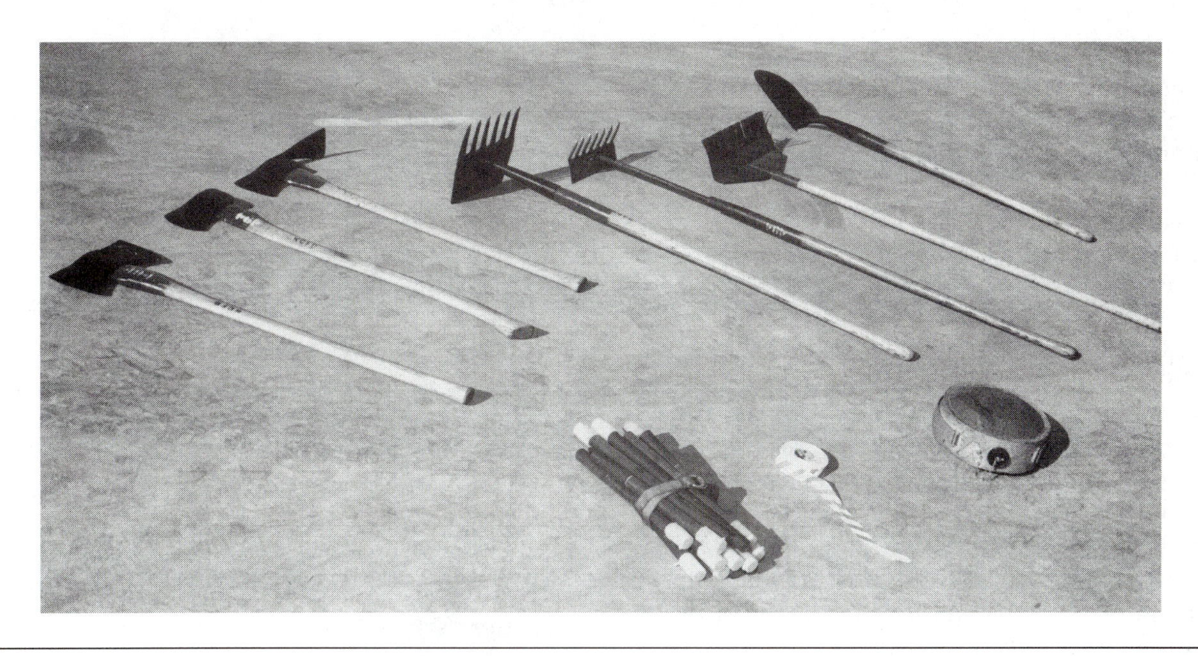

Figure P-10 Wildland firefighting tools, left to right: two axes, Pulaski, two McLeods, fire broom, shovel, fuses, flagging tape, canteen.

carbon dioxide (CO_2), or nitrogen is introduced under pressure and released through the system to the discharge points. Outlets that do not release the contained gas indicate the pipe is obstructed and further investigation is required. Commonly performed on fixed gaseous fire suppression agent systems (CO_2) or where there is a concern of corrosion from frequent application of water to a corrosion susceptible piping system.

Pulaski—A name given for a tool for fighting wildland fires. It has an ax on one side of the head and a grub hoe on the other. It is primarily used for clearing brush and creating firebreaks. The Pulaski, is named for Forest Ranger Ed Pulaski who partially invented the tool. He was also famous for saving members of his firefighting team during a notorious forest fire in 1910 in the northern Rocky Mountains. A tool similar to the Pulaski was originally exhibited by the Collins Tool Company in 1876. After several trials, Ed Pulaski, after earlier discussions by his supervisor (W. G. Wiegel) with his associates (J. Halm and E. Holcomb), constructed a prototype tool in its present form in 1913, which was eventually accepted by the Forest Service in 1920. See Figure P-10.

Pull Station—Terminology that is synonymous with manual pull station or manual call point. *See* **Manual Pull Station (MPS).**

Pump and Roll—An operational firefighting technique used for grass and brush fires that utilizes on-board pumping facilities while the fire protection vehicle is moving. Water is sprayed to extinguish the fire and hose lines are connected to the fire apparatus.

Pump Discharge Pressure (PDP)—The velocity pressure of water leaving the outlet of a pump.

Pump, End Suction and Vertical In-Line—A common type of centrifugal fire pump that has either a horizontal or vertical shaft, with a single suction impeller and single bearing at the drive end. It is useful for adapting to existing systems because of the orientation versatility and minimum installation requirements. Small vertical in-line pumps can be supported with ordinary pipe supports on either side of the pump, eliminating costly foundations or pads. *See also* **Centrifugal Pump; Fire Pump.**

Pump, Fire—*See* **Fire Pump.**

Pump, Horizontal Split Case—A common type of centrifugal fire pump that is characterized by having a double suction impeller with inboard and outboard bearing with a horizontal orientation of the drive shaft of the impeller. It is applied where a positive suction water supply is available. The casing is manufactured in half sections, which gives it a characteristic "split" configuration. The split case for a horizontal unit has the advantage that the rotating element can be removed without disturbing the suction and discharge piping. *See also* **Centrifugal Pump; Fire Pump.**

Pump, Line-Shaft—*See* **Pump, Vertical Shaft, Turbine Type.**

Pump, Positive Displacement—A pump that consists of a rotating member with a number of lobes that move together in a closed casing. Liquid is trapped in the spaces between the lobes, imparted with momentum, and discharged from the pump. In most cases, the liquid is actually discharged in a series of pulses. In the fire protection profession, most positive displacement pumps are used to pump highly viscous fluids, such as additives, to water to enhance its firefighting ability (foam concentrates, wetting agents, etc.).

Pump, Vertical Shaft, Turbine Type—A common type of centrifugal fire pump that has its impeller(s) completely submerged in the supply source. It is suspended from a discharge head by a series of pipes that combine into a column. The column pipe also supports and guides the pump vertical drive (line) shaft and bearings from where it is driven by either a diesel engine or electric motor. This type of fire pump is commonly provided where a suction lift is required, such as from a well, offshore caisson, or water pit. See Figure P-11. *See also* **Centrifugal Pump; Fire Pump.**

Pumper—Terminology used to describe a designated truck used in the fire service that is specifically provided with a water pump, usually a centrifugal type, to supply water in sufficient quantities and pressures from public water supplies for the purpose of fire protection (control, suppression, cooling, etc.). *See also* **Fire Pumper; Triple Combination (Pumper).**

Puncture (Piercing) Nozzle—A specially designed firefighting nozzle which is able to penetrate walls to extinguish a hidden fire when it is forced though by a firefighter. It may also be called a bayonet or piercing nozzle. See Figure P-12.

Pung—A hose wagon that was used in the snowy climates of New England circa 1870s. It had sled rails underneath to facilitate its use in winter conditions. It was commonly used for a low, one-horse box sleigh. Shortened from dialectal *tom-pung,* from the Algonquian native American Indian language of southern New England. *See also* **Hose Cart.**

Purging—The removal of a combustible vapor atmosphere or any residue capable of producing combustible vapors in an enclosure or piping. It is accomplished so that subsequent natural ventilation will not result in the reinstatement of a combustible atmosphere. Purging is commonly accomplished by the use of a protective or inert gas (nitrogen, carbon dioxide, etc.) at a sufficient flow and positive pressure to reduce the concentration of any flammable gas or vapor

Figure P-11 Vertical in-line turbine fire pump being removed from test pit at factory showing end strainer, two-stage pump, column pipes, and discharge head.

initially present to an acceptable level. *See also* **Inerting.**

Purple-K—*See* **Potassium Bicarbonate.**

Pyrex™—A trademark for a type of glass and glassware that is resistant to heat, chemicals, and electricity. It is used to make chemical apparatus, industrial equipment, including piping and

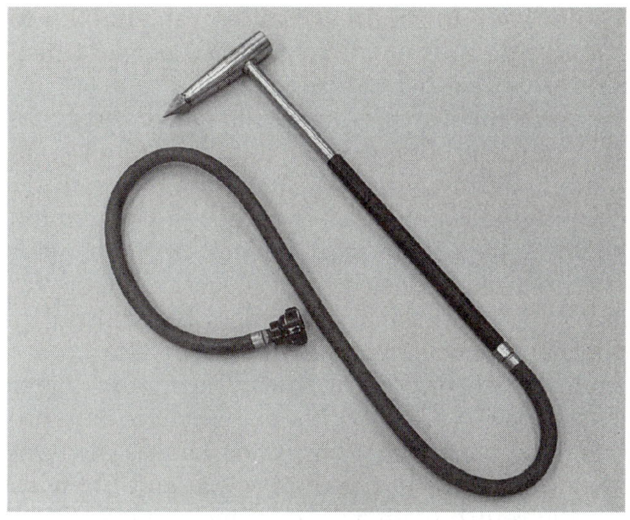

Figure P-12 Puncture nozzle.

thermometers, and ovenware. Chemically, Pyrex contains borosilicate and expands only about one-third as much as common glass (silicate) when heated. Consequently, it is less apt to break when subjected to rapid temperature changes. It is resistant to many chemicals and is an electrical insulator. Fibers and fabrics made of it possess excellent heat insulation and fire-resistant qualities. It is sometimes referred to by the generic term, borosilicate glass.

Pyro—A Greek term meaning "fire." Pyro may also refer to a someone who has a pyromania disorder. *See also* **Pyromania.**

Pyrolysis—The transformation, usually decomposition, of a compound into one or more other substances by heat alone. Most combustion gases include products formed by pyrolysis as well as by combustion. Pyrolysis often precedes combustion. *See also* **Combustion.**

Pyromania—An impulse-control disorder of an individual characterized by the recurrent compulsion to set fires. The term refers only to the setting of fires for sexual or other gratification provided by the fire itself, not to arson for profit or revenge. Pyromania is usually a symptom of underlying psychopathology, often associated with aggressive behaviors. Sigmund Freud, the founder of psychoanalysis, noted that the majority of pyromaniacs are males with a history of bed-wetting. He suggested that pyromania is one of many disorders brought on by the denial of instinctual drives, in this case a male desire to control fire by urination. Later psychoanalysts found his explanation too simplistic. Among other suggested causes of pyromania are the feelings of rejection and the wish for the return of an absent father.

Pyromania usually first surfaces in childhood, and only a small percentage of adult fire-setters actually suffer from the disorder. Pyromaniacs fighting an urge to set fires experience increasing tension that can only be relieved by giving in. After repeated failures to control the impulse, they may cease resistance to avoid this tension. The disorder may be treated by family-centered psychotherapy and antidepressant drugs. A pyromaniac is contrasted with an arsonist, who sets fires primarily for monetary gain or crime concealment, or an incendiarist, who sets fires for political or social action or injustice. *See also* **Arson; Firebug; Incendiarist; Pyro.**

Pyrometer—Any instrument for measuring degrees of heat. Most pyrometers work by measuring radiation from the body whose temperature is to be measured. Radiation-type pyrometers have the advantage of not having to touch the material being measured. Optical pyrometers, for example, measure the temperature of incandescent bodies by comparing them visually with a calibrated incandescent filament that can be adjusted in temperature. In an elementary radiation pyrometer, the radiation from the hot object is focused onto a thermopile (a collection of thermocouples), which generates an electrical voltage that depends on the intercepted radiation. Proper calibration permits this electrical voltage to be converted to the temperature of the hot object.

In resistance pyrometers, a fine wire is put in contact with the object. The instrument converts the change in electrical resistance, caused by heat, to a reading of the temperature of the object. Thermocouple pyrometers measure the output of a thermocouple placed in contact with the hot body; by proper calibration, this output yields temperature. Pyrometers are closely akin to the bolometer and the thermistor and are used in thermometry.

Pyrophoric—The property of a material to ignite spontaneously in air under ordinary conditions, generally taken as a chemical with an autoignition temperature in air at or below 130°F (54.4°C). Certain very finely divided powders overall present so much available surface area to the air that they can spontaneously combust and are called pyrophoric substances. An alloy composed of iron and rare earth metals, called misch metal, is pyrophoric. When scratched it gives off sparks capable of igniting flammable gases. It is used in cigarette lighters, miners' safety lamps, and automatic gas-lighting devices. Iron sulfide may form in hydrocarbon tanks or vessels that upon exposure to air will spontaneously combust. Ignition of pyrophoric materials can be prevented by the keeping them damp until they can be safely disposed of (steaming out vessels).

Pyrostat—An automatic heat sensing device that activates an alarm or extinguisher in case of a fire from a predetermined set point.

Pyrotechnics—Explosive commodities that have extremely high flame speeds and heat releases. Pyrotechnics are primarily used for fireworks displays and munitions. *See also* **Fireworks.**

Quad—Fire service terminology applied to a fire truck that is supplied with quadruple level of fire equipment: water tank, pump, ladders, and hose. See Figure Q-1. *See also* **Quint; Triple Combination.**

Qualitative Risk Analysis—An evaluation of risk based on the observed hazards and protective systems that are in place, as opposed to an evaluation that uses specific numerical techniques. *See also* **Quantitative Risk Analysis (QRA); Risk Assessment.**

Quantitative Risk Analysis (QRA)—An evaluation of both the frequency and the consequences of potential hazardous events to make a logical decision on whether the installation of a particular safety measure can be justified on grounds of safety and loss control. Frequency and consequences are usually combined to produce a

Figure Q-1 Quad fire truck.

measure of risk that can be expressed as the average loss per year in terms of injury or damage arising from an accidental event. The risk calculations of different design alternatives can be compared to determine the safest and most economical options. Calculated risk may be compared to set criteria that have been accepted by society or required by law. *See also* **Fire Modeling; Qualitative Risk Analysis; Risk Assessment.**

Quarter Drainage Time—*See* **Drainage Rate, Foam.**

Quench—To cool a fuel to a temperature below its kindling or flash point to effect fire extinguishment. Quenching is commonly accomplished by soaking the fuel with water; however, any noncombustible liquids that vaporize at a relatively low temperature can be used. A material temperature may be also quenched (lowered) by plunging it into liquids or subjecting it to air blasts. *See also* **Blacken; Fire Quenching Pit.**

Quick Opening Device (QOD)—A device used to increase the opening times of dry pipe sprinkler valves. They are either of the "accelerator" or "exhauster" type. Quick opening devices were developed by the Grinnel Company, Inc. and were first available in 1920. An accelerator should open the sprinkler valve in 15 to 20 seconds, an exhauster takes slightly longer. *See also* **Accelerator; Exhauster.**

Quick-Response Early Suppression Sprinkler (QRES)—A type of quick-response sprinkler that has a thermal element with a response time index (RTI) of 50 (meters-seconds) 1/2 or less and that is able to provide fire suppression of specific fire hazards as specified by listing or approval agencies.

Quick-Response Extended Coverage Sprinkler—A type of quick-response sprinkler that has a thermal element with an a response time index (RTI) of 50 (meters-seconds) 1/2 or less and that is able to provide fire suppression for extended areas as specified by listing or approval agencies. *See also* **Sprinkler, Extended Coverage.**

Quick-Response (QR) Sprinkler—A sprinkler that is fitted with a specially designed response element that allows it to react faster to a fire event by rapidly transferring heat to its actuating mechanism. A quick response sprinkler has a thermal element with a response time index (RTI) of 50 (meters-seconds) 1/2 or less. QR sprinklers are used for specific fire hazards. *See also* **Early Suppression Fast Response (ERFS) Sprinkler; Response Time Index (RTI).**

Quint—Fire service terminology applied to a fire truck that is supplied with quintuple (five) levels of fire equipment: water tank, pump, ladders, hose, and aerial ladder. *See also* **Quad; Triple Combination.**

Radiant Heat—Heat energy carried by electromagnetic waves longer than light waves and shorter than radio waves. Radiant heat (electromagnetic radiation) increases the sensible temperature of any substance capable of absorbing the radiation, especially solid and opaque objects.

Radiant Panel Test—A test for the surface flammability of materials. It is outlined under ASTM E-162, *Standard Test Method for Surface Flammability of Materials Using a Radiant Heat Energy Source*. It is considered an alternative test method to NFPA 255, *Standard Method of Test of Surface Burning Characteristics of Building Materials,* ASTM E-84, UL 723 or the Steiner Tunnel Test. *See also* **Flame Spread Index (FSI).**

Radiant Protective Performance (RPP)—A rating of thermal protection provided to protective clothing to indicate the level of heat protection specified in NFPA 1977, *Standard on Protective Clothing and Equipment for Wildland Firefighting*. The radiant protective performance (RPP) test uses a sample of material (outer shell, inner moisture barrier, and thermal liner materials) exposed to a 20 kW/m^2 (5 cal/ cm^2) per second of heat exposure to simulate a moderate fire exposure for firefighting. *See also* **Thermal Protective Performance (TPP).**

Radiation (Thermal Radiation)—The transfer of heat by electromagnetic waves. It is emitted by a heated surface in all directions and travels directly to its point of absorption at the speed of light; thermal radiation does not require an intervening medium to carry it. Thermal radiation ranges in wavelength from the longest infrared rays through the visible-light spectrum to the shortest ultraviolet rays. The intensity and distribution of radiant energy within this range is governed by the temperature of the emitting surface. The total radiant heat energy emitted by a surface is proportional to the fourth power of its absolute temperature (the Stefan-Boltzmann law). The rate at which a body radiates (or absorbs) thermal radiation depends upon the nature of the surface as well. Objects that are good emitters are also good absorbers (Kirchhoff's radiation law). A blackened surface is an excellent emitter as well as an excellent absorber. If the same surface is silvered, it becomes a poor emitter and a poor absorber. Some specialized firefighting protective clothing is provided with a silvered outer coating to prevent heat absorption and reflect the heat radiation. The thermal radiation level and its duration to an exposed surface determines its hazard. See Table R-1.

Rapid Intervention Crew (RIC) or **Rapid Intervention Team (RIT)**—An assemblage of firefighters used to assist other firefighters in need of emergency assistance. A rapid intervention crew is a requirement of most fire departments.

Rapid Intervention Vehicle (RIV)—A fast-response vehicle that is able to reach an incident

Table R-1 Thermal Radiation Exposure Levels

Exposure	Maximum Exposure Time	Radiation Level KW/m^2
Greatest solar radiation	Continuous in light winds	1.0
Working areas where personnel are continuously exposed	Continuous in light winds	1.6
General areas where personnel may be continuously exposed		1.9 to 2.5
Dehydration of wood; will eventually reach 500°F (260°C)		4.1
Emergency action areas; upper limit for working when wearing normal clothes and intermittently sprayed by water or sheltered	2 minutes	4.7
Emergency action areas	30 seconds	6.3
Immediate evacuation required	Few seconds only	9.5
Structures and equipment not specifically designed to accept radiation		9.5
Wood will be heated up to 800°F (427°C) and would ignite		12.6

scene before larger or heavier capacity emergency vehicles. It generally carries a minimal amount of fire suppression agent for initial fire attack operations.

Rate by Area Method—A method of designing and engineering a fixed carbon dioxide (CO_2) fire suppression system. It is based on the capability of carbon dioxide nozzles to discharge a specified amount of agent over a fixed area of coverage.

Rate by Volume Method—A method of designing and engineering a fixed carbon dioxide (CO_2) fire suppression system for local application. It is based on an imaginary volume larger than the specified hazard to account for dissipation and loss of carbon dioxide during discharge.

Rate Compensated (Heat) Detector—A fire detection device that initiates an alarm signal when the temperature of the air surrounding it reaches a predetermined level, rather than the rate of the air temperature rise. See Figure R-1. *See also* **Heat Detector.**

Rate-of-Rise (ROR)—A method of fire detection by devices or equipment sensitive to an abnormal rate of increase of heat, that is, a rapid rate

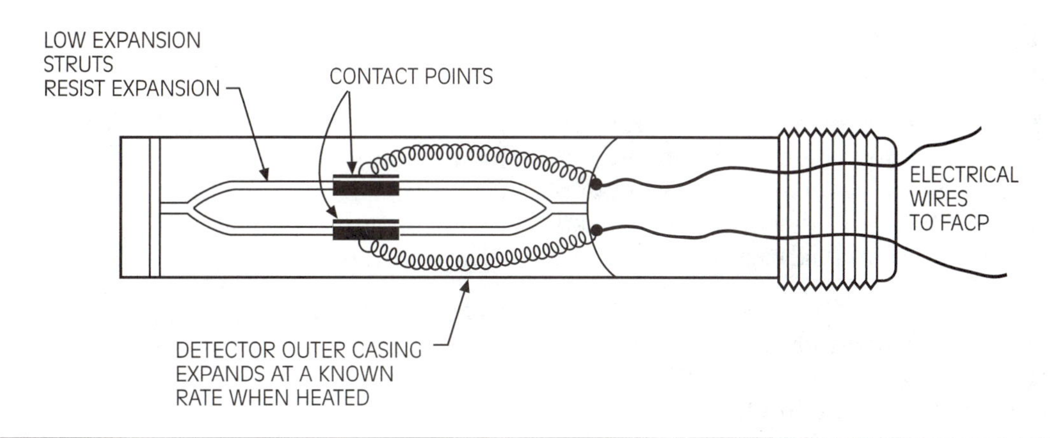

Figure R-1 Spot-type rate compensated heat detector.

increase but not a slow normal rate of increase (from expected ambient conditions).

Rate-of-Rise (Heat) Detector—A fire detection device that initiates a signal when the temperature rises at a rate exceeding a predetermined level, rather than a fixed temperature limit. Most rate-of-rise (ROR) detectors are based on the principle of air expansion when heated. The rate-of-rise heat detector uses confined air in an airtight hollow chamber that causes a pressure increase when heated. This pressure increase is arranged to cause an electrical contact to operate to send an alarm signal. To prevent false alarms from normal heat increases, a compensating vent is normally provided. *See also* **Heat Detector.**

Rating Bureau—An insurance office that determines fire insurance rates for a given locality or area. The insurance rates are primarily based on the historical frequency of fires, an amount of claims, and the level of protection provided. *See also* **Fire Insurance; Insurance Services Office (ISO), Inc.**

Ratio Flow Controller—A foam proportioner that is able to meter the correct amount of foam concentrate into a water stream over a wide range of flows and pressures. The ratio flow controller uses a modified venturi to adjust to flow conditions. As water passes through the inlet jet of the venturi, a reduction in the annular area is created. This reduction allows the metering of foam concentrate into the water stream through the foam concentrate metering orifice. *See also* **Proportioner.**

Rattle Watch—Name for the nighttime patrol established for the city streets in New York (then called New Amsterdam) in 1648 to look out for fires. Upon discovery of a fire, the watchman would swing a large wooden rattle above his head as a warning. Other watchmen on duty would hear the alarm and also swing their rattles until the local church bell was rung, awakening the populace. Because firefighting efforts at that time were mostly manual methods (bucket brigade, hand pumpers), the more personnel available to contribute their manpower, the greater the probability of success in extinguishing the fire. Since the initial fire alarm was given by the twirling of a rattle, the watchmen consequently became known as the Rattle Watch. The creation of the Rattle Watch is considered in some fashion the first organized police and fire department in America.

RECEO-VS—A widely used acronym created by Chief Lloyd Layman to remember strategic priorities. **Rescue, exposures, control, extinguishment,** and **overhaul** are the five prioritized strategies. **Ventilation** and **salvage** occur as and when needed.

Refractive Index—A method of examining produced foam to determine whether it has been accurately proportioned. Light entering a liquid will bend because of refractive properties (change in direction of electromagnetic waves from one medium to another). By measuring the amount of refraction, the density of the material can be determined. A handheld refractometer is used to measure the refractive index of the foam solution samples. This method is not particularly accurate for AFFF or alcohol-resistant AFFFs, since they typically exhibit very low refractive index readings. The refractometer method should also not be used when testing foam percentages of 1 percent or lower because the accuracy, at best, for determining the percent of foam concentrate in a foam solution when using a refractometer is +/– 0.1 percent. For this reason, the conductivity method is a preferable test method where AFFF, alcohol-resistant foam, or foam in 1 percent or less concentration (Class A foams) is to be tested. *See also* **Conductivity Method; Refractometer.**

Refractometer—An instrument used in the fire protection industry to measure the refraction index of produced foam against a predetermined scale to determine the amount of proportioning that has been applied. A small sample of the produced foam is taken for examination to observe the amount of light refraction based on its proportioning percentage. It usually consists of a small, handheld optical device through which the refractive level is viewed from the sample taken. *See also* **Conductivity Method; Refractive Index.**

Refractory—Any material not easily affected by heat, such as firebrick. Refractory materials are commonly used as a lining for furnaces and as fireproofing materials. *See also* **Firebrick; Fireproofing.**

Rekindle—A return to flaming combustion conditions after incomplete extinguishment. Residual heat and hidden embers may reinstate flaming fire conditions after a fire has been declared extinguished if overhauling isn't comprehensive or correctly performed. Rekindling commonly occurs in attics, basements, and walls of structural fires or in wildland fires.

Relay Operation—The transmission of firewater from one pumper to another (in series) to move water to a fire incident. Relay operations are generally required where the water source is a long distance from the fire incident, the water supply is of low pressure, or there is a significant elevation difference to overcome from the water source to the fire location. Pump relays may be open or closed. An open relay consists of one pumper supplying water into a open container from which another pumper is drafting. Closed relays consist of one pumper pumping directly into the suction inlet of another pumper. Excessive pressure and flow are controlled at each pump within the system. Large diameter hose (LDH) is commonly used for relay operations. It may also be called a pump relay or relay pumping. See Figure R-2. *See also* **Water Relay.**

Relief Valve—A pressure control device on a pump to prevent excessive pressures by releasing it to a safe discharge location. Relief valves are arranged to automatically bypass or dump water when the desired pump pressure is exceeded to avoid hazardous conditions.

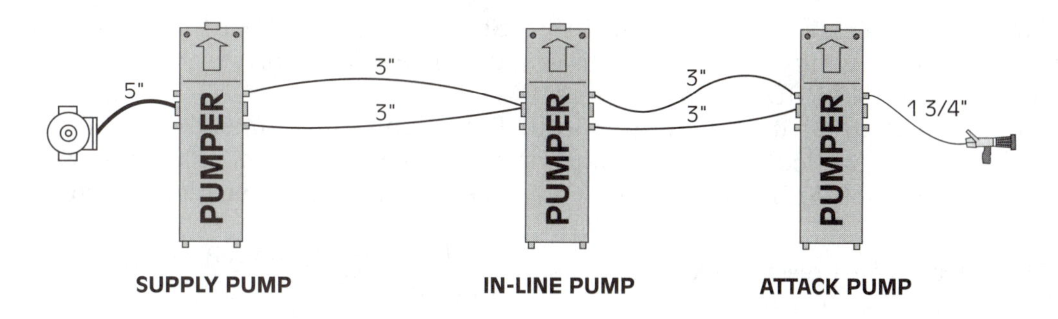

Figure R-2 Example of pump relay operation.

A common arrangement is to direct the excess discharge pressure of the pump via the relief valve back to the intake side of the pump or to atmosphere. *See also* **Pressure Governor.**

Remote Impounding—The diversion and collection of spilled material to a safe location away from the site of the spillage to prevent it from contributing to damage to the site and allow its cleanup/removal. *See also* **Diverting.**

Remote Supervising Station System—A fire alarm system from one or more protected premises that is remotely monitored. See Figure R-3.

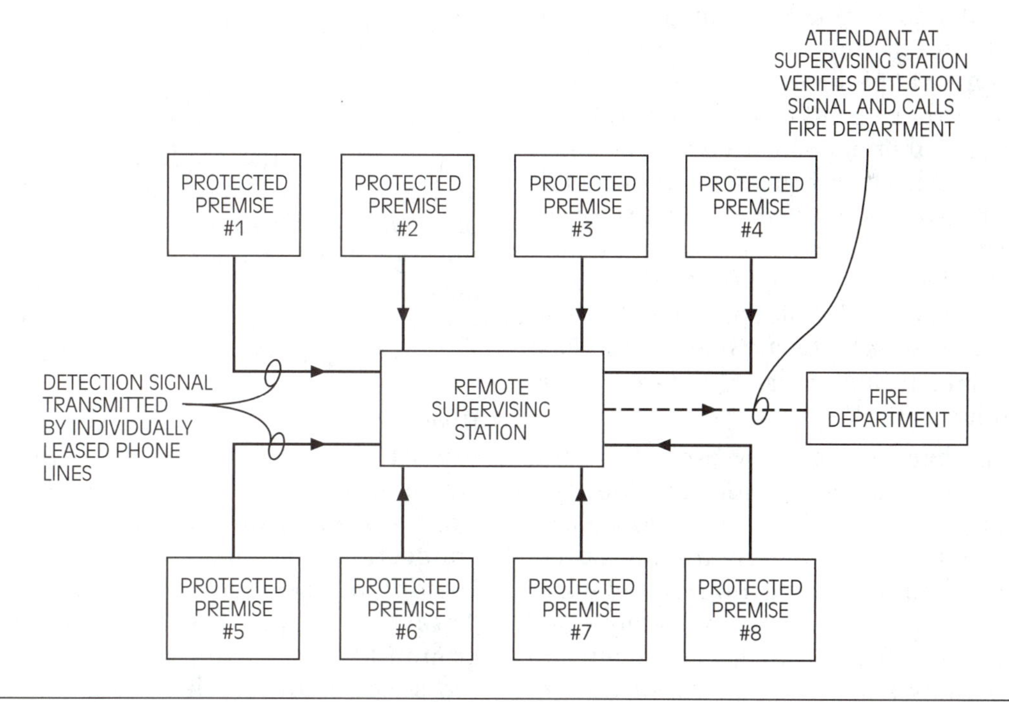

Figure R-3 Remote supervising station system.

Required Delivered Density (RDD)—A sprinkler's delivered density required to achieve fire suppression. The value of RDD depends on the fire size at the time of application of water and is also a measure of a particular hazard's suppressability. *See also* **Actual Delivered Density (ADD).**

Residual Pressure—The pressure existing at a given point in a piping distribution system or hose line at a specific flow rate. The measurement of residual pressure for fire protection applications is necessary to determine if fire protection devices can be adequately supplied with water in an emergency when large volumes of water are being used. *See also* **Static Pressure.**

Response Time Index (RTI)—A relative measure of the sensitivity of an automatic fire sprinkler's thermal element as installed in a specific sprinkler. It is usually determined by plunging a sprinkler into a heated laminar airflow within a test oven. This type of "plunge" test is not currently applicable to certain sprinklers. These sprinklers must have their thermal sensitivity determined by other standardized test methods. A response time index is also used to quantify the responses of heat detectors used in a fire detection system. A normal RTI for a sprinkler is 300. Early suppression fast response (ESFR) sprinklers have an RTI of 50 or less. *See also* **Early Suppression Fast Response (ESFR) Sprinkler; Quick-Response (QR) Sprinkler.**

Retard Chamber—A small vessel provided between the outlet of a automatic sprinkler system alarm check valve and the water flow alarm (pressure switch) for the system. It allows pressure surges and normal fluctuations in the water supply to dissipate their energy before they falsely activate a fire alarm signal. They are designed to allow a specific amount of time to occur before activation of an alarm condition. See Figure R-4. *See also* **Alarm Check Valve; Waterflow Detector.**

Retention—Containment of a spill by the provision of a hole large enough to divert and collect it. *See also* **Diking; Diverting; Remote Impounding.**

Reverse Lay—The laying of fire hose in a direction from a fire incident to a water source (fire hydrant) from a hose cart or fire truck. A reverse lay allows a fire pumper to feed hose lines with adequate flow and pressure from its own pump using a short suction hose to the water supply. The manner in which a fire hose is stored for use should account for the method of hose lays to be used and the type of hose couplings (male or female) available at each end of the hose. It may also be called a fire-to-hydrant lay. See Figure R-5. *See also* **Forward Lay; Hose Lay.**

Rim Seal Fire Protection—The provision of fire protection measures to the flexible seal area of an open floating roof storage tank for combustible or flammable liquids. Fire protection measures usually consist of the application of a fixed foam delivery system for the control and suppression of combustible liquid fires at the seal area through foam application. Guidance in the design and operation of rim seal fire protection systems is provided in NFPA 11, *Standard for Low Expansion Foam*. *See also* **Foam Dam.**

Riser—Main vertical supply piping used in fire protection systems. The riser is the main vertical pipe header of a fire suppression distribution system from which branch lines or connections are made. Sprinkler risers and standpipe risers are common examples that "rise" or supply water vertically from the ground level to the top of a structure to various cross-feed mains or hose connections, respectively. *See also* **Dry Riser; Fire Main.**

R

Figure R-4 Retard chamber (at left side, sprinkler valve inlet to water motor gong).

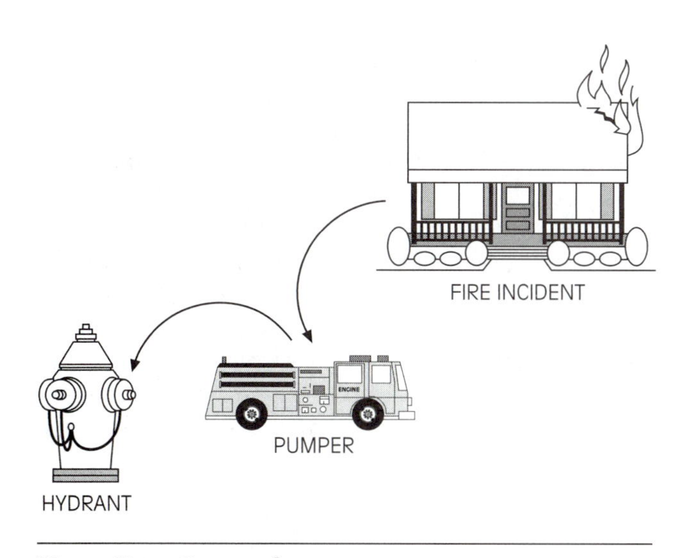

Figure R-5 Reverse lay.

Risk—An assessment (loss of life, property damage, business economic interruption, environmental impact, etc.) of both the probability and consequence of all hazards (explosion, fire, smoke exposure, etc.) of an activity or condition, that is, R = *f* {P, C}. In the insurance industry, "risk" refers to the person or thing insured.

Risk Acceptance or **Risk Retention**—The philosophy of risk management that an entity will bear the responsibility and subsequent consequences for any loss that occurs. *See also* **Risk Management Techniques.**

Risk Area—A defined area where a fire hazard is expected to be contained, due to spacing, fire barriers or breaks, fixed protection, etc. Primarily used for the purposes of estimating firewater flow requirements of a facility in order to design the size of the fixed firewater pumping systems that are required to ensure the risk area is adequately contained, etc.

Risk Assessment—The amount or degree of potential hazard perceived by a given set of parameters and operating conditions. Risk assessment provides a means to identify and rank risks and to obtain information on their extent and nature. By performing a risk assessment, suitable risk control measures can be identified and assessed for adequacy. Risk assessment involves identification of hazards, probability of occurrence, severity, and subsequence consequences. Risk assessment may be either qualitative or quantitative in nature. It may also be called a probabilistic risk assessment (PRA) where probabilities of occurrence are used extensively in the analysis. *See also* **Qualitative Risk Analysis; Quantitative Risk Analysis (QRA).**

Risk Avoidance—The philosophy of risk management that an entity will not enter into an operation that poses an exposure of loss or will reduce the exposure to a potential loss. It is primarily used in risk management practices for corporate cost-benefit analysis for insurance applications. *See also* **Risk Management Techniques.**

Risk Control—The provision of suitable measures or elements to eliminate or control real or potential hazards. *See also* **Risk Management; Risk Management Techniques.**

Risk Engineer—An individual qualified to identify potential hazards and consequences of the hazards, assess the probability of those hazards occurring, and determine appropriate hazard protection measures based on a cost-benefit analysis and regulatory requirements.

Risk Insurance—The provision of financial reimbursement through an agency for an economic loss for specific conditions and periods for the payment of premium (annual fee based on the hazard of the risk involved).

Risk Management—The process that combines information from risk assessment with economic, political, legal, and ethical considerations to make decisions. *See also* **Risk Control.**

Risk Management Techniques—Various methods available to handle a hazard that primarily include risk acceptance, risk avoidance, risk control, and risk reduction. *See also* **Risk Acceptance** or **Risk Retention; Risk Avoidance; Risk Control; Risk Reduction.**

Risk Reduction—The lowering of a loss exposure through the provision of risk avoidance, prevention techniques, loss control measures, or risk financing instruments. *See also* **Risk Management Techniques.**

Room Design Method, Sprinkler—A method of designing and engineering a fire protection sprinkler system based on only the sprinklers in a particular room. Application of a room design method places more emphasis on firewall ratings and protection of openings that exist in the firewalls.

Rotary Pump—A pump designed for positive displacement pumping action. A rotary gear pump consists of two gears that rotate within the walls of a circular housing. The inlet and outlet ports are on opposite sides of where the gear teeth mesh together. Fluid is drawn into the clear space between the meshing gears and its pressure is subsequently increased by the rotation of the gears acting together. Rotary pumps are commonly employed in the fire protection industry where viscous fluids (foam concentrates) are involved. *See also* **Fire Pump; Foam Pump.**

Rubber-Covered Hose—A fire hose constructed with an inner liner, an intermediate reinforced jacket, and an outer rubber cover. The inner liner is typically made from a synthetic rubber or special thermoplastic compound. The intermediate jacket is a synthetic fiber weave, which is then covered by a rubber cover. *See also* **Double-Jacketed Hose.**

Rundown—The capability of a fire protection water spray applied to a vertical surface (typically a vessel) to flow downward and cover those areas that do not have a direct water spray applied to them, or to supplement water sprays that are applied at the bottom surfaces of a vessel. In some cases, the most vulnerable portion of a pressure vessel from a fire is the area that does not have a liquid quantity in it to absorb heat and may be highly vulnerable to rupture. Therefore, only the top portion of the vessel may be provided with immediate protection, and rundown of the water may be left to provide protection of the lower portions that may be considered less vulnerable. NFPA 15, *Standard for Water Spray Fixed Systems for Fire Protection,* does not consider protection from rundown for spherical or horizontal vessels adequate to provide a wettable surface for protection on their lower half, and recommends direct water spray nozzles where such protection is required.

Running Fire—A fire from a burning liquid fuel that flows by gravity over surfaces to a lower elevation. The fire characteristics are similar to a pool fire, except that it is moving or draining to a lower level. *See also* **Pool Fire.**

Runoff—Water from a fire protection applications that drains outside the fire area and has not absorbed sufficient heat from the fire to vaporize. Runoff may cause additional water damage, and if it contains hydrocarbon liquids that have a lower density than water (they float on its surface), the runoff may cause the fire incident to spread outside the initial fire area.

Safe Refuge—A location free of risk from an incident of concern. An area designated as safe refuge has the capability to provide protection against the designated emergency, such as fire, explosion, toxic gases, etc. *See also* **Evacuation; Shelter-In-Place (SIP).**

Safety Blowout (Backfire Preventer)—A protective device that provides a bursting disc for excessive pressure release, a means for stopping a flame front, and an electric switch or other release mechanism for actuating a built-in or separate safety shutoff normally located in the discharge piping of large mixing machines. A check valve, signaling means, or both may also be incorporated. *See also* **Backfire Preventer (Arrester).**

Safety Can—A small metallic waste disposal can, usually with a 5 gallon (18.9 liter) capacity with a spring closing lid, that will safely relieve pressures when exposed to a fire.

Safety Lamp or **Davy Lamp**—A lamp used in areas of combustible gas that prevents ignition of the gas by the provision of a wire screen that encloses the lamp flame. The wire screen absorbs the heat of the (oil lamp) light source before it can contact a gas, thereby preventing its ignition. It was invented by the British chemist Sir Humphry Davy (1778–1829) in 1815 for use by coal miners where firedamp was present. George Stephenson (1781–1848), a British inventor and engineer, also independently invented a similar miner's safety lamp at about the same time, but shared credit for this invention with Sir Humphry Davy. *See also* **Flame Arrester.**

Saint Elmo's Fire—The glow accompanying the brushlike discharges of atmospheric electricity that usually appears as a tip of light on the extremities of such pointed objects as church towers or the masts of ships during stormy weather. It is commonly accompanied by a crackling or fizzing noise. St. Elmo's fire, or corona discharge, is commonly observed on the periphery of propellers and along the wing tips, windshield, and nose of aircraft flying in dry snow, in ice crystals, or near thunderstorms. Various flight procedures, in addition to mechanical and electrical devices designed to reduce the charge accumulation, are utilized as safeguards in preventing or minimizing discharges. The name St. Elmo is an Italian corruption, through Sant Ermo, of St. Erasmus, the patron saint of Mediterranean sailors, who regard St. Elmo's fire as the visible sign of his guardianship over them.

Salamander—A mythical animal having the power to endure fire without harm. Also referred to as an elemental being, in one of the theories of Paracelsus, that inhabits fire. Paracelsus declared that when wood burns, "that which burns is sulfur, that which vaporizes is mercury, and that which turns to ashes is salt." *See also* **Fire; Theory of Paracelsus.**

Salamander Fire Protection Engineering Honor Society—An honorary fire protection engineering organization dedicated to the promotion of the original investigation and the objective of fraternal association within the profession. It was originally established at Illinois Institute of Technology in 1922. Salamander provides for three classes of membership. They consist of active, alumni, and honorary members. The active members are selected from students enrolled in Fire Protection Engineering studies. These members automatically obtain alumni membership upon graduation. Honorary membership is bestowed only on those individuals who have distinguished themselves in the discipline of Fire Protection Engineering through academic excellence.

Salvage—The methods by which firefighters protect items of value and the interiors of buildings from fire, smoke, and water damage. Objects are covered with waterproof covers, and water is removed by water vacuums, mops, squeegees, water chutes, and portable pumps. Almost all fire departments carry salvage equipment in their apparatus. Fire departments in some large cities maintain special salvage companies. *See also* **Loss Control; Smoke Curtain.**

S

Sanborn Map—A scale drawing, formerly used by insurance agencies, that indicated all structures on an insured property, depicting construction characteristics, occupancy, sources of fire protection, and exposures, or "COPE." They are no longer used by insurers. They were prepared to help visualize and explain the features of a property for insurance underwriting purposes and provided a permanent record of these features for future reference. Standard symbols and abbreviations were used to facilitate efficiency in their preparation. They were named after the Sanborn Map Company. In Europe, Goad maps were used that were prepared on a larger scale. Some insurance agencies prepared their own maps for their insured properties.

Schedule System—A method of design for fire protection sprinkler systems in which the pipe sizes are predetermined by the occupancy classification and defined in a table. A given number of sprinklers are allowed to be fed from a specified pipe size. NFPA 13, *Standard for the Installation of Sprinkler Systems,* delineates scheduled sprinkler systems.

Scorch—The affects to a material surface, such as discoloring (browning or blackening) by limited carbonization or texture damage (drying or shriveling), from heat exposure. Removal of the heat exposure stops further destruction.

Seal Fire—A fire that occurs at the flexible sealing membrane of the floating roof of a combustible open-top floating roof storage tank. Most seal fires are ignited by lighting strikes. Seal fires can be extinguished with portable fire extinguishers, foam hand lines, foam monitors, and fixed foam application systems. *See also* **Seal Protection.**

Seal Protection—A method of floating roof storage tank fire protection where fire suppression foam is provided to the storage tank seal area from a foam delivery system. A foam dam is provided on the floating roof in parallel with the seal area, and low expansion foam is provided to the fill the volume created by the dam and seal area for fire extinguishment. *See also* **Foam Dam; Seal Fire.**

Search (Primary, Secondary) by Firefighters—Primary: quick search for viable victums, routine function on all fires, before under control; secondary: thorough search of buildings by firefighters after fire is under control.

Seat of a Fire—The location of the main body of a fire (flames, heat, combustion reaction, etc.).

Secondary Exit—An alternative escape route from a structure or building. It is provided in addition to the required exits and is used in case some unforeseen condition prevents the use of the primary designated exits. *See also* **Exit.**

Secondary Fire—*See* **Spot Fire.**

Self-Expelling Extinguisher—A portable fire extinguisher in which the firefighting agent has sufficient vapor pressure at normal operating temperatures to expel the agent when it is released through the extinguisher nozzle. A carbon dioxide (CO_2) portable fire extinguisher is an example of a self-expelling extinguisher. *See also* **Fire Extinguisher, Portable.**

Self-Extinguishing—The inherent characteristic of a material undergoing a combustion process to extinguish itself once the source of ignition is removed, without the available fuel or oxidizer being depleted.

Self-Heating—The result of exothermic reactions, occurring spontaneously in some materials under certain conditions, whereby heat is liberated at a rate sufficient to raise the temperature of the material.

Self-Igniting Rock—Rocks that contain minerals that cause the rocks to self-heat and ignite due to chemical oxidation and spontaneous combustion.

Self-Ignition—Ignition resulting from self-heating. Synonymous with spontaneous ignition. *See also* **Spontaneous Heating.**

Self-Ignition Temperature—The minimum temperature at which the self-heating properties of a material lead to ignition. *See also* **Ignition Temperature; Kindling Temperature.**

Self Insurance or **Self Insured**—An arrangement whereby an entity assumes its own financial risk.

Self-Sustained Burning—Combustion or burning that continues to propagate itself without an outside ignition source.

Semi-Subsurface Foam Injection—The protection of combustible liquid storage tank through the application of foam from a floating hose contained at the bottom of the tank, that rises to the surface and distributes foam at the time of application. *See also* **Subsurface Foam Injection.**

Setchkin Test—A test to determine the flash ignition temperature of polymeric materials in ASTM D1929, *Standard Test Method for Ignition Properties of Plastics,* or BS 476, Part 4. A specimen is placed in an electrically heated tubular furnace in a controlled airflow. The temperature at which the gases ignite in the presence of a spark is called the flash ignition temperature. Named for N. P. Setchkin, who developed the test. *See also* **Flash Ignition Temperature.**

Sewer Gas—*See* **Gas, Sewage.**

Shell Hemispherical Vapor Cloud Explosion Model—A mathematical model of a vapor cloud explosion blast. It provides a plot of peak pressure wave duration as well as a plot of peak overpressure with a series of curves to account for various deflagration strengths as well as detonation. Developed by the research office of Shell Petroleum. The Shell model provides practical advice for the selection of an appropriate overpressure for congested regions of a process plant by means of a flowchart rather than as a separate series of overpressure curves, usually encountered in other explosion models. *See also* **TNO Multi-Energy Vapor Cloud Explosion Model.**

Shelter-In-Place (SIP)—A method of protection in which, instead of escaping from a fire risk (because avenues of escape are unavailable or time consuming), an individual protects the immediate vicinity to avoid injury. Shelter-in-place cannot be utilized where available oxygen supplies are insufficient, cannot be isolated from contaminants or barriers between the occupants and the incident are not adequate to withstand the exposure (i.e., fire, blast, etc.). *See also* **Fire Shelter; Safe Refuge.**

Shock Wave—A pressure pulse formed by an explosion in which a sharp discontinuity in pressure travels as a wave at a supersonic velocity through any elastic medium such as air, water, or a solid substance. Shock waves differ from sound waves in that the wave front, in which compression takes place, is a region of sudden and violent change in stress, density, and temperature. Shock waves travel faster than sound, and as their speed increases, the amplitude is raised. The intensity of a shock wave decreases very quickly as it travels because some of the energy of the shock wave is expended to heat the medium in which it travels. The amplitude of a strong shock wave, as created in air by an explosion, decreases almost as the inverse square of the distance until the wave has become so weak that it obeys the laws of acoustic waves. *See also* **Blast.**

Shotgun (Sprinkler) System—A wet pipe sprinkler system with an electronic alarm system in place of a water motor gong, retard chamber, etc.

Shutoff Pressure—The pressure at the discharge of a centrifugal fire pump with the pump running at its rated speed in rpm but with no flow of water. The shutoff pressure should not exceed 140 percent of total pressure at its rated capacity. *See also* **Fire Pump Curve.**

S

Siamese Connection—A twin hose, Y-shaped connection provided to connect two hoses to a single outlet. A siamese connection is commonly connected to water-based piping networks to allow the connection of a mobile fire pumper. The connection of a mobile fire truck pump allows for a boost in the supply pressure to the system. For buildings, these are provided at the outside of the exterior wall at a location distant from the building or on a standpipe sidewalk connection at a location convenient to fire truck access to the building. They are connected to the beginning of the fire protection system that it supports. It is commonly called a fire department connection and is configured in a Y shape, resulting in the term siamese connection. They commonly have female connections on the two inlets and a male outlet connection. *See also* **Fire Department Connection (FDC).**

Sight Flow Connection or **Drain**—A metal fitting with a viewing window provided in a water line to visually verify the flow of water in the system. Provided when a closed connection to a drain is provided from a fire protection system and verification water flow is needed that otherwise cannot be readily observed. They may also be required at cooling water drain lines for diesel engines, for water-cooled right angle gear drives supporting fire water pumps, or at an inspector's test connection for a sprinkler system. *See also* **Inspector's Test (Valve) Connection.**

Single-Jacketed Hose—A fire hose constructed with an inner liner and an outer protective reinforced jacket. The inner liner is typically made from a synthetic rubber or special thermoplastic compound. The outer jacket is a synthetic fiber weave. The two are bonded together. *See also* **Double-Jacketed Hose.**

Single Point Failure (SPF)—A location in a system that if failure occurs it will cause the entire system to fail because backup or alternative measures to accomplish the task are not available.

Single Station Alarm Device—A smoke detector that incorporates its alarm and control circuitry in the same housing.

Siren—A high-pitched wailing sound readily distinguishable from whistles, horns, or other monotone audible devices and easily discernable from the confusion of other sounds. Sirens are commonly employed to announce an emergency condition.

Skin Fire—A flammable liquid fire, such as a spill on a solid surface, where the liquid is not present in a depth exceeding 1.0 in. (2.5 cm). *See also* **Pool Fire.**

Slow Burning—A characteristic of the burning rate when related to a particular fire test.

Small Orifice Sprinkler—A standard spray sprinkler fitted with a small orifice that generally discharges less water at the same water pressure than the standard sprinkler. It is normally used when a lower density is required or when high residual water pressure may be present.

Smoke—The gaseous products of the burning of carbonaceous materials made visible by the presence of small particles of carbon (the small particles that are of liquid or solid consistencies produced as a by-product of insufficient air supplies to a combustion process). Smoke particles are usually less than one micrometer, or one-millionth of a meter, in size, so small as to not be individually visible to normal human observation.

Smoke produced by different fires will vary enormously. Entrained air is usually the largest component of produced smoke, and provided some assumptions of a fire can be made, estimates on the rate of smoke production are possible.

Smoke travel is affected by the combustion particle rise, spread, rate of burn, coagulation, and ambient air movements (ventilation, obstacles, wind). Combustion gases tend to rise due to the heat of the fire, since they are lighter than the surrounding air. They will spread out when encountering objects such as a ceiling or structural member. Smoke particulates will readily penetrate every available opening, such as cracks, crevices, stairways, etc.

Rate of burn is the amount of material consumed during the fire incident for any given time period. Particle coagulation is the rate at which combustion particles gather into groups large enough to precipitate out of the air. Coagulation occurs continuously because of the mutual attraction of the combustion particles. Air movement will direct smoke particles in a particular pattern or direction. *See also* **Cold Smoke; Combustion, Incomplete; Fire Gases.**

Smoke Barrier—A continuous surface (wall, floor, HVAC damper, or ceiling assembly) that is designed and constructed to restrict the movement of smoke. A smoke barrier may or may not also have a fire resistance rating. Such barriers might have protected openings. *See also* **Damper, Smoke.**

Smoke Bomb—A device for generating smoke from a chemical source (a pyrotechnic device) to simulate fire conditions. Smoke bombs are used in confined spaces for testing and training purposes (testing smoke detection, smoke management systems, or firefighter training). They usually produce smoke at a standard rate and quality and can be supplied in various durations. Smoke bombs are sometimes called smoke candles. *See also* **Chemical Smoke; Cold Smoke.**

Smoke Chaser—Terminology used for a forest firefighter who is lightly equipped to enable him or her to get to a fire quickly.

Smoke Compartment—An area enclosed by smoke barriers on all sides, including the top and bottom.

Smoke Condensate—The condensed residue of suspended vapors and liquid products of incomplete combustion.

Smoke Control—The control of smoke movement by the use of the airflow by itself, if it is of sufficient velocity and application of air pressure differences of sufficient strength across a barrier. Dilution of a smoke environment by only supplying air and extracting air is not considered a method of smoke control within an enclosure for fire safety concerns. *See also* **Smoke Management System (SMS).**

Smoke Control System—A system to limit and direct smoke movement within a building to protect occupants and assist with evacuation measures. It consists of mechanical fans that are engineered to produce airflows and pressure differences within the building compartments to achieve smoke control. *See also* **Smoke Management System (SMS).**

Smoke Control Zone—The subdivision of a building to inhibit the movement of smoke from one area to another for the purpose of life safety, evacuation, and property protection. A smoke control zone can consist of one or more floors, or a floor can consist of more than one smoke control zone. Each zone is separated from the others by partitions, floors, and doors that can be closed to prevent the spread of smoke. *See also* **Smoke Management System (SMS).**

Smoke Curtain—Salvage covers placed around an area by the fire service to prevent the spread of smoke that may cause further damage. *See also* **Salvage.**

Smoke Damage—The harmful effects to property from the occurrence of unwanted smoke exposure and combustion gases, consisting of stains, odors, and contamination. Exposure of combustion gases and smoke in some locations may cause property damages higher than physical fire damages. High technology clean rooms and the food processing industry, for example, require high cleanliness standards for their products and smoke damage may cause harmful chemicals to deposited on the products, making them unsuitable for use or salvage.

Smoke Damper—*See* **Damper, Smoke.**

Smoke Density—The relative quantity of solid and gaseous airborne products of combustion in a given volume. *See also* **Flame Spread Index (FSI).**

Smoke Density Index (SDI)—*See* **Smoke Developed Rating.**

Smoke Detection—The sensing of the products of combustion and sending a signal or an alarm for the purpose of safeguarding life or property. Various devices are available that can sense the presence of smoke, which is considered evidence of unwanted combustion. *See also* **Linear (Beam) Smoke Detection (LSD) System; Smoke Detector.**

Smoke Detector—A device that senses visible or invisible particles of combustion. They are very effective for a slow smoldering fire and will generally provide an alarm before a heat detector. They may cause a false alarm or be ineffective if not sited and installed where air currents from ventilation or air conditioning systems are likely to carry smoke and other products of combustion away from the detectors.

Usually used to warn occupants of a building of the presence of a fire before it reaches a rapidly spreading stage and inhibits escape or attempts to extinguish it. On sensing smoke, the detectors emit a loud, high-pitched alarm tone, usually warbling or intermittent, and are usually accompanied by a flashing light. There are two types of smoke detector: photoelectric and ionization. Photoelectric smoke detectors utilize a light-sensitive cell in either of two ways. In one type, a light source (a small spotlight) causes a photoelectric cell to generate current that keeps an alarm circuit open until visible particles of smoke interrupt the ray of light or laser beam, breaking the circuit and setting off the alarm. The other photoelectric detector widely used in private dwellings employs a detection chamber shaped so that the light-sensitive element cannot ordinarily "see" the light source (usually a light-emitting diode, LED). When particles of smoke enter a portion of the chamber that is aligned with both the LED and the photocell, the particles diffuse or scatter the light ray so it can be "seen" by the photocell. Consequently, a current is generated by the light-sensitive cell, producing an alarm. See Figures S-1 and S-2.

Ionization detectors employ radioactive material—in quantities so tiny they are believed to pose no significant health hazard—to ionize the air molecules between a pair of electrodes in the detection chamber. This enables a minute current to be conducted by the ionized air. When smoke enters the chamber, particles attach themselves to ions and diminish the flow of current by attaching themselves to the ions in the air from the radioactive source. The reduction in current sets off the alarm circuit. See Figure S-3.

Photoelectric smoke detectors are relatively slow to respond and are most effective in sensing the larger smoke particles generated by a smoldering, slow-burning fire. Ionization detectors are much faster to respond and are best at sensing the tiny smoke particles released by a fast-burning fire. Ionization smoke detection is also more responsive to invisible particles (smaller than

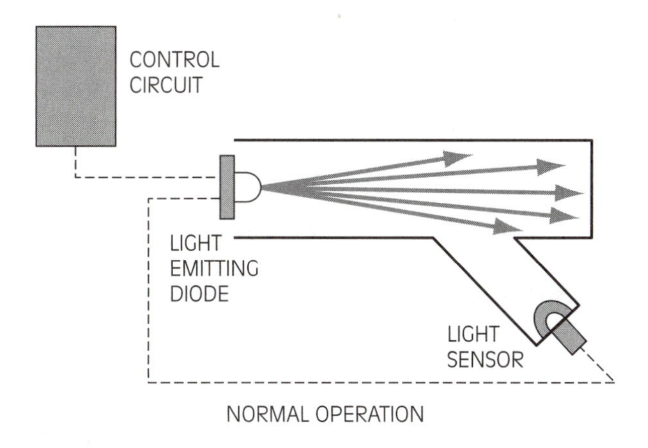

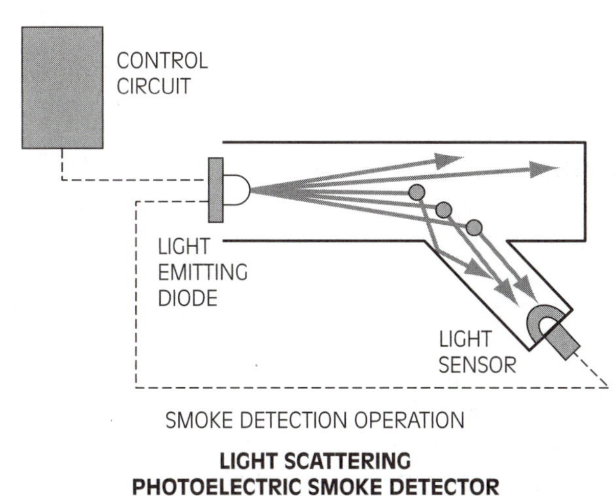

Figure S-1 Light-scattering photoelectric smoke detector.

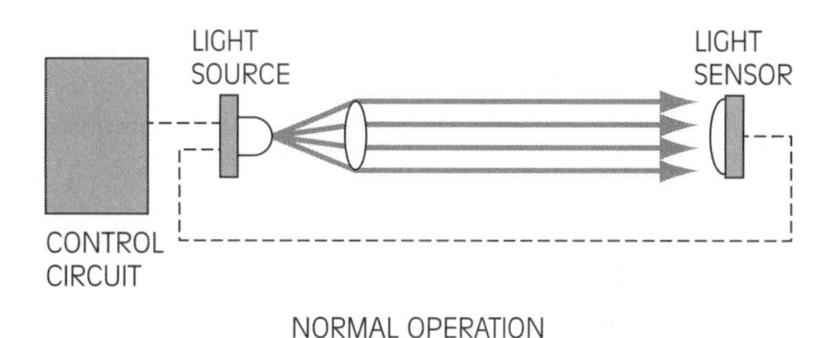

Figure S-2 Photoelectric light obscuration smoke detector.

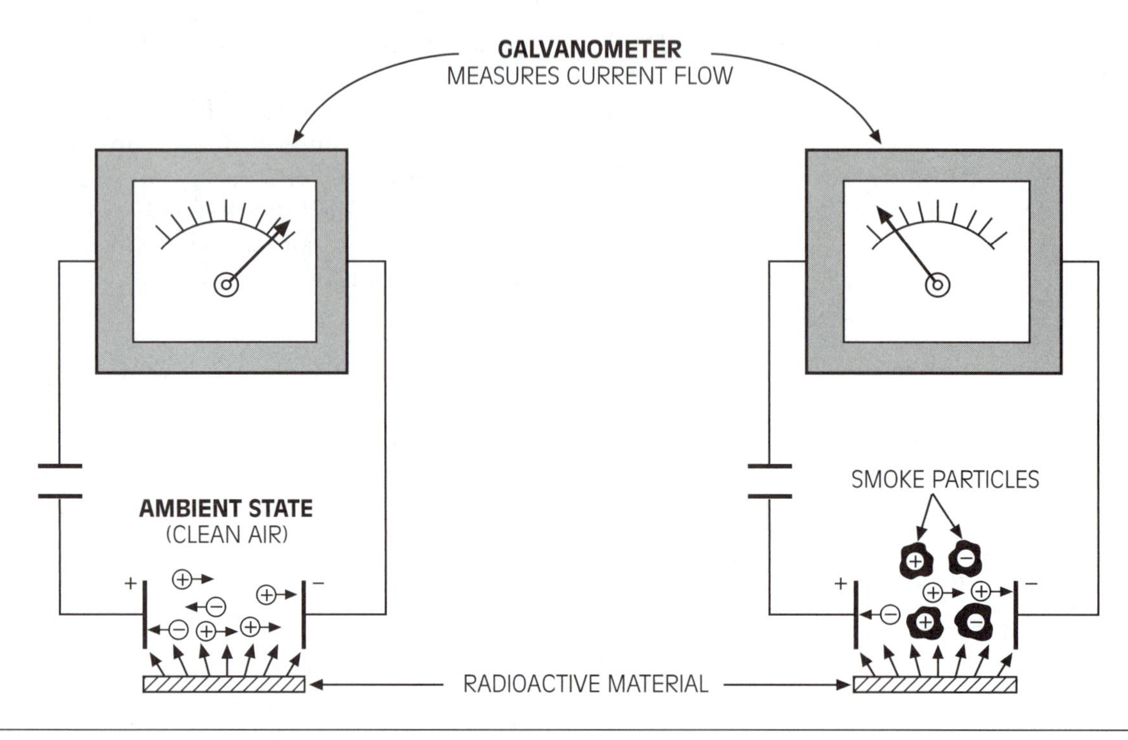

Figure S-3 Ionization smoke detectors, principle of operation.

1 micron in size) produced by most flaming fires. It is somewhat less responsive to the larger particles typical of most smoldering fires. For this reason, some manufacturers produce combination versions of detectors. Many fire prevention authorities recommend the use of both photoelectric and ionization types in various locations in a private home. Either type of detector can be powered by batteries or by household current.

Air sampling smoke detectors is a fire detection system where smoke is transported to and into a sampling port but they aspirate smoke into the detector sensing chamber rather than relying totally on outside forces (fire plume strength or environmental airflows). These systems actively draw smoke into the sensing chamber using suction fans. Smoke in the immediate vicinity of the sampling ports is drawn into the detector sensing chamber. Air sampling smoke detectors have been employed in zero gravity environments (space vehicles) to detect the presence of smoke.

The first smoke alarm was invented by W. Jaeger and E. Meili of Switzerland in 1941. The alarm was part of a project named Minerva Fire Alarm System. It was battery powered and had a flashing light and audible alarm when activated. It was also capable of sending a signal to the local fire station. *See also* **Aspirating Smoke Detection (ASD); Fire Alarm Signal; Fire Detection; IR Smoke Detection; Photoelectric**

(Light Scattering) Smoke Detector; Projected Beam Smoke Detector; Smoke Detection; Smoke Detector, Duct.

Smoke Detector, Duct—A device located within a building air-handling duct, protruding into the duct, or located outside the duct that detects visible or invisible particles of combustion flowing within the duct. Actuation of the device may allow operation of certain control functions. National and local fire codes recognize the hazard posed by building air-handling systems to spread smoke, toxic gases, and flame from one area to another unless they are shut down. The primary purpose of duct smoke detection is to prevent injury, panic, and property damage by preventing the spread (recirculation) of smoke. Duct smoke detection can also assist in protecting the air-handling system itself as well as sensitive equipment such as computer hardware. Duct smoke detectors may also be used to activate smoke exhaust dampers. Duct smoke detectors are not rated to be used as general area protection nor are general area detectors listed as duct smoke detectors. It may also be called a duct detector (DD). See Figures S-4 and S-5. *See also* **Smoke Detector; Very Early Smoke Detection and Alarm (VESDA) System.**

Smoke Developed Index (SDI)—A relative index of smoke produced during the burning of a material as measured by a recognized test. The smoke developed rating of materials is determined by NFPA 255, *Standard Test of Surface Burning Characteristics of Building Materials;* ASTM E-84, *Surface Burning Characteristics of Building Materials;* and UL 723, *Tests for Surface Burning Characteristics of Building Materials.*

Smoke Developed Rating—A relative index for the smoke produced from a building material test sample as measured and calculated by the Steiner Tunnel Tests (NFPA 255, *Standard Method of Test of Surface Burning Characteristics of Building Materials;* ASTM E-84, *Surface Burning Characteristics of Building Materials;* or UL 723, *Tests for Surface Burning Characteristics*

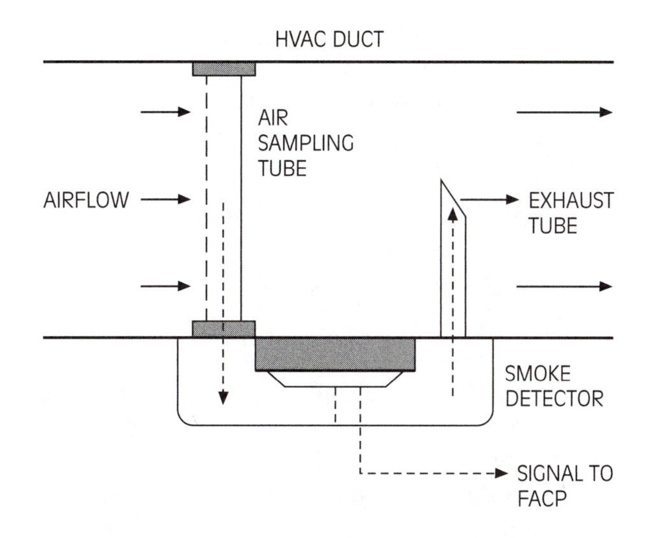

Figure S-4 Air sampling duct detector.

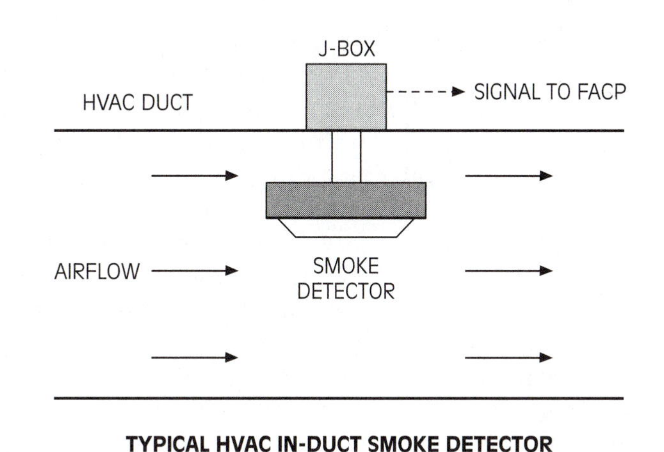

Figure S-5 In-duct detector.

S

of Building Materials). Red oak has a rating of 100, whereas cement board has a rating of zero. It may also be called a smoke density index (SDI). *See also* **Flame Spread Index (FSI); Fuel Contributed Rating.**

Smoke Eater—Slang terminology referring to a firefighter. It has been applied due to the consequences of firefighters inhaling smoke during a fire incident.

Smoke Ejector—A device similar to a fan used to exhaust heat, smoke, and harmful combustion gases from a post-fire enclosed environment and to induct fresh air to the affected enclosure. Smoke ejectors are usually carried as part of the complement of equipment on a firefighting vehicle and are commonly electrically powered. *See also* **Smoke Extractor.**

Smoke Exhaust System—Natural (chimney) or mechanical (fans) ventilation for the removal of smoke from an enclosure to its exterior. The provision of a tenable environment for human life is not considered within the capability of a smoke exhaust system. *See also* **Smoke Management System (SMS).**

Smoke Explosion—*See* **Backdraft.**

Smoke Extraction—The removal of smoke from an enclosed structure to aid firefighting operations. It is generally acknowledged that correct ventilation in fire conditions reduces (lateral) fire spread and resultant damage, enables firefighters to enter a building fire more easily, and provides greater visibility for firefighting activities. Manual efforts at the time of the fire may be employed, either through rapidly cut openings or the use of portable smoke extraction fans. Where smoke production may be anticipated in buildings, they are provided with automatic smoke extraction devices. *See also* **Venting, Fire.**

Smoke Extractor—Machine or fan blower device for extracting or removing smoke, heat, and gases from a building or an enclosure. See Figure S-6. *See also* **Smoke Ejector.**

Smoke Fans—*See* **Smoke Extractor; Hydraulic Ventilation.**

Smokehouse—A structure used to provide simulated smoke conditions for the training of firefighters. Training is provided in smoke environments where conditions can be monitored and observations made to improve performance. Simulated smoke is used as a safety measure and real fires are avoided. *See also* **Burn Building.**

Smoke Indicator Hole or **Inspection Hole**—A small triangular hole cut into the roof of a facility by the fire service during a response to a fire incident for the purposes of accurately determining the condition of the roof.

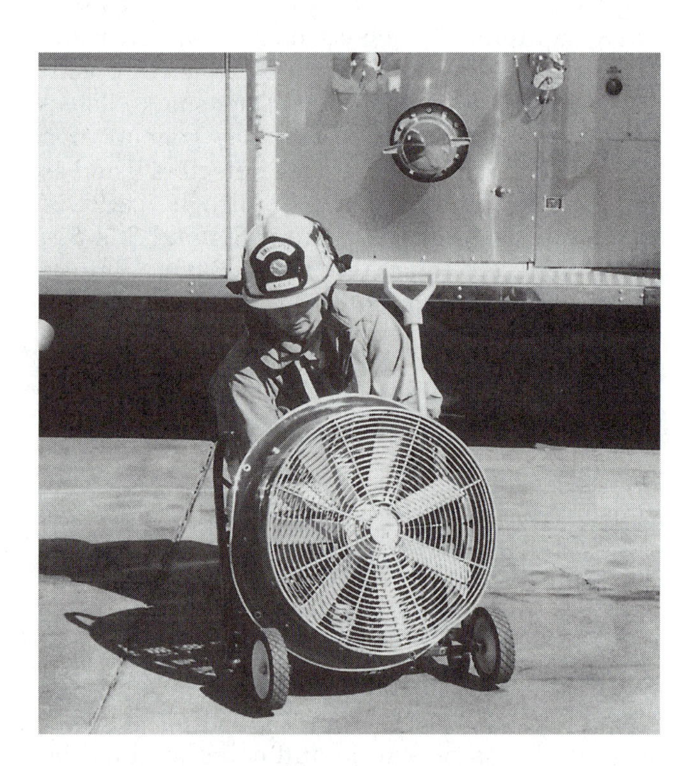

Figure S-6 Gasoline-powered ventilation fan.

S

Smoke Inhalation—The breathing of the combustion products into the lungs. It is considered an injury that damages the respiratory system. The main dangers of smoke inhalation to the lungs are the presence of narcotic gases, principally carbon monoxide (CO), hydrogen cyanide (HCN), carbon dioxide (CO_2), and the asphyxiating effects of an oxygen-depleted atmosphere. Inhalation of narcotic gases often leads to hyperventilation, leading to an increase in the amount of narcotic gases taken into the lungs. Narcotic gases also cause incapacitation by attacking the central nervous system. A low level of oxygen in the blood results in low oxygen levels to the brain, which causes impaired judgment and concentration. These effects may confuse, panic, or incapacitate an individual. Incapacity occurs in less than 10 minutes with a 0.2 percentage concentration of carbon monoxide (CO) if heavy activities are being performed. Carbon monoxide combines with the hemoglobin of the blood, preventing oxygen from binding with hemoglobin, which will cause death. Carbon monoxide has an affinity to hemoglobin 300 times that of oxygen. The degree of poisoning depends on the time of exposure and concentration of the combustion gases. If the percentage of carbon monoxide in the blood rises to 70 to 80 percent, death is likely to ensue.

Hydrogen cyanide is also referred to as hydrocyanic acid. The cyanides are true protoplasmic poisons, combining in the tissues with enzymes associated with cellular oxidation. They therefore render oxygen unavailable to the tissues and cause death through asphyxia. Inhaling concentrations of more than 180 ppm of HCN leads to unconsciousness in a matter of minutes, but the fatal effects would normally be caused by carbon monoxide poisoning after HCN has made the victim unconscious. Exposure to concentrations of 100 to 200 ppm for periods of 30 to 60 minutes can also cause death. Inhalation of hot smoke gases into the lungs will also cause tissue damage (burns) such that fatal effects could result in 6 to 24 hours after the exposure. Whenever the effects of smoke may affect individuals, protective measures must be provided, such as smoke management systems, smoke barriers, or fresh air supplies. *See also* **Combustion Gases; Fire Injury.**

Smoke Interface—The layer in a compartment that separates the smoke layer from the non-smoke layer. A smoke layer will gradually increase as the fire increases if the smoke layer is not vented, which lowers the smoke interface and may eventually fill the compartment. Fully developed fires have a smoke interface several centimeters (inches) above the floor. Cooling, or a decrease in the fire, may allow the smoke layer to dissipate and the interface will rise. May also be called smoke layer interface. *See also* **Neutral Plane.**

Smoke Jumper—A forest firefighter who parachutes to locations otherwise difficult to reach to combat forest fires.

Smoke Layer—The accumulated thickness of smoke in an enclosure.

Smoke Machine—A device used to create artifical non-toxic smoke for firefighter, industry or public evacuation training or drills.

Smoke Management System (SMS)—Natural or mechanical ventilation for the control or removal of smoke from an enclosure. Smoke management systems provide for smoke control to assist in personnel evacuation and firefighting activities. They provide pressurized areas within a building to prevent the entrance of smoke or to direct smoke to the outside of the building.

Smoke control systems can be designated as dedicated or nondedicated. A dedicated smoke control system is provided for smoke control only within an enclosure. It is a separate air moving and distribution system that does not function under normal building operating conditions. It is specifically designated for smoke control functions. A nondedicated smoke control system shares its components with other building systems such as the heating, ventilation, and air conditioning (HVAC) system. Activation of it causes the system to change its mode of operation from normal building HVAC requirements to that of smoke control. See Figure S-7. *See also* **Firestopping; Pressure Sandwich; Smoke Control; Smoke Control System; Smoke Control Zone; Smoke Exhaust System; Zoned Smoke Control.**

Smoke Pencil—A chemical solid that is ignited to produce smoke for testing purposes, primarily for the integrity testing of enclosures that are protected by fixed gaseous fire suppression systems.

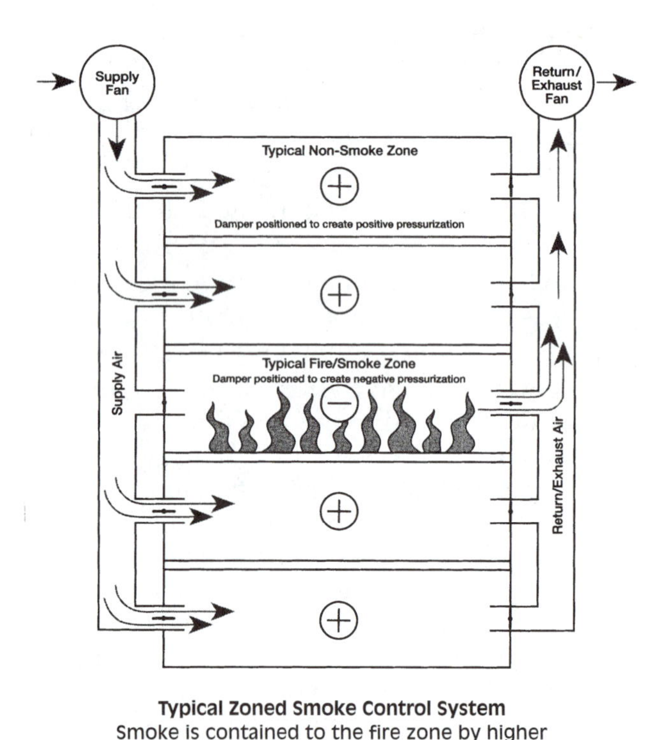

Figure S-7 Example of smoke management system. (*Reprinted with permission of Greenheck Fan Corp.*)

Smokeproof—Resistant to the spread of smoke.

Smoke Seal—A flexible membrane provided around the edge of a rated fire door frame. It is used to prevent the passage of smoke particles and combustion gases through the door seam surrounding a fire-rated door when it is closed.

Smoke Stop—A barrier provided to stop the spread of smoke to another area. *See also* **Curtain Boards; Draft Curtain.**

Smoke Test—A method for confirming the integrity of a chimney and for detecting any cracks in a masonry chimney flue, or deterioration or breaks in the seal or joints of a factory-built or metal chimney flue. Smoke is generated in a fireplace or solid fuel-burning appliance while simultaneously covering the chimney termination. Smoke leakages are then checked for through the chimney walls or suspected openings.

Smoke Visibility—The ability to perceive objects through smoke at a specific distance. Smoke visibility is necessary during fire conditions for evacuation of occupants, rescue operations, and firefighting activities. The ability to see through smoke is a measure of smoke visibility and can be related to the mass optical density (the yield of solid and liquid particulates of smoke generation). Smoke reduces visibility by a reduction in available light through the absorption and scattering of light by the smoke particulates. *See also* **Mass Optical Density.**

Smokey Bear—Symbolic character for the prevention of forest fires. The character is the property of the government of the United States of America, whose integrity is maintained by the Secretary for the United States Department of Agriculture (USDA). The Forest Service of the USDA operates the Smokey Bear program. Besides the Forest Service, the character is used by many US governmental agencies concerned with the prevention of careless forest fires, including the National Association of State Foresters, the Bureau of Land Management, and the Bureau of Indian Affairs. The character was originally created in 1944 by the USDA Forest Service and the Advertising Council to help prevent the useless destruction of timber resources needed for the WWII effort by reminding Americans to be careful with fire in national forests. The name Smokey was reputed to be taken from "Smokey Joe" Martin, a firefighter for the city of New York. In 1950, state and national forest rangers found a very badly burned black bear cub clinging to a charred tree in New Mexico's Lincoln National Forest, whom they thought would be appropriate to name Smokey Bear. The orphan bear cub was flown by Forest Rangers to Santa Fe, New Mexico. He was nursed back to health at the home of a game warden and was later sent to the National Zoo in Washington, DC. He was referred to as "the living symbol" of Smokey Bear. Posters and advertisements commonly promoted forest fire prevention showing Smokey Bear with the message, "Only You Can Prevent Forest Fires." The living symbol Smokey Bear was retired by the Forest Service in May 1975 and a second, younger Smokey Bear replaced him. The first Smokey Bear died in late 1976 and was buried in Capitan, New Mexico. The second bear died in 1991 and the living symbol phase was discontinued. The Smokey Bear character has never been retired and is continually used to emphasize to the public careless forest fire prevention practices in North America.

Smoking—The human use or possession of materials that are used for recreational inhalation of burning substances (cigarettes, cigars, pipes, etc.). Smoking materials represent an ignition source for unwanted fires if carelessly handled.

Smoldering—To burn and smoke without flame. It is a relatively slow combustion process (low rate of heat release) that uses very little oxygen. The hotter the environment, the less oxygen is required. Smoldering combustion occurs between available oxygen and the surface of a solid fuel. The fuel surface undergoes charring and glowing. A smoldering fire can develop into a diffused flaming fire due to a change in conditions, such as an increase of airflow to the smoldering surface, or after sufficient total energy has been produced. It is generally associated with substances with a high surface-to-weight ratio, such as saw dust, coal dust, and some foams, such as neoprene and polyurethane. Cigarettes carelessly discarded into upholstered furniture or bedding are a common source of smoldering fires in homes and offices. Smoldering fires in typical dwelling enclosures can produce a significant amount of carbon monoxide and can cause incapacitation in one to two hours.

Smoldering fires can continue to burn for many weeks, for example, in bales of cotton and jute and within heaps of sawdust. Smoldering is hazardous, as it produces more toxic compounds than flaming combustion per unit of mass burned, and it provides a chance for flaming combustion from a heat source too weak to directly produce flame. See Figure S-8. *See also* **Cold Smoke; Combustion; Fire Stages; Glowing Combustion; Holdover Fire.**

Smother—To suppress a fire by the exclusion of oxygen or by reducing the oxygen concentration below that required for combustion though the introduction of a barrier between the fuel source and oxygen source. Some materials may not be able to be smothered because they contain sufficient oxygen for self-combustion, such as cellulose nitrate. Common methods of smothering include a cover (blanket), layer of foam liquid, blanket of carbon dioxide (CO_2), steam, and other inert gases. The effectiveness of gaseous agents for smothering depends on

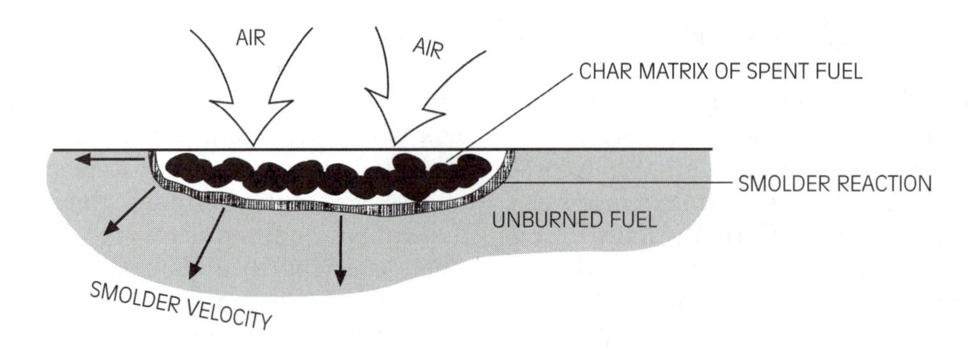

Figure S-8 Schematic of smoldering.

applying them at a high enough rate to completely fill the entire flame space at one time. If a hot surface is in the vicinity of a fuel smothered with a gaseous agent, the fuel will immediately re-ignite if the gas is removed or blown away.

Snorkel or **Snorkel Truck**—The snorkel truck is a vehicle specially equipped with a hydraulically operated elevating boom mounted on a turntable, for use in either firefighting or rescue work. A large capacity firewater nozzle is commonly provided at the end of the boom, providing a characteristic "snorkel" look to the arrangement. The nozzle is used to deliver large quantities of water for firefighting close to the upper levels of a building. It was introduced by the Chicago Fire Department in 1958 and named by Fire Commissioner Robert Quinn. It may also be called an elevating nozzle. *See also* **Elevating Nozzle; Ladder Truck; Water Tower.**

Snuffing Steam—Pressurized steam (water vapor) used to smother and inhibit fire conditions generally in the firebox of a heater or furnace due to a tube leak. It is commonly utilized in the process industries due to readily available steam supplies in the facility for process support. Steam extinguishes fire by the exclusion of free air and the reduction of oxygen content to the immediate area. It is most effective in relatively small confined areas. A standard on the use of snuffing steam has never been published, but Appendix F of NFPA 86, *Standard for Ovens and Furnaces,* provides some limited information on the general requirements in the design of a snuffing steam system. Snuffing steam presents a personal burn hazard from the superheated steam if directed on unprotected skin. *See also* **Steam Smothering.**

Society of Fire Protection Engineers (SFPE)—An international US-based organization that was originally established as a section of the National Fire Protection Association (NFPA) in 1950, but was later independently organized in 1971. Its stated purpose is to advance the art and science of fire protection engineering and its allied fields, to maintain high ethical standards among its members, and to foster fire protection engineering education. Chapters of the society are located in the United States, Canada, Europe, and Australia. Membership is available to individuals possessing engineering, physical science, or engineering technology degrees with appropriate experience and to those in associated fields and to individuals supporting SFPE goals and objectives. See Figure S-9. *See also* **Institution of Fire Engineers (IFE).**

Figure S-9 SFPE logo. *(Reprinted with permission of Society of Fire Protection Engineers).*

Soda-Acid Extinguisher—*See* **Chemical Extinguisher.**

Sodium Bicarbonate—A dry chemical fire suppression agent. It was the first dry chemical agent to be formulated and is effective on Class B and C fires. It has about 50 percent greater effectiveness than water when applied to the same fire. It is usually compounded with other particular materials to render the mixture water repellent and therefore capable of flowing from a pressurized container (portable fire extinguisher). *See also* **Dry Chemical; Multipurpose Dry Chemical; Potassium Bicarbonate.**

Soot—Fine black particles, chiefly composed of carbon, produced by incomplete combustion of coal, oil, wood, or other fuels. During combustion, a cracking or separation of molecules occurs due to the high temperatures involved. This cracking separates the carbon from other portions of the molecules. If a cold object is introduced into the outer portions of a flame, the temperature of that part of the flame is lowered below the point of combustion, and unburned carbon and carbon monoxide are given off. For example, if a noncombustible object is passed through a candle flame, it receives a deposit of carbon in the form of soot. The tendency to form soot is also partly dependent on the molecular structure of the fuel. Aromatic hydrocarbons show a greater tendency to form soot than do aliphatic materials. Soot formation increases the burning rate because the soot increases the flame radiation, which increases the burning rate. Soot will also contribute to the amount of smoke damage from a fire. *See also* **Ash.**

Spalling—Generally, the breaking away or explosion of concrete materials from a fire exposure. It occurs due to stresses set up by steep temperature gradients onto aggregates in the concrete that expand, or moisture that is trapped and vaporizes without any means of venting safely. Lightweight aggregates have a relatively steep temperature gradient and yield readily to the stress gradient; therefore, spalling rarely occurs in this type of concrete. Recent experimental research has indicated four types of spalling may occur: aggregate spalling, corner spalling, surface spalling, and explosive spalling. Each type has its own causative factors. Some of these factors have opposing influences on the different types of spalling, thereby causing confusion in the understanding of spalling. Aggregate spalling is thought to be a manifestation of sheer failure of individual aggregate particles local to the heated concrete surface. The susceptibility of concrete to aggregate spalling is related to the coefficients of thermal expansion of the aggregate and surrounding mortar, the size and thermal diffusivity of the aggregate, and the rate of surface heating. Corner spalling is sometimes referred to as sloughing-off. It is a consequence of concrete losing its tensile strength at elevated temperatures. Surface spalling generally arises because of excessive pore pressure generation within heated concretes. It is observed to increase under conditions of increased rate of heating, increased degree of pore saturation, and reduced permeability of concrete. Surface spalling may be alleviated by controlling the build up of pore pressures. Explosive spalling is the catastrophic failure of concrete elements and may be caused by either of two separate mechanisms. Excessive pore pressures within heated concrete can cause explosive spalling, but is normally limited to unloaded or small specimens. Explosive spalling may also occur as a result of the superimposition of thermal stresses induced by a fire on the ambient applied load stresses. Spalling reduces the fire resistance of concrete, since it reduces the amount of protective material available against a fire exposure.

Concretes are rated into two grades on the basis of fire resistance affected by aggregates—Grade 1 and Grade 2—when tested in accordance with ASTM E-119, *Standard Methods of Fire Tests of Building Construction and Materials* (Similar to NFPA 251, UL 263).

Spanner, Hydrant (Wrench)—A wrench or tool specifically designed and used to open and close a fire hydrant, remove or attach outlet caps, connect or loosen hose couplings, or as a small prying tool. See Figure S-10.

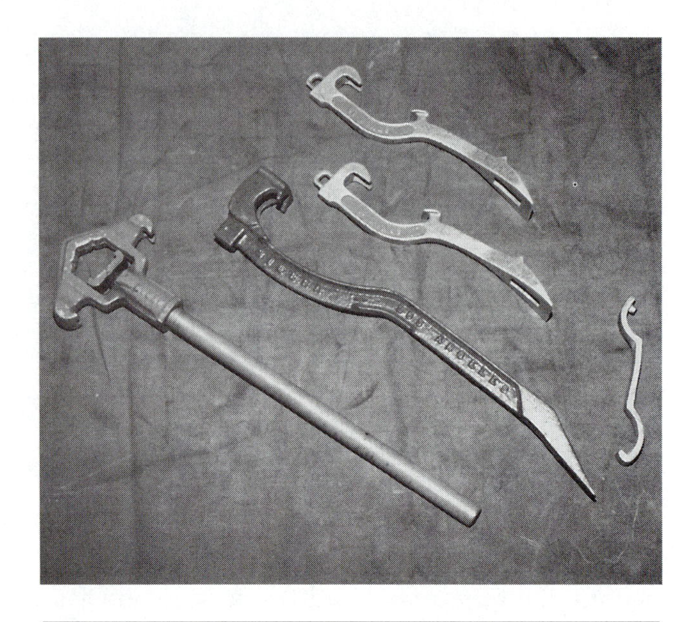

Figure S-10 Hydrant wrench spanners.

Spark—A small, incandescent particle from electrical failures or a moving ember from a fire source. During electrical fire investigations, the term spark is reserved for particles thrown out by arcs, whereas an arc is a luminous electrical discharge. If copper and steel are involved in arcing, the released spatters of melted metal begin to cool as they sail through the air. When aluminum materials are involved in the faulting, the particles may burn as they sail in the air and continue to be extremely hot until they burn out or they are extinguished when they fall on a material. Therefore, aluminum sparks have a greater ability to ignite fine fuels than do sparks from copper or steel. Sparks from arcs are generally an inefficient ignition source. Besides adequate temperature, the size of the particles is important for the total heat content of the particles and the ability to ignite a fuel. *See also* **Arc; Ignitable Liquid; Ignition Energy; Incendive Spark; Incendive Arc; Non-Incendive; Spark Ignition.**

Spark Arrester—Screening material or a screening device attached to an opening, such as an internal combustion engine, to prevent the passage of sparks or a chimney/burner stack termination to prevent emission of brands to the outside atmosphere. Arresters cannot eliminate spark hazards, but they do provide abatement measures. NFPA 211, *Standard for Chimneys, Fireplaces, Vents, and Solid Fuel-Burning Appliances,* states that openings should be greater than 0.5 in. (1.27 cm) in diameter and block the passage of spheres less than 0.375 in. (0.95 cm) where spark arrester screens are required on chimney outlets. *See also* **Fire Screen.**

Spark or **Ember Detector**—An optical radiant energy fire detector that is designed to detect sparks or embers, or both. These devices are normally intended to operate in dark environments and in the infrared (IR) part of the spectrum using a photodiode to sense a small amount of radiant energy (using an IR beam). They are intended to sense a spark or ember before the flaming stage of a fire. They are commonly employed in the dust collection systems in silo storage systems for protection against the initiation of a fire.

Spark Extinguishing System—A method of protection against sparks in which the radiant energy of a spark is detected and the spark is immediately quenched or extinguished.

Spark Ignition—The initiation of a fire or explosive event through the application of spark energy. The ability of a spark to produce ignition is governed largely by its amount of energy. Tests have shown that saturated hydrocarbon gases and vapors require approximately 0.25 millijoule of discharge energy for spark ignition of optimum mixtures with air. Unsaturated hydrocarbons can have lower minimum ignition energies. It has also been demonstrated that sparks arising from potential differences of less than 1,500 volts are unlikely to be hazardous in saturated hydrocarbon gases because of the short gap and heat loss to the terminals. Tests have demonstrated that dusts and fibers usually require discharge energy of one or two magnitudes greater than that of common gases and vapors for spark ignition of optimum mixtures with air (the ignition energy requirement diminishes rapidly with decreased particle size of dusts). Ignition energy can be reduced when an increase in oxygen concentration occurs relative to that of air. Tests have demonstrated that a mixture of a flammable gas to a dust suspension can lower the ignition energy of the dust, even if the gas is present at a concentration below its lower flammable limit. The mixture can be flammable even if both components are below their respective lower flammable limits. In such instances, a mixture can be ignited at an energy level approaching that of the most easily ignited component.[9] *See also* **Arc; Ignition, Direct Electric; Ignition Energy; Incendive Spark; Incendivity; Spark; Static Ignition Hazard.**

Spark Protected—Electrical equipment that will not emit a spark or will contain any spark generated by means of its container, insulation, shield, or screen.

Spark Resistant—Materials treated, made, or constructed to prevent the generation of sparks that could act as an ignition source. Spark-resistant or nonferrous materials have a flame spread index of 25 or less when tested in accordance with NFPA 255, *Standard Method of Test of Surface Burning Characteristics of Building Materials.* Air Movement and Control Association (AMCA) Standard 99-0401-86, *Classifications for Spark Resistant Construction,* provides informative material regarding spark-resistant fan construction.

Sparky the Fire Dog™—Symbolic character for the prevention of fires, primarily directed toward childhood education. He is a registered trademark and mascot of the National Fire Protection Association (NFPA). The character originated in 1951. He is shown as an adult Dalmatian dog appearing in firefighter protective clothing, helmet, and boots. A Dalmatian dog was chosen because they were traditionally kept by firehouses during the 19th century to direct horse drawn wagons and guard the firehouse. The name Sparky was commonly given to a firehouse dog because of the correlation of sparks from a fire. See Figure S-11. *See also* **Firehouse Dog.**

Special Services Fire Apparatus—A multipurpose fire service vehicle that provides support

S

[9] NFPA 77 Static Electricity.

Sparky the Fire Dog®!

Figure S-11 Sparky® the fire dog. (Sparky is a registered trademark of the National Fire Protection Association, Quincy, MA 02269. Reprinted with permission of the National Fire Protection Association, Quincy, MA 02269.)

services at emergency scenes. These services include rescue, command, hazardous material containment, air supply, electrical generation and floodlighting, or transportation of support equipment and personnel.

Specific Heat—The ratio of the quantity of heat required to raise the temperature of a body one degree to that required to raise the temperature of an equal mass of water one degree. The term is also used in a narrower sense to mean the amount of heat, in calories, required to raise the temperature of one gram of a substance by one degree Celsius. In the 18th century, the Scottish scientist Joseph Black (1728–1799) noticed that equal masses of different substances needed different amounts of heat to raise them through the same temperature interval. From this observation, he established the concept of specific heat.

Splash Shield or **Partition**—A metal shield or plate formerly provided between the fire pump and its electric motor driver to prevent water from leaking or spraying from the fire pump and damaging the electric motor. The National Fire Protection Association (NFPA) currently recommends that motors supporting firewater pumps and liable to water exposure be constructed and rated for such an environment.

Spontaneous Combustion—The outbreak of fire without application of heat from an external source. Spontaneous combustion may occur when combustible matter, such as hay or coal, is stored in bulk. It begins with a slow oxidation process (bacterial fermentation or atmospheric oxidation) under conditions not permitting ready dissipation of heat, such as in the center of a haystack or a pile of oily rags. Oxidation gradually raises the temperature inside the mass to the point at which a fire starts. Crops are commonly dried before storage or, during storage, by forced circulation of air, to prevent spontaneous combustion by inhibiting fermentation. For the same reason, soft coal in small sizes is wetted to suppress aerial oxidation.

The ignition points of some vegetable and animal oils are low. They oxidize so quickly that they generate a great deal of heat. If kept in a confined place, they may burst into flame. Fires may be caused by the spontaneous combustion of heaps of rags, paper, and similar materials that are soaked with oil. Coal and charcoal stored in large piles sometimes generate enough heat to set themselves on fire. Certain bacteria in moist hay may cause the temperature of the hay to rise rapidly and start a fire. A form of spontaneous combustion, hypergolic ignition, is used to fire a liquid-fuel rocket. Two liquids are pumped into the rocket combustion chamber: a chemical oxidizer and a fuel with which it reacts. On contact, they rise to ignition temperature. Through oxidation they burst into flame. Burning at a high temperature, the pressure they create provides the jet thrust that propels the rocket. *See also* **Fire; Hypergolic.**

Spontaneous Heating—The process whereby a material increases in temperature without drawing heat from its surroundings. Spontaneous heating of a substance to its ignition temperature results in spontaneous ignition or combustion. The process results from oxidation and is often aided by bacterial action where agricultural products are involved. *See also* **Self-Ignition; Spontaneous Ignition.**

Spontaneous Ignition—The initiation of combustion of a material by an internal chemical or biological reaction that has produced sufficient heat to ignite the material. White phosphorous will ignite when exposed to air, as will sodium and potassium if added to water. Spray paint and oil varnishes are capable of self heating when in contact with cellulosic materials such as rags and may lead to self ignition. Coal stacks and piles of grain can also oxidize and thus heat up raising the possibility of self-ignition. *See also* **Autoignition; Spontaneous Heating.**

Spot Burning—The localized burning of waste timber (slash). *See also* **Broadcast Burning.**

Spot Detector—A device whose detecting element is concentrated at a particular location. Examples are bimetallic detectors, fusible alloy detectors, local rate-of-rise and smoke detectors, and thermoelectric detectors. Spot-type detectors have a defined area of coverage. *See also* **Fire Detector.**

Spot Fire—Term used by the forest service to describe a secondary fire that is located away from the main fire area. It is usually started by embers or sparks that are carried by convection currents or natural wind and deposited away from the fire area.

Spray Pattern—The pattern produced by a divergent flow, the pattern varying with the nozzle pressure and adjustment to the spray creating device. The water spray pattern for a nozzle is important to ensure adequate coverage for fire protection.

Spray Sprinkler—The standard sprinkler for fire protection purposes, generally manufactured after 1953. Water is discharged from the deflector in a solid spray pattern in a horizontal direction (180-degree angle). Little or no water is directed upward to the ceiling and consequentially, heat absorption directly into the water spray is achieved. Upper-level cooling and area coverage per sprinkler was enhanced with the introduction of the spray sprinkler as opposed to the "old style" sprinklers being used. The discharge from a standard sprinkler with a standard orifice will provide coverage to a circle 16 ft. (4.9 m) in diameter, 4 ft. (1.2 m) below the sprinkler with a discharge of 15 gpm (0.95 liters/second). A standard sprinkler is available in an upright or pendent designation. The spray sprinkler was primarily developed and tested at the laboratories of Factory Mutual under the direction of Norman J. Thomson from 1947 to 1952, following development of fog and spray nozzles by the US Navy and Coast Guard during WWII. The NFPA standard on sprinkler systems was amended in 1953 to recognize the acceptance by the industry of the spray sprinkler. *See also* **Old Style Sprinkler; Sprinkler; Sprinkler Head; Sprinkler, Standard Orifice; Sprinkler, Upright; Standard Sprinkler Pendent (SSP); Standard Sprinkler Upright (SSU).**

Sprinkler—A water deflector spray nozzle device used to provide distribution of water in specific characteristic patterns and densities for the purpose of cooling exposures exposed to unacceptable heat radiation, and controlling and suppressing fires or combustible vapor dispersions. Water droplet size from a discharging sprinkler is one key factor in determining the effectiveness of its water spray. Water droplets penetrate a fire plume to reach a burning commodity by two modes: gravity and momentum. In the gravity mode, the downward velocity of the water droplets falling through a fire plume must be greater than the upward velocity of the fire plume for it to reach the base of the fire. Gravity action alone cannot accomplish this. Increased system pressure provides water droplets with greater downward thrust (momentum) to overcome the upward thrust of the fire plume. *See also* **Old Style Sprinkler; Spray Sprinkler.**

Sprinkler Alarm—A device arranged so that it is able to detect the flow of water from one or more sprinklers and will result in an alarm signal to be produced on the premises. It may also alert a constantly manned monitoring station of the sprinkler system activation. *See also* **Waterflow Alarm; Water Motor Alarm** or **Gong.**

Sprinkler, Concealed—A sprinkler that is recessed in the surrounding surface and provided with an unobtrusive cover plate to conceal the sprinkler from view for aesthetic reasons. The cover plate is usually soldered to the periphery of the sprinkler housing at three points. A small spring is also inserted to assist with separation of the cover plate from the sprinkler housing at the time of the fire. *See also* **Sprinkler, Flush; Sprinkler, Recessed.**

Sprinkler Control Valve—A valve provided on the riser, header, or feed main to a sprinkler system to isolate the system from the supply source. Sprinkler control valves are to be the indicating type, so they can be readily seen if the system has been isolated. Tamper switches are sometimes required to be fitted in order to alarm when a sprinkler system control valve has been closed.

Sprinkler, Cornice—Sprinklers provided for the fire protection of building cornices (overhanging decorative framing provided for windows) or for general building exposure protection. They are generally of the open type with a deflector plate that directs the water spray to a certain location on the outside surface of a building or structure.

Sprinkler Deflector—The portion of a sprinkler head that deflects the water stream into a spray pattern of desired shape, density, and distance. For common sprinklers, it consists of a circular metal plate attached perpendicularly to the sprinkler pipe connection, although for special sprinkler applications, unique sprinkler deflectors are provided (side wall sprinklers).

Sprinkler, Dry (Pendent)—An automatic sprinkler that is not provided with water continuously at its inlet. It is provided where freezing conditions are a concern if the sprinkler system is seasonally drained down. A seal is provided at the main supply pipe to prevent water from entering the sprinkler assembly until the sprinkler is activated from fire conditions. Typically, it is designed so that the fusible element opens the sprinkler and releases a spring-loaded tube that breaks the glass inlet water seal. This allows water to flow to the sprinkler.

Sprinkler Escutcheon—A circular decorative plate provided around the drop nipple to a sprinkler head, which extends below a ceiling. They cover the gap in the finished ceiling and afford a more aesthetic appearance to the fire sprinkler installation.

Sprinkler, Extended Coverage—A sprinkler that provides a larger area of coverage than do standard sprinklers. Normally, these sprinklers have higher flow rates, which allow them a greater spacing distance between sprinklers. *See also* **Quick-Response Extended Coverage Sprinkler.**

Sprinkler, Flush—A sprinkler in which all or part of the body, including the shank thread, is mounted above the lower plane of the ceiling. Mainly provided for aesthetic reasons or for protection against external impacts. *See also* **Sprinkler, Concealed; Sprinkler, Recessed.**

Sprinkler, Foam-Water—An open-type sprinkler that is designed to air aspirate the foam-water solution and deliver it in a spray pattern. It consists of an open barrel body with a deflected sprinkler mounted at the end.

Sprinkler Guard—A wire or sheet steel frame that is provided around a sprinkler to provide protection against mechanical injury to a sprinkler but that does not impact the water spray pattern of the sprinkler.

Sprinkler Head—A water spray device for the application of water onto a fire hazard in defined densities over a specific area. They are usually designed to activate individually by the provision of a heat-sensing element (fusible solder or glass bulb) that releases the contained water supply source to the sprinkler it supplies. A sprinkler may also be open without a heat-esponsive element. Sprinklers are specified for amount of coverage, orifice size, type of configuration (upright, pendent, sidewall, etc.), and temperature rating. Standard sprinklers have an orifice of 0.5 in. (1.27 cm). See Figure S-12. *See also* **Spray Sprinkler.**

Sprinkler, In-Rack—A sprinkler that is provided with a water shield to prevent its heat-responsive element from being wetted and cooled from water release from sprinklers located at the roof of the facility. It may also be called intermediate level or rack storage sprinklers.

Sprinkler, Old Style—*See* **Old Style Sprinkler.**

Sprinkler, On-Off—A cycling (on-off), self-actuating, snap-action, heat-actuated sprinkler. Water flow automatically shuts off from the sprinkler when the fire has been extinguished (no heat is available to activate the sprinkler head) and it is automatically reset for later operations. This type of sprinkler requires a water supply that is free of contaminants (potable) that could interfere with its operation. It does not have to be replaced after operation. It is provided to avoid water damage by eliminating the need to shut off the water supply after a fire has been extinguished. Typical applications include areas containing high-value inventories, materials, or equipment highly sensitive to water, areas subject to flash or repeat fires, and where the water supply is limited.

Sprinkler, Open—A sprinkler device that has a permanent open orifice and is not actuated by a heat responsive element. Instead, an upstream device controls waterflow from the sprinkler. Its primary purpose is to provide adequate distribution of water in a prescribed pattern.

Sprinkler, Pendent—A sprinkler designed for and installed with the head in a downward fashion from the piping, rather than placed in an

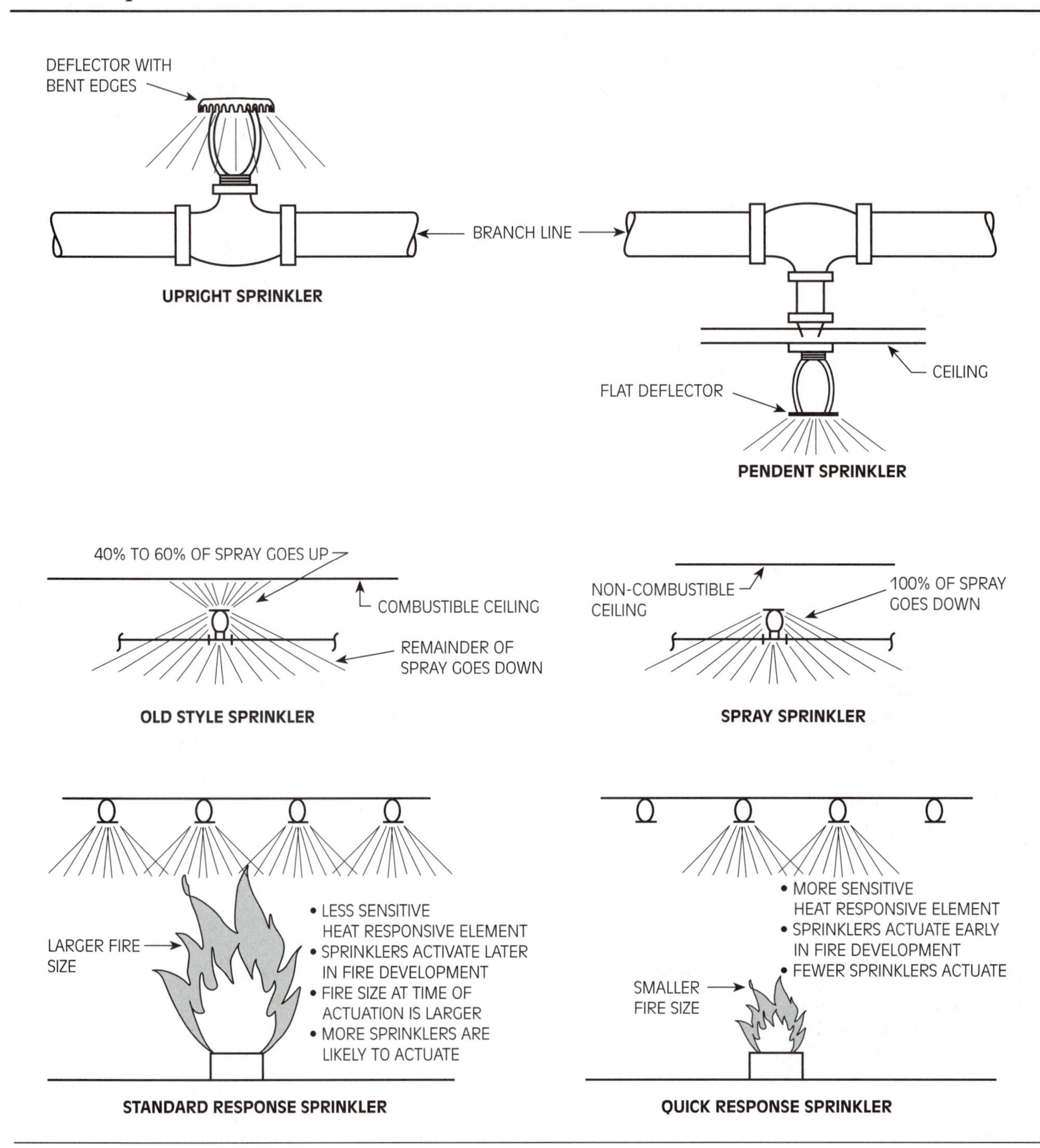

Figure S-12 Sprinkler differences.

upward position above the supply pipe. They are primarily used where upright sprinklers cannot be used because of lack of space (headroom) or where concealment of sprinkler piping above a false ceiling is desired because of aesthetic reasons (office areas). *See also* **Sprinkler, Upright; Standard Sprinkler Pendent (SSP).**

Sprinkler, Pilot—An automatic sprinkler head or thermostaic fixed temperature device used in a pneumatic or hydraulic fire detection system, normally connected to an actuating valve that releases when the pilot device is activated. *See also* **Pilot Head Detection System.**

Sprinkler Pintle—An indicating device on sprinklers that have small and large orifices and a standard 0.5 in. (1.27 cm) pipe thread. A pintle highlights the sprinkler orifice size difference compared to standard orifices; that is, 0.5 in. (1.27 cm) sprinklers with 0.5 in. (1.27 cm) pipe threads. It consists of a small, short cylinder centrally mounted and perpendicular to the deflector plate, on the side opposite the water discharge.

Sprinkler, Recessed—Sprinklers in which all or part of the body, other than the shank thread, is mounted within a recessed housing. Recessed sprinklers are mainly provided for aesthetic reasons, although protection of the sprinkler against external impacts may also be a consideration. They also allow a sprinkler system to be installed, tested, and evaluated before the installation of the finished ceiling. See Figure S-13. *See also* **Sprinkler, Concealed; Sprinkler, Flush.**

Figure S-13 Recessed sprinkler.

Sprinkler, Residential—A type of fast-response sprinkler that is well known for its ability to enhance human survivability in the room of fire origin and is used in the protection of dwelling units as specified by listing or approval agencies. The first effective fast-response sprinkler for residential use was developed by the Factory Mutual Research Corporation (under contract to the United States Fire Administration) and was demonstrated in 1979. *See also* **Early Suppression Fast Response (ESFR) Sprinkler.**

Sprinkler Riser—The vertical portion of a sprinkler system piping from the ground main to the horizontal cross main that feeds the branch lines. See Figure S-14.

Sprinkler, Sidewall—Sprinkler designed to be installed on piping along the sides of a room instead of the normal sprinkler spacing requirements. The sprinkler is made with a special deflector that deflects most of the water away from the nearby walls in a pattern similar to a quarter of a sphere. A small portion of the water is directed at the wall behind the sprinkler. Sidewall sprinklers are generally used because of aesthetic concerns, building construction arrangements, or installation economy considerations.

Sprinkler Spacing—Distribution of automatic sprinklers to provide the area coverage specified for light, ordinary, and extra hazardous occupancies.

Figure S-14 Sprinkler riser.

Sprinkler, Standard—*See* **Spray Sprinkler.**

Sprinkler, Standard Orifice—A sprinkler with an orifice size of 0.5 in. (1.27 cm). *See also* **Spray Sprinkler.**

Sprinkler Stopper—One of several devices to stop the flow from individual sprinklers.

Sprinkler System—For fire protection purposes, an integrated system of underground and overhead piping designed in accordance with fire protection engineering standards. The installation includes one or more automatic water supplies. The portion of the sprinkler system above ground is a network of specially sized or hydraulically designed piping installed in a building, structure, or area, generally overhead, and to which sprinklers are attached in a systematic pattern. The valve controlling each system riser is located in the system riser or its supply piping. Each sprinkler system riser includes a device for actuating an alarm when the system is in operation. The system is usually activated by heat from a fire and discharges water over the fire area.

The first recorded patented sprinkler system was developed in London in 1806 by John Carey. It consisted of a pipe fed by a gravity tank with a number of valves held closed by counterweights on strings; when a fire burned the strings, the valves were opened. The sprinkler head consisted of an outlet similar to a water can perforated nozzle that faced downward. A refined sprinkler system was patented (British Patent No. 3201) by William Congreve in 1809. His system used fusible metal on the wires controlling water supply valves and had various water distribution devices including perforated pipes, devices similar to sidewall sprinklers. Many manually operated systems were installed in 19th-century buildings. The first system in America was installed in a plant in Lowell, Massachusetts in about 1852. A number of perforated pipes were fed by a main riser that could be turned on in an adjoining area. James B. Francis improved the distribution of this system by using pipe with perforations about 0.1 in. (0.25 cm) in diameter and spaced 9 in. (22.86 cm) apart, alternately on different sides, to provide a spray at water at an angle slightly above the horizontal. Insurance companies of the time continued to improve on the design. These systems resulted in frequent water damage in parts of a room or building untouched by fire. An improvement was sought and found in the Parmelee sprinkler head, which was introduced in the United States in the 1870s. The Parmelee head had a normally closed orifice that was opened by heat from a fire. The first sprinkler successfully used over a long period was the Grinnel "glass button," which appeared in 1890 (previous Grinnel types were developed from 1884 to 1888). Since about 1900, most changes to sprinklers have been refinements in the design (deflector or activating mechanism improvements) rather than conceptual changes. Modern versions use a fusible link or a bulb containing chemicals that breaks at about 160°F (70°C) to open the orifice. Modern sprinkler heads are designed to direct a spray downward. Most sprinkler systems are wet-head; that is, they use pipes filled with water. Where there is danger of freezing, however, dry-head sprinklers are used, in which the pipes are filled with air under moderate pressure; when the system is activated, the air escapes, opening the water-feeder valves. An improved version has air under only atmospheric pressure and is activated by heat-sensing devices. Another special type, used in high-hazard locations, is the deluge system, which delivers a large volume of water quickly.

The definitions of several types of sprinkler systems follow:

Wet Pipe System—A sprinkler system that uses automatic sprinklers installed in a piping system containing water and connected to a water supply. Individual sprinklers discharge immediately when they are affected by the heat of a fire. Sprinklers that are not affected by the heat remain closed. It is used where there is no danger of the pipes freezing and where no other conditions require the use of a special system. See Figure S-15.

Antifreeze System—A wet pipe sprinkler system that uses automatic sprinklers installed in a piping system containing an antifreeze solution and connected to a water supply. The antifreeze solution is discharged (followed by water) immediately upon operation of the sprinklers, which are opened from the heat effects from a fire.

Dry Pipe System—A sprinkler system that uses automatic sprinklers installed in a piping system containing air or nitrogen under pressure. A release of pressure on the system (as from the opening of a sprinkler) permits the water pressure to open a valve known as a dry pipe valve. The water then flows into the piping system and out the opened sprinklers. Dry pipe systems operate more slowly than do wet pipe systems and are more expensive to install and maintain, therefore they are only used where there is an absolute necessity, such as freezing conditions. See Figure S-16.

Pre-Action System—A sprinkler system using automatic sprinklers installed in a piping system containing air that may or may not be under pressure, with a supplemental detection system installed in the same areas as the sprinklers. Actuation of a detection system opens a valve that permits water to flow into the sprinkler piping system and to be discharged from any sprinklers that have opened from the effects of a fire. Sprinklers that are not affected by heat from a fire remain closed. They are designed to counteract the operational delay of dry pipe systems and eliminate the damage from a broken pipe or sprinkler head. See Figure S-17.

Combined Dry Pipe and Pre-Action System—A sprinkler system that uses automatic sprinklers installed in a piping system containing air under pressure, with a supplemental fire detection system installed in the same areas as the sprinklers. Operation of the detection system actuates tripping devices that open dry pipe valves simultaneously without a loss of air pressure in the system. Operation of the fire detection system also opens air exhaust valves at the end of the system feed main, facilitating the filling of the system

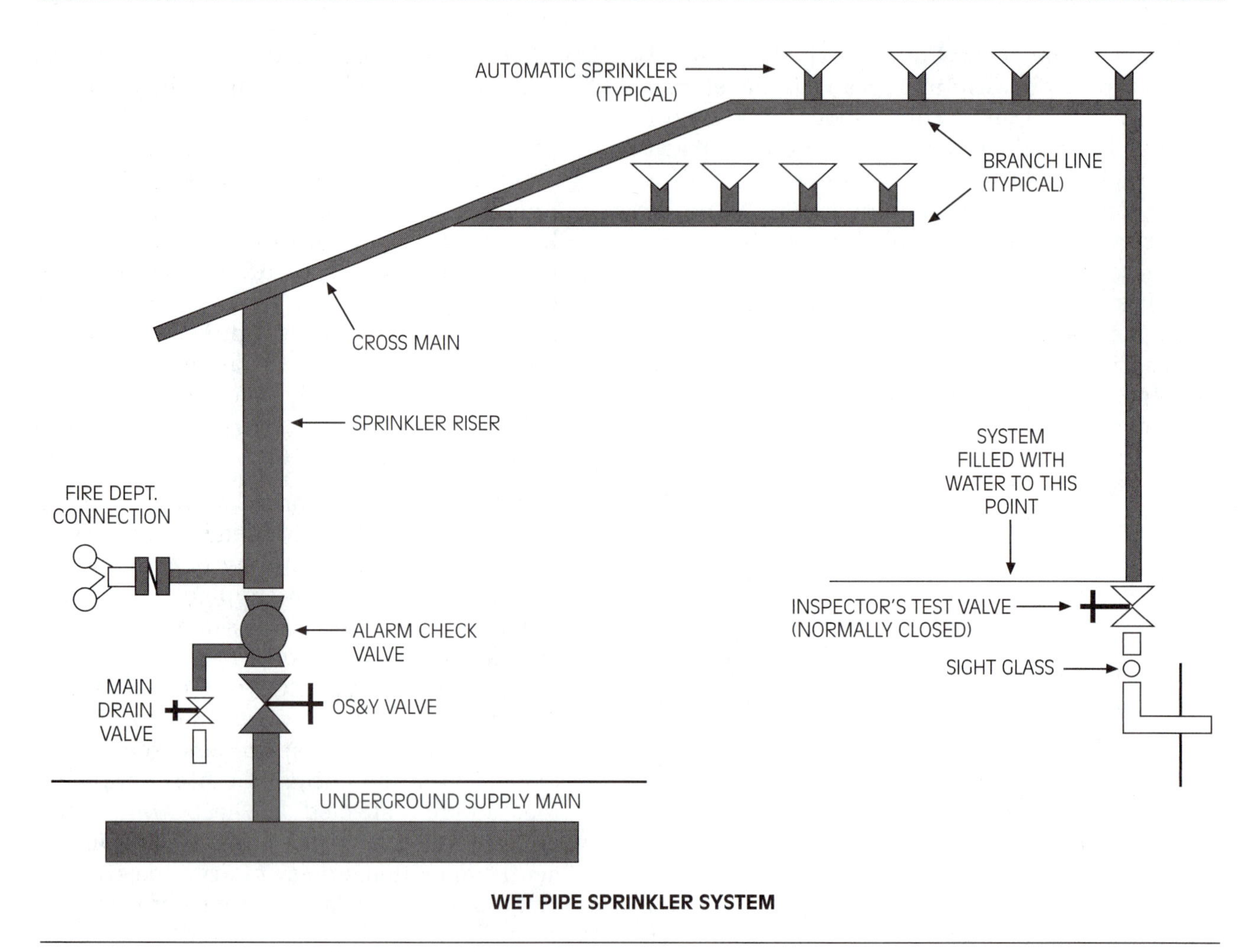

Figure S-15 Wet pipe sprinkler system schematic.

with water, which normally occurs before any sprinklers open. The detection system also serves as an automatic fire alarm system for the area. Only sprinklers that are affected by heat from a fire are opened; others remain closed.

Deluge System—A sprinkler system using open sprinklers installed in a piping system connected to a water supply through a valve that is opened by the operation of a detection system installed in the same areas as the sprinklers. When the deluge valve opens, water flows into the system piping and discharges from all sprinklers. There are no closed sprinklers in a deluge system. Its objective is to deliver the most amount of

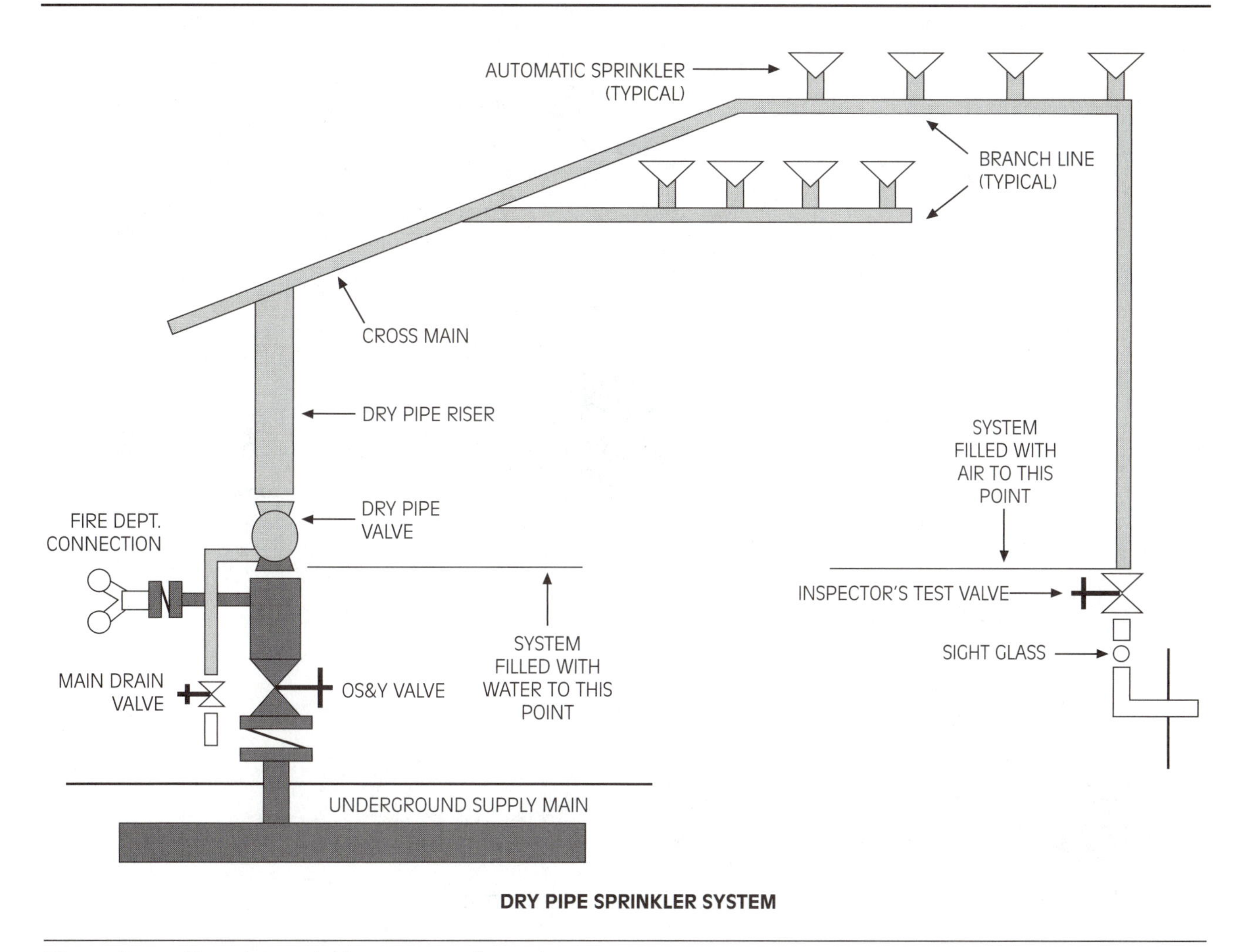

Figure S-16 Dry pipe sprinkler system schematic.

water in the least amount of time. Deluge systems are specified for high hazard locations where a fire occurs quickly and reaches very high temperatures, such as from highly flammable fuels. See Figure S-18.

Water Spray System—A fixed pipe system connected to a water supply and equipped with spray nozzles for specific discharge and distribution over the surface or area to be protected. The piping system is connected to the water supply through an automatic or manually activated valve that initiates the flow of water. The control valve is actuated by the operation of automatic fire detection devices installed in the same area as the water spray nozzles (in special cases the

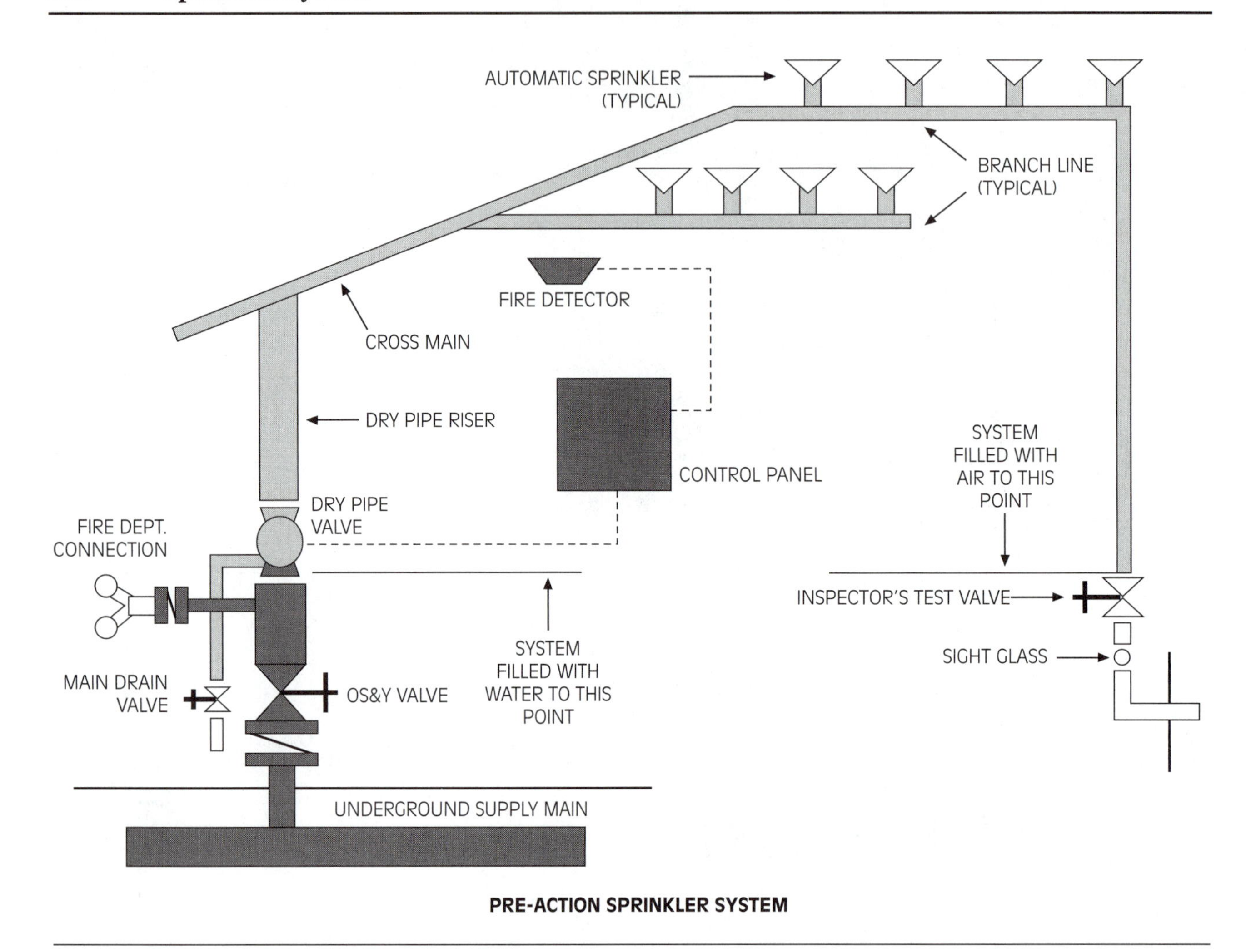

Figure S-17 Preaction system schematic.

automatic detection devices may be located in another area).

Foam-Water Sprinkler System—A fire protection piping system that is connected to a source of air-foam concentrate and a water supply and is equipped with appropriate discharge devices for extinguishing agent discharge and for distribution over the area to be protected.

Foam-Water Spray System—A fire protection piping system connected to a source of air-foam concentrate and a water supply and equipped with foam-water spray nozzles (aspirating or non-aspirating) for extinguishing agent

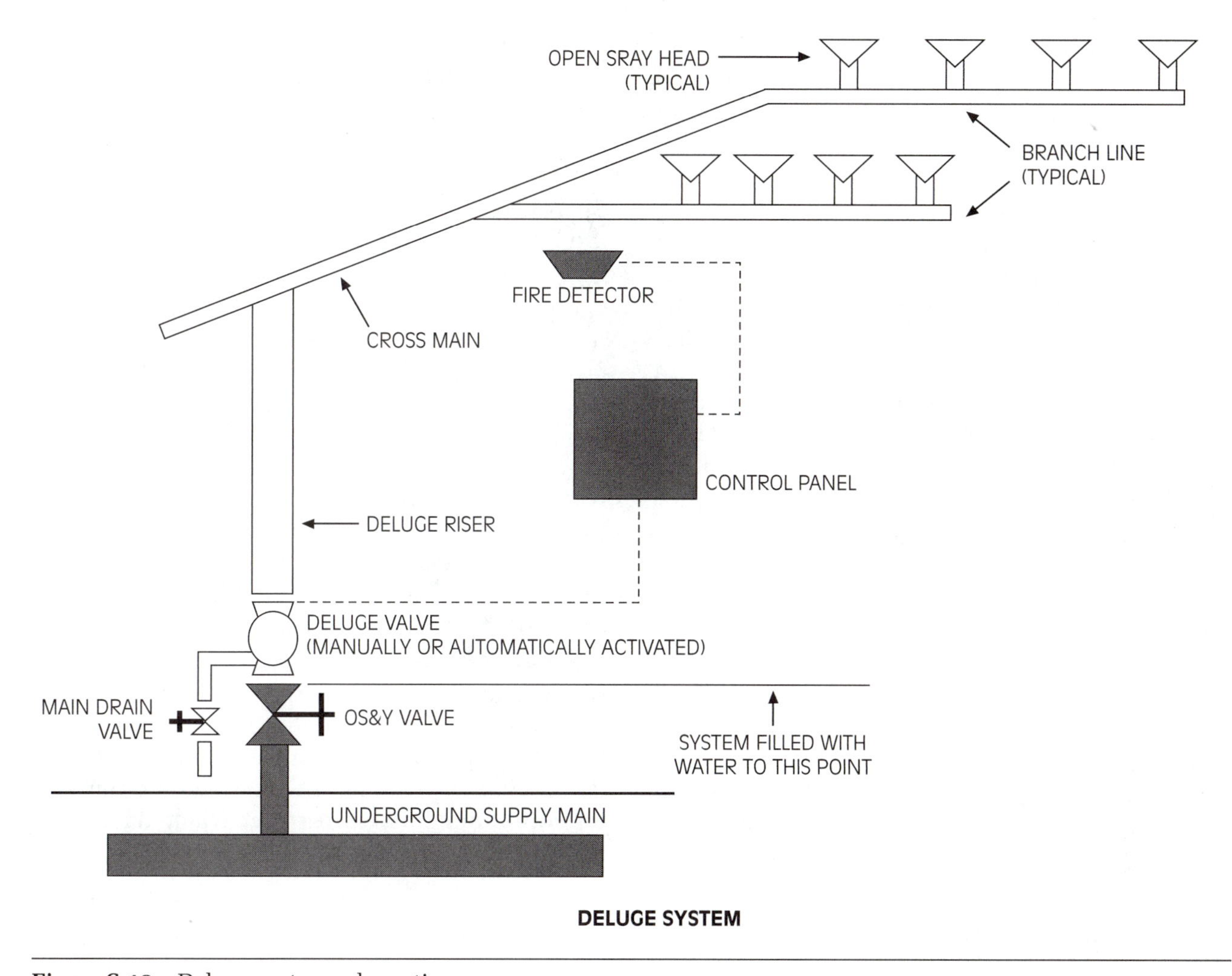

Figure S-18 Deluge system schematic.

S

discharge and for distribution over the area to be protected.

Fire protection sprinkler systems may also have several different piping arrangements.

Gridded System—A sprinkler system piping arrangement where parallel cross-mains are connected by multiple branch lines. An operating sprinkler receives water from both ends of its branch line while other branch lines help transfer water between cross-mains.

Looped System—A sprinkler system piping arrangement where multiple cross-mains are connected to provide more than one path for water to flow to an operating sprinkler, and branch lines are connected to each other.

Circulating, Closed-Loop System—A wet pipe sprinkler system that has a non-fire-protection connection to an automatic sprinkler system in a closed-loop piping arrangement. This allows the sprinkler piping to conduct water for heating or cooling in an economical fashion without impacting the ability of the sprinkler system to support its fire protection purpose. Water is not removed or used from the system, but is only circulated through the piping system. *See also* **Alarm Check Valve; Dry Pipe Valve; Dry Riser; Dry Sprinkler; Open Head System; Perforated Pipe.**

Sprinkler Temperature Classes—Sprinklers are designated to operate at specific fire temperatures and are segregated into temperature classes. The actual temperature rating of sprinklers may be less important than is popularly perceived. Where ceiling temperatures rise rapidly, the difference between 165°F (74°C) and 212°F (100°C) for the first sprinkler to operate may not be important, but it may affect the number of heads that operate. Higher temperature sprinklers are used where the ambient temperatures may be higher than ordinary temperature ratings.

These temperature classes are defined in Table S-1.

Sprinkler Tong—A portable tool used to stop the flow from a sprinkler head. See Figure S-19.

Sprinkler, Upright—A sprinkler designed for and placed in an upward position above the supply pipe, rather than installed in a downward fashion from the supply pipe. It directs 100 percent of its water toward the floor. A sprinkler designated for upright installation cannot be used in the downward position because the water will be directed to the ceiling instead of toward the fire incident and will not achieve its density pattern for fire control and extinguishment. *See also* **Spray Sprinkler; Sprinkler, Pendent; Standard Sprinkler Upright (SSU).**

Sprinkler Wedge—An improvised triangular block (usually wooden), which is used to interrupt the flow from an individual sprinkler head by inserting it between the deflector plate and the orifice. It provides an intermediate measure to stop water flow to prevent water damage in

Table S-1 Sprinkler Temperature Classes

Class	Color Code*	Color Glass Bulb	Temperature Rating °F (°C)
Ordinary	Uncolored or Black	Orange or Red	135°F to 170°F (57°C to 77°C)
Intermediate	White	Yellow or Green	175°F to 225°F (79°C to 107°C)
High	Blue	Blue	250°F to 300°F (121°C to 149°C)
Extra high	Red	Purple	325°F to 375°F (163°C to 191°C)
Very extra high	Green	Black	400°F to 475°F (204°C to 246°C)
Ultra high	Orange	Black	500°F to 575°F (260°C to 302°C)
Ultra high	Orange	Black	650°F (343°C)

*Used for fusible link sprinklers and chemical pellet sprinklers but not for glass bulb sprinklers.

Figure S-19 Sprinkler tong.

Figure S-20 Sprinkler wedge.

the immediate area after a fire is extinguished before the main isolation valve for the system is closed. See Figure S-20.

Squib-Activated System—A fire protection system that utilizes a small explosive charge, a "squib," to enhance rapid operating of the activation valve to release the suppressant agent. Squib-activated systems are used where an extremely fast activation time for agent release is desired. The squib is activated by electronic fire detection systems.

Squirrel Tail Suction—Name given to the provision of a hard suction hose permanently attached to the front suction connection of a fire pumper, where the remainder of the hose is curved around the front until connected to a hydrant. The arrangement saves times when making a connection to a fire hydrant for water supplies. *See also* **Hose, Hard Suction.**

Squirt or **Fire Squirt**—A medieval device similar to a syringe used to apply water to a fire. Water was pushed from a cylindrical shape out through a nozzle at one end by a plunger placed at the opposite end. During the Great Fire of London in 1666, the citizens only had handheld fire squirts or buckets to apply water to the fire. See Figure S-21. Also used as a trade name for an articulating aerial device that can provide an elevated master stream without a platform to hold firefighters. Similiar devices also may be called a "snozzie". *See also* **Snorkel.**

Stack Effect—*See* **Chimney Effect.**

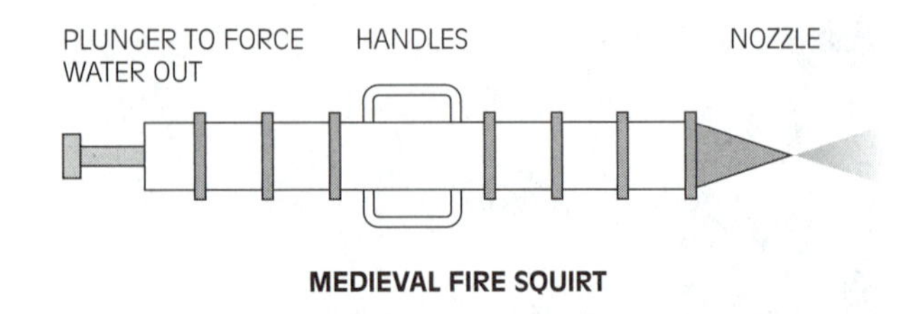

Figure S-21 Medieval fire squirt.

Standard Fire Prevention Code—A fire prevention code issued by the Southeastern Association of Fire Chiefs. *See also* **Fire Code.**

Standard Fire Test—A furnace fire test using a time temperature curve that simulates a standard fire and that results in an "A" rating for successful resistance to the fire exposure. The fire test is intended to simulate a fire from ordinary combustible materials. *See also* **Fire Test.**

Standard Sprinkler Pendent (SSP)—A sprinkler designated to be installed with its outlet oriented to allow its spray to be directed upward. *See also* **Spray Sprinkler; Sprinkler, Pendent.**

Standard Sprinkler Upright (SSU)—A sprinkler designated to be installed with its outlet orientated to allow the spray to be directed downwards. *See also* **Spray Sprinkler; Sprinkler, Upright.**

Standard Time Temperature Curve (STTC)—*See* **Standard Fire Test.**

Standpipe Hose Cabinet—A cabinet provided for the provision of standpipe outlets and/or fire hose storage. Fire hoses are usually preconnected and stored in a rack with release pins. See Figure S-22. *See also* **Hose Rack.**

Figure S-22 Standpipe hose cabinet.

Standpipe, Manual—A standpipe system that relies on the fire service to supply water to it to meet its demands.

Standpipe, Semi-Automatic—A standpipe system that is connected to an adequate water supply, but requires a control device activation to supply water to the hose outlets.

Standpipe System—The provision of piping, riser pipes, valves, firewater hose connections, and associated devices for the purpose of providing or supplying firewater hose applications in a building or structure by the occupants or fire department personnel. Standpipe systems are classified according to their intended use by the building occupants, the fire department, or both.

Many high-rise or other large buildings have an internal system of water mains (standpipes) connected to fire hose stations. Trained occupants or employees of the building management operate the hoses until the fire department arrives. Firefighters can also connect their hoses to outlets near the fire.

The National Fire Protection Association (NFPA) classifies standpipe systems based on their intended use as Classes I, II, or III. Class I is provided for fire service use or other personnel trained in handling heavy fire streams. It is distinguished by the provision of 2.5 in. (6.35 cm) hose stations or hose connections. A Class II standpipe system is provided for use by the building occupants and by the fire service during initial attack operations. It is characterized by the provision of 1.5 in. (3.81 cm) hose stations. A Class III standpipe system combines both the features of Class I and Class II systems. See Figure S-23. *See also* **First Aid Standpipes** or **Hoses.**

Standpipe, Wet—A standpipe system that is permanently charged with water for immediate use.

State Fire Marshal (SFM)—*See* **Fire Marshal.**

Figure S-23 Standpipe outlet connection for fire hose.

Static Ignition Hazard—An electrical charge build-up of sufficient energy to be considered an ignition source. For an electrostatic charge to be considered an ignition source, four conditions must be present: (1) a means of generating an electrostatic charge, (2) a means of accumulating an electrostatic charge of sufficient energy to be capable of producing an incendiary spark, (3) a spark gap, and (4) an ignitable mixture in the spark gap. Removal of one or more of these features will eliminate a static ignition hazard. Static charges can accumulate on personnel and metallic equipment. If the static accumulation is separated by materials that are electrically nonconducting, a dangerous potential difference may occur. These nonconducting materials or insulators act as barriers to inhibit the free movement of electrostatic charges, preventing the equalization of potential differences. A spark discharge can occur only when there is no other available path of greater conductivity by which this equalization can be affected (bonding or grounding). *See also* **Ignitable Liquid; Ignition Energy; Ignition Source; Incendive Spark; Incendivity; Spark Ignition.**

Static Pressure—The pressure created by a confined liquid at rest (no flow in a system). Static pressure may be developed through the difference in elevation between two points of a confined liquid or it can be created artificially through the application of force applied to a confined fluid. *See also* **Head; Residual Pressure.**

Static Source Hydrant—*See* **Hydrant, Dry** (or **Suction) Fire.**

Steamer—A large outlet or connection provided on a fire hydrant (steamer connection) or a fire pumper. It may also refer to a steam-driven fire pumper.

Steamer Connection—A hose connection on a pumper with 4.5 in. (11.43 cm) diameter or larger, compatible with the 4.5 in. (11.43 cm) outlet on a fire hydrant. *See also* **Hose, Hard Suction; Steamer Outlet.**

Steamer Outlet—The large connection on a hydrant that is 4.5 in. (11.43 cm) in diameter or larger. It was originally provided for steam fire engines and came to be called the steamer outlet due to its large size. *See also* **Steamer Connection.**

Steam Pumper or **Steamer**—*See* **Fire Steamer.**

Steam Smothering—A purging method used in the process industries to reduce the oxygen content of an enclosure to prevent a hazardous atmosphere from forming or as a method to suppress a fire incident in enclosed spaces or of small size. Most process industries use vast quantities of steam to power drivers or for their process heating requirements. Since steam is a readily available resource in these plants, it is commonly used for purging and firefighting of enclosed process heaters when convenient. High pressure steam is directed onto a fire to smother a fire or inhibit combustion conditions. It is not effective for cooling or protection from fire damage. *See also* **Snuffing Steam.**

Steiner Tunnel Test—*See* **Flame Spread Rating (FSR).**

Stoichiometric Combustion—The complete burning of a mixture, where both the fuel and the oxidizer are completely consumed. *See also* **Perfect Combustion.**

S

Stoichiometric Mixture—A mixture of a fuel and oxidizer such that complete combustion may just occur (the oxidizer concentration is just sufficient to completely consume all of the available fuel). *See also* **Perfect Combustion.**

Stop, Drop and Roll—*See* **Drop and Roll.**

Stored Pressure Extinguisher—A portable fire extinguisher in which the firefighting agent and expelling medium are stored in the same container. A pressurized multipurpose dry chemical portable fire extinguisher is an example of a stored pressure extinguisher. *See also* **Fire Extinguisher, Portable.**

Stratification—The effect of heated smoke particles or gaseous combustion products becoming less dense than the ambient "cooler" air, and rising until there is no longer a difference in temperature between it and the surrounding atmosphere. Stratification can also be caused by forced ventilation.

Street Box—*See* **Municipal Fire Alarm System.**

Structural Firefighting Protective Clothing—*See* **Turnout Clothing** or **Gear.**

Submergence Time—The amount of time required by a high expansion foam application to ensure the extinguishment of a fire. The submergence time for a location is dependent on the type of hazards, fuel type, and its geometry.

Subsurface Foam Injection—The introduction of firefighting foam beneath the surface of certain flammable liquids to secure extinguishment of a fire burning on its surface. It is normally provided for the protection of atmospheric and low pressure storage tanks. Foam is produced though a high backpressure foam maker and is forced into the bottom of the protected storage tank. The injection line may be an existing product line or a dedicated foam injection line specifically designed for subsurface injection. The foam becomes highly buoyant due to the entrainment of air and travels up through the tank contents to form a vapor-tight blanket on the surface of the liquid in the tank. It can be applied to any of the various types of atmospheric pressure storage tanks, but is generally not recommended for storage tanks with a floating roof since distribution to the seal area from the internal dispersion points may be difficult to achieve. See Figure S-24. *See also* **Foam; High Backpressure Foam Maker; Overhead Foam Injection; Semi-Subsurface Foam Injection.**

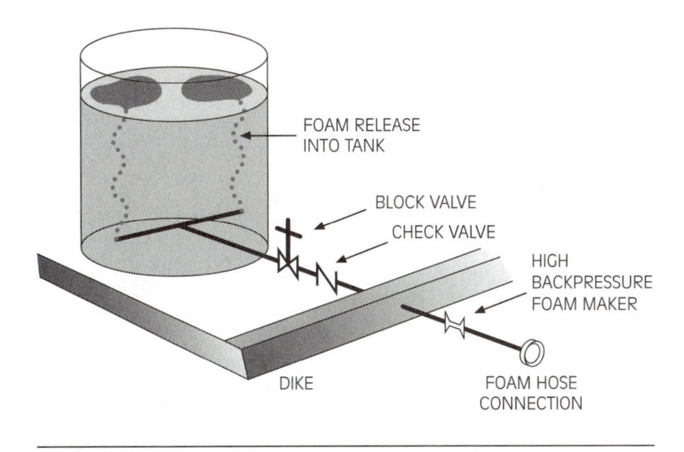

Figure S-24 Schematic of cone-roof flammable liquid storage tank with subsurface low expansion foam system.

Supervised Flame—A flame whose presence or absence is detected by a primary safety control. Supervised flames generally occur in furnaces or boilers where their absence may lead to conditions that may be unsafe (buildup of combustible vapors leading to an explosion condition). *See also* **Combustion Detector.**

Supply Curve—A graphical representation of the amount of water pressure and quantity available for a water-based fire protection system, or a facility based on hydraulic calculations of the system requirements, or actual water flow test of the system. A supply curve is compared against a demand curve to verify the facility water supplies and requirements can meet the system requirements. *See also* **Demand Curve.**

Supply Line—A firewater hose line used to provide water to an apparatus.

Suppressant, (Fire)—A substance that reduces flaming or glowing phases of combustion by either removing heat from a flame, chemically interfering with a combustion process, or diluting a flammable mixture.

Suppression, (Fire)—The sum of all the work performed to extinguish an unwanted fire from the time of its discovery until fire extinguishment.

Surface Burn—Combustion limited to the exposed surface of a material.

Surface Emissive Power (SEP)—The heat that is radiated outward from a flame per unit surface area of the flame. Its measurement and calculation units are normally kW/m^2.

Surface Fire—*See* **Forest Fire.**

Surge—*See* **Water Hammer.**

Tamper Switch (TS)—An electrical switch provided to a main isolation supply valve on a water-based fire protection system to indicate its unauthorized use (closure). Tamper switches are usually arranged to alarm the local fire alarm control panel to indicate an unexpected impairment to the system. They are commonly fitted to the riser isolation valve or the post indicator valve (PIV) of the system where there is a concern the supply valve may be inappropriately closed.

Tanker—An aircraft capable of carrying and dropping water or fire retardants on a fire incident (usually a wildfire). May also still be used to describe a land-based water carrying apparatus. *See also* **Mobile Water Supply Apparatus.**

Temperature—The intensity of sensible heat of a body as measured by a thermometer or similar instrument. It is expressed in terms of any of several arbitrary scales and indicates the direction in which heat energy will spontaneously flow, that is, from a hotter body (one at a higher temperature) to a colder one (one at a lower temperature), using the Fahrenheit, Celsius, Rankine, or Kelvin scales. The lowest possible temperature is absolute zero on the Kelvin temperature scale (−273° on the Celsius scale). At absolute zero, it is impossible for a body to release any energy.

Tempered Glass—A type of glass that has been treated to improve its stability and resistance to heat, impact, and distortion. Glass sheets are tempered at about 1,200°F (650°C) followed by a sudden chilling. This treatment increases the strength of the glass sheets approximately six times. When such glass does break (due to impact, expansion, etc.), it shatters into blunt granules. *See also* **Fire Window Assembly.**

Theoretical Air—The quantity of air necessary to completely burn a substance. The amount can be calculated by the chemical reaction of the substance for the combustion process. For example, in ideal combustion $CH_4 + 2O_2 = CO_2 + 2(H_2O)$. In other words, 1 lb. (0.45 kg) of methane combines with 4 lbs. (1.8 kg) of oxygen to produce 2.75 lbs. (1.2 kg) of carbon dioxide (CO_2) and 2.25 lbs. (1.02 kg) of water vapor. Theoretical air may also be called stoichiometric air or theoretical combustion air.

Theoretical Critical Fire Area (TCA)—Terminology used in aircraft firefighting. It is an area to be protected in any post-accident situation that would permit the safe evacuation of the aircraft occupants. The purpose of the critical area concept is not to define fire attack procedures but to allow a basis for calculating the quantities of extinguishing agents necessary to achieve protection within an acceptable period of time. The theoretical critical fire area (TCA) is a rectangle, the longitudinal dimension of which is the overall length of the aircraft, and the width includes the fuselage and extends beyond it by a predetermined set distance that is dependent on the overall width. An aircraft length multiplied by the calculated width equals the size of the TCA. *See also* **Practical Critical Fire Area (PCA).**

Theoretical Lift—For fire protection applications, it is defined as the maximum amount of lift (or height) that can be produced by a pump, producing a vacuum condition in its suction inlet. *See also* **Draft.**

Theory of Paracelsus—A theory of fire perpetuated in the Middle Ages in which it was theorized that an elemental being inhabited fire. The theory was developed by Paracelsus (1493–1541), who was an medieval German-Swiss physician and alchemist. *See also* **Salamander.**

Thermal Analysis—A calculation in which the results are temperature distributions for a given heat input. The heating may be described in terms of either temperature or radiation levels.

Thermal Barrier—A heat protective barrier. Thermal barriers are provided in firefighting protective clothing and in building construction. For building construction, it is defined as a material that will limit the average temperature rise of the unexposed surface to not more than 250°F (121°C) after 15 minutes of fire exposure, complying with the standard time temperature curve of NFPA 251, *Standard Methods of Fire Tests of Building Construction and Materials.*

T

Thermal Burn—A burn caused by radiant heat. *See also* **Burn Injury.**

Thermal Capacity—The ability of a material to store heat. It is often referred to as the heat required to raise the temperature of the material by one degree.

Thermal Column—*See* **Plume.**

Thermal Conductivity—The measure of heat transfer through a medium via random molecular motion. Thermal conductivity is the exchange of energy between adjacent molecules and electrons in the conducting medium. The rate of heat flow in a rod of material is proportional to the cross-sectional area of the rod and to the temperature difference between the ends and inversely proportional to the length, expressed as $H = -k(A/l)(T_2 - T_1)$. The minus sign arises because heat always flows from higher to lower temperature. A substance of large thermal conductivity k is a good heat conductor, whereas one with small thermal conductivity is a poor heat conductor or good thermal insulator. Metals are generally a better thermal conductor than are other materials. A fire that impinges on metal components within a building will transmit the heat effects of a fire much faster than concrete or plaster will, and they may aid in the spread of a fire. *See also* **Conduction; K-Factor.**

Thermal Convection—A process by which heat is transferred by the movement of a heated fluid such as air or water. Natural convection results from the tendency of most fluids to expand when heated (to become less dense and to rise as a result of the increased buoyancy). Circulation caused by this effect accounts for the uniform heating of water in a kettle or air in a heated room; the heated molecules expand in the space they move in through increased speed against one another, rise, and then cool and come closer together again, with an increase in density and a resultant sinking. Forced convection involves the transport of fluid by methods other than that resulting from variation of density with temperature. Atmospheric convection currents can be set up by local heating effects such as solar radiation (heating and rising) or contact with cold surface masses (cooling and sinking). Such convection currents primarily move vertically and account for many atmospheric phenomena, such as clouds and thunderstorms.

See also **Convection; Fire Plume; Heat Transfer; Plume.**

Thermal Detection—The detection of a specific set point temperature by a thermal element in a thermal device. *See also* **Heat Detection.**

Thermal Element—A device that responds to specified heat input. It employs either a bimetal or a melting alloy joint to initiate the opening action. They are commonly employed in automatic sprinkler heads, fire alarm heat detectors, or overload relays (for the protection of electrical circuits against overloading). *See also* **Fixed Temperature Detector; Fusible Element; Fusible Metal.**

Thermal Expansion—The general increase in the volume of a material as its temperature is increased. It is usually expressed as a fractional change in dimensions or volume per unit temperature change; a linear expansion coefficient is usually employed in describing the expansion of a solid, whereas a volume expansion coefficient is more useful for a liquid or a gas. The amount of increase per degree temperature is called the coefficient of thermal expansion. Thermal expansion coefficients are specific for each material because the magnitudes of the bonding forces within each material are specific. The expansion of building materials during a fire incident may lead to structural failure. The collapse of buildings from structural failure during a fire has led to the death or injury of firefighters and additional property damage.

Thermal Ignition—The ignition of a combustion process in a material by the application of sufficient heat until the ignition point of the material is attained. See also **Hot Surface Ignition.**

Thermal Injury—*See* **Burn Injury.**

Thermal Insulation—One or more layers of noncombustible or fire-resistant, high-density material to reduce the passage of heat for protection against exposure of an ignition source (hot surface), burn injuries, or heat damage. See also **Thermal Protective Clothing.**

Thermal Lag—When a fixed temperature device senses a rise in ambient temperature, the temperature of the surrounding air will always be higher than the operating temperature of the device itself. This difference between the operating temperature and the actual air temperature is commonly referred to as thermal lag, and is proportional to the rate at which the temperature is rising. See Figure T-1.

Thermal Layering—The process of gases to form layers based on temperature where the hottest layers in a confined space are located at the highest elevations (due to lower densities) and the lowest temperature gases are located at the lowest elevations (due to the highest densities).

Thermal Plume—A column of heat rising from a thermal source. *See also* **Fire Plume; Plume.**

Thermal Protective Clothing—The protective apparel provided for and used by firefighters and other individuals as protective insulation against the adverse effects of heat. Generally consisting of helmets, boots, gloves, hoods, coats, and pants. *See also* **Thermal Insulation; Turnout Clothing** or **Gear.**

Thermal Protective Performance (TPP)—A rating of thermal protection provided to protective clothing to indicate the level of heat protection specified in NFPA 1971, *Protective Ensemble for Structural Fire.* The thermal protective performance (TPP) test uses a sample of material (outer shell, inner moisture barrier, and thermal liner materials) exposed to a brief thermal environment produced by radiant heat and flames from laboratory burners. Approximately 80 kW/m^2 (2 cal/cm^2) per second of heat exposure is

T

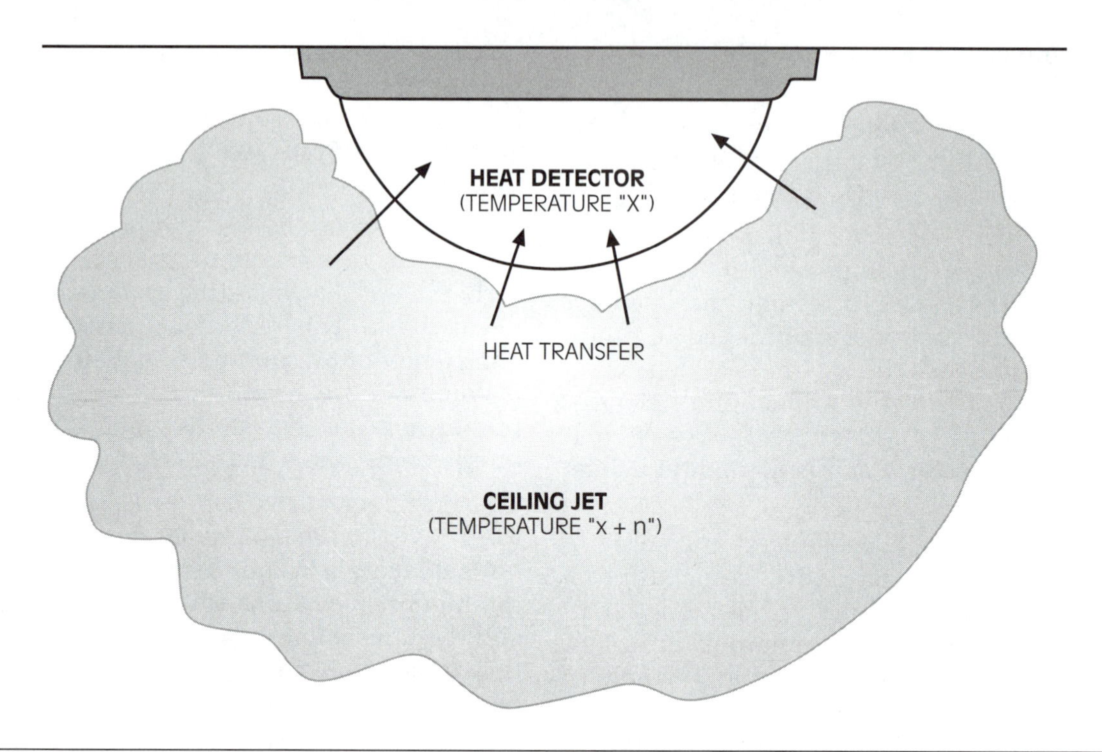

Figure T-1 Example of thermal lag.

produced to simulate a moderate level flash fire exposure. *See also* **Radiant Protective Performance (RPP).**

Thermal Radiation—A process by which energy, in the form of electromagnetic radiation, is emitted by a heated surface in all directions and travels directly to its point of absorption at the speed of light. Thermal radiation does not require an intervening medium to carry it. Thermal radiation ranges in wavelength from the longest infrared rays through the visible light spectrum to the shortest ultraviolet rays. The intensity and distribution of radiant energy within this range is governed by the temperature of the emitting surface. The total radiant heat energy emitted by a surface is proportional to the fourth power of its absolute temperature (the Stefan-Boltzmann law). The rate at which a body radiates (or absorbs) thermal radiation depends upon the nature of the surface as well. Objects that are good emitters are also good absorbers (Kirchhoff's radiation law). A blackened surface is an excellent emitter as well as an excellent absorber. If the same surface is silvered, it becomes a poor emitter and a poor absorber. The heating of a room by an open-hearth fireplace is an example of transfer of energy by radiation. The flames, coals, and hot bricks radiate heat directly to the objects in the

T

room with little of this heat being absorbed by the intervening air. Most of the air that is drawn from the room and heated in the fireplace does not reenter the room in a current of convection, but is carried up the chimney together with the products of combustion.

Thermal Sensitivity—The speed at which a specific response is initiated following exposure to a heat source. Thermal elements are evaluated for the thermal sensitivity with which the thermal element operates as installed in a specific sprinkler or sprinkler assembly. The response time index (RTI) is one measure of thermal sensitivity used to measure the speed of response. Fast response sprinklers have a thermal element with an RTI of 50 (meters-seconds) 1/2 or less. Standard response sprinklers have a thermal element with an RTI of 80 (meters-seconds) 1/2 or more. *See also* **Early Suppression Fast Response (ESFR) Sprinkler.**

Thermal Shield—*See* **Thermal Insulation.**

Thermistor—An electrical resistance element made of a semiconducting material consisting of a mixture of oxides of manganese and nickel; its resistance varies with temperature. Thermistors (temperature-sensitive or thermal resistors) are used as temperature-measuring devices. *See also* **Fixed Temperature Detector.**

Thermocouple—A temperature measuring device composed of two pieces of different metals welded or soldered together at one end. When temperature changes at the welded junction, an electromotive force is generated and appears at the free ends. This is translated into a temperature reading. Thermocouples are commonly employed to measure temperature during fire testing experiments and material fire-resistive tests.

Thermo Decomposition—When most combustible substances burn or are heated, some of the molecules can break down into simpler substances rather than the oxidation products of combustion. Some polymers, such as polymethylmethacrylate, will regenerate the monomer (methyl methacrylate) when heated, whereas others, such as polyvinylchloride (PVC), will produce hydrochloric acid gas (HCl) on thermal decomposition plus a range of carbon-containing materials.

Thermowell—A cavity within a vessel or line, but sealed off from it, for the purpose of inserting a thermocouple or thermometer for temperature measurements. Primarily used in the chemical or hydrocarbon process industries.

Time Temperature Curve—*See* **Cellulosic Fires; Fire Curve; Hydrocarbon Fire; Time Temperature Curve, Cellulosic Fire.**

Time Temperature Curve, Cellulosic Fire—A standardized graphical model used for the simulation of a fire exposure from ordinary combustibles based on temperatures experienced during the fire. The curve is referenced in various standards (NFPA, ISO, ASTM, DIN, SOLAS, etc.) and is defined by the following equation:

$$T - T_0 = 345 \log_{10} (8t + 1),$$

Where:

T = Temperature at time t

T_0 = Temperature at time $t = 0$

t = time in minutes from the start of fire

The time temperature curve was proposed in 1903. The curve was later developed in 1916 by ASTM as a criterion for fire endurance testing and was adopted in 1917 as ASTM C-19. It was based on the best judgment of the time of a structural fire. In 1922, the US National Bureau of Standards (now NIST) conducted ten fire tests to validate the time temperature curve. The conclusion of these tests indicated the standard time temperature curve was substantiated.

Contemporary fire testing and research has indicated a much more rapid temperature rise than the ASTM E-119 fire curve for structural fires. Some of the difference may be accounted for by the change and amount of materials used in modern structures as compared to those used earlier in the century. NFPA has used a more realistic fire exposure temperature for the testing of early suppression fast response (ESFR) sprinklers, that is, the ultra fast T-squared fire growth model and a high-rise fire exposure test that is used in the petrochemical industry, UL Standard 1709, *Standard for Safety, Rapid Rise Fire Tests of Protected Materials for Structural Steel.* See Figure T-2. *See also* **Fire Resistance Rating; Hydrocarbon Fire.**

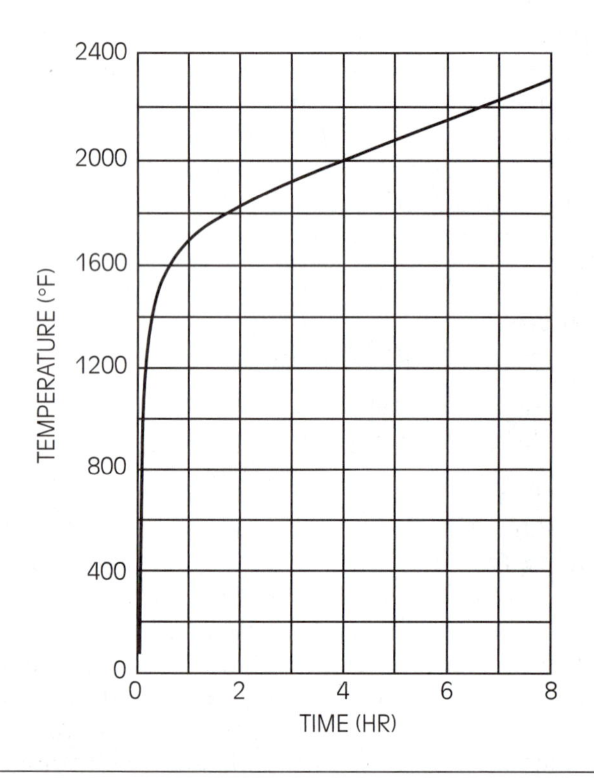

Figure T-2 Standard time-temperature curve.

Time Temperature Curve, Hydrocarbon Fire—A standardized graphical model used for the simulation of a fire exposure from hydrocarbon materials based on temperatures experienced during the fire. The curve indicates a rapid rise in fire temperature within the first few minutes of the fire as opposed to a cellulosic fire, wherein the fire temperature is shown as a relative gradual increase. UL 1709, *Rapid Rise Fire Tests of Protection Materials for Structural Steel,* is generally used in the industry as the standard time temperature curve for hydrocarbon fire exposures.

TNO Multi-Energy Vapor Cloud Explosion Model—A mathematical and graphical model of a vapor cloud explosion blast. It provides a plot of peak pressure wave duration as well as a plot of peak overpressure with a series of curves to account for various deflagration strengths as well as detonation. It was developed by the Dutch Research Organization, TNO. The parameters of the TNO multi-energy model are not well defined, but are intended to correlate with the degree of confinement and congestion of process plant structures and arrangements. *See also* **Shell Hemispherical Vapor Cloud Explosion Model.**

TNT Equivalence Vapor Cloud Explosion Model—A mathematical and graphical model of a vapor cloud explosion blast. It provides a plot of peak pressure wave duration as well as a plot of peak overpressure with a series of curves to account for various deflagration strengths as well as detonation. The model is based on the expected blast damage from an equivalent force from the explosion of TNT (trinitrotoluene) at a specific point. The model has been questioned for its accuracy due to inaccurate comparison to a large gas volume that is dispersed and then explodes. The TNT equivalence model was one of the first mathematical models used to predict blast overpressures from vapor cloud explosions; however, newer models have been developed to overcome the deficiencies of the TNT equivalence model (Baker-Strehlow, TNO multi-energy, Shell hemispherical), as well as flame acceleration, vapor dispersion, congestion, etc. *See also* **Baker-Strehlow Vapor Cloud Explosion Model.**

Torch Fire—*See* **Jet Fire.**

Total Flooding—The provision of a fire extinguishing agent in an enclosed area that completely fills the volume to effect extinguishment or prevent a fire incident. They may be gaseous, liquid, or solid. Most common agents are gaseous, such as carbon dioxide (CO_2) or Halon; liquid types use foaming agents, such as high expansion foam, and solid form may use dry chemical systems. Gaseous agents require a particular concentration of agent to be achieved within the volume before extinguishment can be achieved. Total flooding is used where it may be difficult to immediately reach the seat of a fire, such as in machinery spaces, a computer room, and engine compartments. Because some agents (CO_2) may deplete oxygen from the enclosure to extinguish the fire, special precautions (pre-alarm, evacuation notification, reentry precautions, etc.) must be implemented where personnel may also be present. Application of a total flooding system to an enclosure requires the enclosure to be adequately sealed to prevent the release of the agent out of the enclosure once it is applied. *See also* **Carbon Dioxide (CO_2) Fire Suppression System; Enclosure Integrity Test (EIT).**

Tower Ladder—Term used to describe a telescoping aerial platform of a fire apparatus.

Transfer Valve—A pumper control valve provided to configure a multistage pump arrangement to either volume or pressure operation. By positioning the valve, water flow is directed to flow in parallel (for volume) or in series (for pressure) through the pump stages. When pumping in the volume mode, each impeller is providing partial (half) flow, while generating the same pressure; while in the pressure mode, one impeller pumps to the next impeller, producing the same flow but double the pressure. The transfer valve is usually located on the pumper pump panel.

Transitional Attack—A situation when firefighters are in the offensive attack mode and are preparing to go to a defensive attack or vice-versa. Firefighters should never operate in both modes at the same time. *See also* **Defensive Attack; Offense Attack.**

Trash Line—Slang terminology for a small diameter fire hose line provided on a fire truck that is pre-connected and available for immediate use. It is primarily intended for use on small fires, such as trash fires.

Travel Distance—The distance from a point on the floor of a building to a vertical exit, horizontal exit, or outside exit measured along the line of travel, except that usually in one story, low, or moderate hazard industrial or storage units, travel distance may be considered as the distance from any point to an aisle, passageway, or other exit connection. Travel distance is also referred to as the distance to a fire extinguisher.

Trench Cut—A type of roof ventilation cut from one wall to another to create a fire break along the surface of a roof.

Trench Effect—A fast fire growth phenomenon that is the result of confining and concentrating combustion materials, flames, and products of combustion in an inclined channel. It was identified during the investigation into the fire at London's King's Cross underground subway station in 1987, in which fire rapidly spread up escalator steps and into the public hall above. Four factors have been identified as important parameters for the occurrence of a trench effect fire: the slope of the trench, the trench's geometrical profile, the combustible nature of the material lining the trench, and the ignition source. *See also* **Chimney Effect.**

Triple Combination (Pumper)—Terminology for a fire service pumper that is equipped with three main fire protection support features: a water pump, a fire hose bed, and a water storage tank. *See also* **Pumper; Quad; Quint.**

Trouble Signal—A signal indicating trouble of any nature, such as a ground or wiring circuit failure, occurring in the devices or wiring of a protective signaling system. Trouble signals are provided on electrical fire protection systems (fire alarm systems) to indicate a fault on the system.

Turbulent Burning Velocity—The burning velocity of a flame when turbulence is present in the flammable mixture.

Turbulent Flame—A flame burning in a turbulent flammable mixture.

Turnout Clothing or **Gear**—Protective garments used for fighting a fire beyond its incipient stage. Turnout clothing includes a helmet with face shield, coat, trousers, gloves, and insulated firefighters' boots. NFPA Standard 1971, *Protective Clothing for Structural Fire Fighting,* specifies performance, testing, and fire resistance of protective coats and trousers. NFPA 1972, *Standard for Helmets for Structural Fire Fighting,* specifies helmet performance requirements for impact, penetration, flammability, thermal endurance, retention, and limited electrical insulation. NFPA 1973, *Standard on Gloves for Structural Fire Fighting,* provides specifications for gloves to protect firefighters against adverse environmental effects to the hands and wrists during structural firefighting and skin exposure to blood or other liquid-borne pathogens and exposure to limited common liquids. NFPA 1974, *Standard on Protective Footwear for Structural Fire Fighting* provides specifications to mitigate adverse environmental effects to the foot and ankle during structural firefighting. It may also be called structural firefighting protective clothing or bunker gear. See Figure T-3. *See also* **Entry Clothing; Thermal Protective Clothing.**

Figure T-3 Firefighter in full structural firefighting PPE.

T

Turret Pipe—A large master water stream appliance normally mounted on a pumper truck or trailer that is directly connected to the discharge of a pump. It may also be called a deck gun or deck pipe.

Twin Agent Unit or **Twinned Agent Systems**—A manual firefighting unit assembly of both foam and dry chemical applicators for simultaneous application of both agents. Twin agent units are effective in quick fire control and sealing the surface of liquid fuels. They apply a stream of potassium bicarbonate dry chemical (Purple K) for knockdown and aqueous film forming foam (AFFF) for vapor coverage. They were primarily developed and used for aircraft crash fires. Some arrangements are available that can be used by one person. They are usually mounted on a small truck, as a turret assembly on a crash rescue truck or at fixed locations where a frequent fire hazard occurs (e.g., rotary aircraft landing/takeoff heli-decks), or other areas where flammable liquids might be handled. *See also* **Foam; Dry Chemical; Potassium Bicarbonate.**

Twin Fluid Media Systems—A water delivery system that produces water mist for fire protection applications. The system produces water mist by impingement of two fluids delivered from separate piping systems. One set of piping provides water to the nozzle and the second piping network provides an atomizing fluid or medium. *See also* **Dual Agent System; Water Mist.**

Two In/Two Out Rule—The procedure of having a crew standing by completely prepared to immediately enter a structure to rescue the interior crew should a problem develope.

Type I (Foam) Discharge Outlet—A discharge outlet of a fixed foam system that conducts and delivers air aspirated foam gently onto the surface of the liquid to be protected, without submergence of the foam or agitation of the liquid surface, as defined by NFPA 11, *Standard for Low Expansion Foam*. Type I discharge outlets are generally considered obsolete because nearly all currently manufactured foams are suitable for use with Type II discharge outlets (tolerant to submergence). Porous tubes (Moeller tubes) and foam troughs along the inside of tank are common Type I foam discharge outlets. *See also* **Foam Chamber; Foam Chute; Foam Trough; Moeller Tube.**

Type II (Foam) Discharge Outlet—A discharge outlet of a foam chamber for a fixed foam system that does not deliver air aspirated foam gently onto the surface of the liquid to be protected, but is designed to lessen submergence of the foam and agitation of the liquid surface, as defined by NFPA 11, *Standard for Low Expansion Foam*. Most modern foams use Type II foam discharge outlets as they are more resistant to submergence. It may also be called a foam chamber. *See also* **Foam Chamber.**

Ultra High-Speed Water Spray System—A fire control and suppression system that is designed to discharge water within 100 milliseconds after fire detection. Ultra high-speed water spray systems are provided with materials with an extremely high flame spread and heat release may be present such as munitions, propellants, or pyrotechnics. *See also* **Explosion Suppression System; Point Protection; Ultra High-Speed Water Spray System, Solenoid Operated.**

Ultra High-Speed Water Spray System, Solenoid Operated—A fire suppression system that is designed to discharge water within 100 milliseconds after fire detection but is activated by a pilot head detection system and discharge nozzles controlled by solenoid valves. Release of the pilot line pressure causes water flow activation to all discharge nozzles. *See also* **Ultra High-Speed Water Spray System.**

Ultraviolet (UV)—Wavelengths of the electromagnetic spectrum that are shorter than the blue-violet light of the visible spectrum and larger than x-rays. Descriptive of invisible radiation between wavelengths of 10^5 to 10^6 centimeters. UV radiation is emitted by the combustion process and fire detection devices are available to sense these emissions. *See also* **Ultraviolet (UV) Fire Detection.**

Ultraviolet (UV) Fire Detection—An optical flame-sensing device used to detect a fire by the presence of radiation within ultraviolet (UV) wavelengths (wavelengths of 4,000 angstroms or less). They use a Geiger-Mueller tube to detect the UV radiation. UV detectors may be sensitive to lightning and welding operations and should be shielded from these effects. A UV detector should be positioned accurately and within its specified range to operate effectively (they are subject to the inverse square law of optics). UV detectors are not affected by sunlight. The lens of the device must also be kept clear for optimum sensing ability. *See also* **Ultraviolet (UV); Ultraviolet-Infrared (UV/IR) Fire Detection.**

Ultraviolet-Infrared (UV/IR) Fire Detection—An optical-sensing device used to detect fire by the presence of both ultraviolet (UV) and infrared (IR) radiation by the use of both UV and IR fire detectors. Both radiation sources must be present simultaneously for the detector to activate. UV/IR detectors are used where false alarms may be a concern by using either a UV or an IR detector by itself. *See also* **IR Fire Detector; Ultraviolet (UV) Fire Detection.**

Unburned Island—An area of unburned vegetation within a fire perimeter. *See also* **Fire Perimeter; Forest Fire.**

Unconfined Vapor Cloud Explosion (UVCE)—An explosion in which the cloud of vapor (gases or mist) ignites in air, resulting in a detonation accompanied by a blast wave and intense heat. UVCE explosions generally occur in the process industries (chemical and petroleum) and may cause vast devastation to the facility due to the blast effects of the incident. Unconfined vapor cloud explosions occur in the open air, but have some degree of confinement, which allows the flame front to accelerate to achieve explosion parameters. The term vapor cloud explosion (VCE) is therefore sometimes used to describe the same phenomenon and may be considered more appropriate.

The vapor cloud explosion is the result of a flame front propagating through a premixed volume of air and flammable gas or vapor. The flame front must propagate with sufficient velocity to create a pressure wave.

For a VCE to occur, certain conditions must be met. These include (1) release of flammable material, (2) sufficient mixing of the flammable material in air to allow rapid flame front propagation, (3) ignition, and (4) confinement of the flame path, which tends to accelerate the flame front. The blast effects of a VCE can vary greatly and are determined by the flame speed. The flame speed of the explosion is affected by the turbulence created within the vapor cloud as it passes through congested or confined areas where the vapor is released. Normal process industry plant design and equipment arrangements have been shown to create enough congestion and confinement to produce turbulent vapor cloud conditions and allow rapid flame propagation to occur. The timing of the vapor cloud ignition, whether immediate or delayed, will determine the amount of flammable material released and the magnitude of the vapor cloud explosion. *See also* **Explosion.**

Uncontrolled Chemical Reaction—An increased reaction rate or rapid decomposition, usually involving materials that undergo exothermic chemical reactions or that have high heat of decomposition.

Underwriter's Laboratories (UL)—A worldwide recognized independent fire and safety testing laboratory. Underwriter's Laboratories developed as a result of the concern by the Western Underwriters Association over the first practical installation and exhibit of electric lighting at the 1893 World's Colombian Exposition held in Chicago. They hired an electrical engineer to evaluate the electrical problems. A laboratory was arranged in a room over a local fire station, which formally evolved into Underwriter's Laboratories (UL) in 1896. *See also* **Approved; Classified; Labeled; Listed.**

Underwriters Playpipe—Portable firewater hose nozzle used in the late 19th century for firefighting. It was later applied in the evaluation of water systems supporting fire protection systems or equipment. It consists of a smooth bore pipe approximately 30 in. (76.2 cm) long that provides a laminar hose stream for measuring the quantity of water supplies by using a pitot tube with 1.0 or 1.125 in. (2.54 or 2.86 cm) tips. It was commonly fitted with swivel handles at the base and wound with cord. The handles and cord assist in holding and directing the nozzle. It was designed by John R. Freeman (1855–1932) in 1889. It is mostly considered obsolete, and was mainly used by underwriters for testing private water supplies rather than for firefighting.

Underwriting—The selection of subjects for insurance in such a manner that general company objectives are met. The main objective of underwriting is to see that the risk accepted by the insurer corresponds to that assumed in the rating structure. There is often a tendency toward adverse selection, which the underwriter must try to prevent. An adverse selection occurs when those most likely to suffer loss are covered in greater proportion than are others. The insurer must decide upon certain standards, terms, and conditions for applicants;

project estimated losses and expenses through the anticipated period of coverage; and calculate reasonably accurate rates to cover these losses and expenses. Since many factors affect losses and expenses, the underwriting task is complex and uncertain. Poor underwriting has resulted in the failure of many insurers.

In some types of insurance, major underwriting decisions are made in the field; in other types, they are made at the home office. In the field of property insurance, the contract is cancelable if the home-office underwriter later finds the risk unacceptable. It is not uncommon for a property insurer to accept large risks only to cancel them later after the full facts are analyzed.

An important initial task of the underwriter is to try to prevent adverse selection by analyzing the hazards that surround the risk. Three basic types of hazards have been identified as moral, psychological, and physical. A moral hazard exists when the applicant either may want an outright loss to occur or may have a tendency to be less than careful with property. A psychological hazard exists when an individual unconsciously behaves in such a way as to engender losses. Physical hazards are conditions surrounding property or persons that increase the danger of loss. Physical hazards include such things as woodframe construction in buildings, particularly in areas where such properties are densely concentrated. Earthquake insurance rates tend to be high where geologic faults exist (as in San Francisco, which is built almost directly over such a fault). In fire insurance, the physical hazards are analyzed according to four major factors: type of construction, the protection rating of the city in which the property is located, exposure to other structures that may spread a conflagration, and type of occupancy. *See also* **Fireline.**

Uniform Building Code (UBC)—A building construction code that is concerned with the construction of buildings for fire, life, and structural safety. It is maintained and published by the International Council of Building Officials (ICBO). It is provided for adoption by city and state governments, mostly in the western United States.

Uniform Fire Code (UFC)—A regulation concerning fire safety or prevention measures for equipment, operations, material storage and handling, and the care or maintenance of fire protection systems. The Uniform Fire Code was originally developed by southern California fire inspectors and adopted by the California Fire Chiefs Association 1961. It was published by the International Conference of Building Officials and the Western Fire Chiefs Association in 1971, partly to coordinate it with the Uniform Building Code. In 1991, the International Fire Code Institute assumed responsibility for its publication and development.

United States Fire Administration (USFA)—An office of the US Federal Emergency Management Agency (FEMA) that develops fire service educational and training programs. It operates the National Fire Academy (NFA) in Emmitsburg, Maryland and the National Fire Data Center (NFDC), through which it performs research on fires and firefighter and residential safety through incident data collection from the National Fire Incident Reporting System. Its objective is to reduce losses from fire through enhanced fire prevention and control. It was established by the Federal Fire Prevention and Control Act of 1974 (US Public Law 93-498). *See also* **Federal Emergency Management Agency (FEMA); National Fire Incident Reporting System (NFIRS); National Fire Protection Association (NFPA).**

Unprotected Opening—An opening in a fire barrier that will allows the passage of fires, combustion gases, or smoke. *See also* **Fire Stop; Penetration Seal.**

Updraft—An upward current of air or gases. Updrafts normally occur in fires due to the combustion process generating heated gases that immediately rise from hot convection currents. *See also* **Backdraft** or **Smoke Explosion; Indraft.**

Upper Explosive Limit (UEL)—The maximum proportion of vapor or gas in air above which propagation of a flame does not occur. It is the upper limit of the flammable or explosive range. UELs are determined in accordance with ASTM E-681, *Standard Test Method for Concentration Limits of Flammability of Chemicals. See also* **Detonation Limits; Explosive Limits; Lower Explosive Limit (LEL); Upper Flammable Limit (UFL).**

Upper Flammable Limit (UFL)—Synonymous with upper explosive limit. *See also* **Flammable Limit** or **Flammability Limits; Lower Flammable Limit (LFL); Upper Explosive Limit (UEL); Vapor.**

Vamps—Slang terminology for volunteer firefighters. The term started due to the association of gaily colored socks, which were called vamps, that were worn by early volunteer firefighters.

Vapor—The material that emanates from a substance that is liquid at standard conditions. Vapors form the fuel source for free-burning fires or explosions. *See also* **Lower Flammable Limit (LFL); Upper Flammable Limit (UFL).**

Vapor Cloud Explosion (VCE)—*See* **Unconfined Vapor Cloud Explosion (UVCE).**

Vapor Pressure—The pressure exerted by a volatile liquid as determined by ASTM D-323, *Standard Method of Test for Vapor Pressure of Petroleum Products (Reid Method),* commonly referred to as Reid Vapor Pressure (RVP).

Velocity Pressure—The pressure water exerts within a stream due to its velocity.

Vent—An opening for the passage of, or dissipation of, fluids such as gases, fumes, and smoke. Vents may be permanent (a chimney) or provided during emergency conditions by firefighters using manual tools to cut an opening in a roof. *See also* **Venting, Fire.**

Vented Suppressive Shield (VSS)—A barrier designed to protect personnel and equipment from explosive incidents. Developed by the US Army, it consists of a barrier wall or container made of perforated metal sheets, spaced layers of metal angles, or "I" beams and screening specifically designed and engineered for the blast that may be encountered. It prevents the passage of blast pressure, fragments, or flame.

Vent, Heat, and Smoke—An assembly rated for the release of heat or smoke from a fire event. Heat and smoke vents are commonly provided in the roofs of buildings. They may be activated by means of automatic detection or constructed of materials that cause the material to melt from the heat of the fire and create an opening for venting. The sizing of heat or smoke vents should be based on the anticipated fire event. *See also* **Venting, Fire.**

Ventilation—*See* **Venting, Fire.**

Ventilation, Horizontal—Channeled pathway for fire ventilation through horizontal openings.

Ventilation-Limited Fire or **Ventilation-Controlled Fire**—A fire in which the combustion rate is controlled by the availability of oxygen rather than by the supply of fuel. Fires in confined spaces, such as building occupancies, are generally considered ventilation controlled fires because the available air supply is limited. The maximum production of carbon monoxide, smoke, and energy occurs when the ventilation limitation occurs for a compartment fire. *See also* **Fuel-Controlled** or **Fuel-Limited Fire.**

Ventilation, Mechanical—The use of exhaust fans, blowers, air conditioning systems, or smoke ejectors to remove products of combustion (smoke, heat, gases) from an area affected by a fire event.

Ventilation, Vertical—*See* **Venting, Fire.**

Venting, Fire—The escape of smoke, noxious or toxic fumes, and heat through openings in a building provided as part of the structure (a chimney) or instituted during emergency firefighting actions for the removal of hot gases and smoke particles. In fire conditions, it is generally accepted that the efficient venting of heat, hot gases, and smoke reduces the lateral spread and subsequent damage and enables firefighters to more easily enter a building on fire and begin fire protection measures. In order for roof or ceiling vents to operate efficiently, there has to be an adequate source of low level replacement air. Ventilation through the top of a structure or its roof vents or similar devices (skylights) is called vertical ventilation or top ventilation. *See also* **Smoke Extraction; Vent; Vent, Heat, and Smoke.**

Very Early Smoke Detection and Alarm (VESDA®) System—A registered company trademark name for a smoke detection and alarm system that samples air for smoke particles using air suction tubes to specific points for the protected hazard. An air sampling smoke detection system such as VESDA is considered 100 to 1,000 times more sensitive for fire detection than spot ionization smoke detectors. *See also* **Active Smoke Detection System; Aspirating Smoke Detection (ASD); Projected Beam Smoke Detector; Smoke Detector; Duct.**

V

Void(s)—Spaces within a collapsed area that are open and may allow someone to survive a building collapse.

Volunteer Fire Department—A firefighting organization that is made up of volunteer firefighters. *See also* **Fire Department (FD); Firefighter; Volunteer Firefighter.**

Volunteer Firefighter—An individual who contributes his or her time and manpower for the purpose of fire protection activities within an organized fire department, either paid or volunteer. A volunteer firefighter is responsible for the same activities as paid firefighter. In some localities, a volunteer is "on-call," and responds to a notification of a fire incident. An individual becomes a volunteer firefighter due to personal concern for civic responsibility and interest in fire protection. Liabilities for injuries and death benefits to a volunteer firefighter are the responsibility of the governmental organization to whom the volunteer firefighter is providing his or her services. *See also* **Firefighter; Volunteer Fire Department.**

Vortex Plate—A plate provided around the intake of a firewater pump suction bell to prevent the formation of vortices. Formation of vortices at the intake of a firewater pump can cause damage to the pump impeller due to erosion caused by

Figure V-1 "V" pattern.

air created by the vortices in the pump intake stream.

V Pattern—The characteristic cone-shaped pattern left on a vertical surface at or near a fire's point of origin. Fire patterns are useful in arson investigation or fire recreations. See Figure V-1. *See also* **Burn Pattern.**

Vulcan—The Roman god of fire, Vulcan was representative of fire worship in Roman civilization. Originally an old Italian deity who seems to have been associated with volcanic fire, Vulcan was identified with the Greek god Hephaestus. In classical time in Rome, his festival, the Volcanalia, was celebrated on August 23. *See also* **Hephaestus.**

Wall (Post) Indicator Valve (WPIV or WIV)—*See* **Post Indicator Valve (PIV).**

Wash Down—To hose down or apply water to an area for the removal of debris or cleaning.

Water—The most efficient, cheapest, and most readily available medium for extinguishing fires of a general nature. It is an odorless, colorless, virtually incompressible liquid that freezes at 32°F (0°C). Water boils at 212°F (100°C), and therefore is highly useful as a firefighting agent. It takes 7.48 gals. or 1 cu. ft. (28.3 liters) of water to generate 1,650 cu. ft. (46,722 liters) of steam. Water may be used to cool locations or equipment exposed to a fire to prevent damage or spread of the fire. It is also combined with chemical agents to form foam to combat petroleum and chemical fires. Its primary function as a fire-extinguishing medium is as a coolant. Maximum cooling effect is achieved if the water is in the form of a fine spray or fog, which cools the flames and burning surfaces. If the temperature of the burning surface is cooled below its flash point, the fire will be extinguished. Oils with very low flash points, such as gasoline, cannot be cooled sufficiently by water spray, but some diminution of the intensity of the flames is achieved by the cooling effect of the water droplets. Conversely, heavy fuels and oils with high flash points can be effectively and easily extinguished by water spray or fog application. *See also* **Fire Extinguisher, Portable; Fire Suppressant Agent; Firewater; Water Fog; Water Spray.**

Water Additive—A variety of chemicals may be added to water to improve its ability to extinguish fires. Wetting agents added to water can reduce its surface tension. This makes the water more penetrating and facilitates the formation of small drops necessary for rapid heat absorption. By adding foam-producing chemicals and liquids to water, fire-blanketing foam is produced. Foam is used to extinguish fires in combustible liquids, such as oil, petroleum, and tar, and for fighting fires at airports, refineries, and petroleum distribution facilities. Chemical additives can expand the volume of foam 1,000 times. This high expansion foam-water solution is useful in fighting fires in basements and other difficult-to-reach areas because the fire can be smothered quickly with little water damage. *See also* **Foam; High Expansion Foam; Low Expansion Foam; Wetting Agent.**

Water Bombers—*See* **Fire Bombers.**

Water Cannon—A large water application nozzle that cannot be handheld due to its size and water reaction forces. *See also* **Heavy Stream.**

Water Columning—A condition in a dry pipe sprinkler system in which the weight of water in the riser prevents the operation of the dry pipe valve.

Water Curtain—A screen or wide angle spray of water that is set up and used to protect exposures from fire effects mainly from radiated heat, smoke, and billowing flames. It normally consists of open or closed sprinkler heads or perforated pipes installed on the exterior of a building at eaves, cornices, window openings, or peaked roofs under manual control, or installed around the openings in floors or walls of a building with the water supply under thermostatic control. Manual firefighting operations may also provide and position water spray nozzles to provide a water curtain to protect exposures. It may also be called a water screen.

Water Damage—The damage sustained to a property as a direct result of water-based firefighting efforts or because of leakage from a fixed water-based suppression system (sprinkler system).

Water Deluge System—A network of small diameter piping and open spray nozzles connected to a control valve from a water supply source that is capable of delivering the designed water spray to the protected area. When the control valve is opened, water is simultaneously released from all spray nozzles.

Water Distribution System—A system provided to transport adequate water supplies for residential, industrial, commercial, and fire protection purposes. It consists of a network of distribution pipes and control valves. Where flow or pressure requirements cannot be met, gravity feed (elevated water tanks) or mechanical pumps are provided.

Lead pipes were in use in London in 1236 to convey water to the city, but mostly open aqueducts or conduits were provided. London's first pumped water supply was built in 1581. The Thames River current turned a water wheel that enabled water to be pumped after a design of similar devices in Germany. Wooden pipes were used in 1613 to direct water to different parts of London. The wooden pipes were made from logs of elm with bores of 2 to 10 in. (5 to 25.4 cm). They were made by using a water wheel that bored into the trunk of the log. Joints were made by driving tapered ends into larger ends. At one time, London had 400 miles (643 km) of wooden pipes in use. Philadelphia replaced its wooden mains with cast iron pipes in 1818 and New York City laid its first cast iron water mains in 1829. *See also* **Fire Main; Firewater Distribution System; Water Distribution System.**

Water Flooding—The introduction of water into a process vessel or tank containing liquid with a lower density than water in order to displace its contents above the introduced water. It is used when a leak occurs on the vessel or tank to allow water to drain out and prevent process material from being released and causing a hazard. Water flooding connections are primarily provided in the process industries to prevent the release of combustible materials. They are intended for use by industry fire brigade pumpers. Care must be taken in the application of water flooding connections that the piping material chosen and process conditions are not the source of failure in the system. In practice, the logistics of performing such an operation while simultaneously conducting firefighting or prevention activities may preclude the usefulness of water flooding, although for very large volume containers it might be useful.

Waterflow Alarm—A device that indicates the flow of a fluid in a liquid (water-based fixed fire suppression system). It alerts occupants that the system has activated or has been damaged and water flow is occurring, requiring immediate action. Most waterflow alarms are connected to the local fire alarm control panel to sound an alarm throughout the entire premises or relay to a manned monitoring and control station. *See also* **Alarm Check Valve; Sprinkler Alarm; Water Motor Alarm** or **Gong; Waterflow Detector.**

Waterflow Detector—A device that detects any flow of water from a fixed fire protection water-based system equal to or greater than that from a single automatic sprinkler of the smallest orifice size installed on the system. It is required to produce an audible alarm on the premises within five minutes after such flow begins, and will continue until the flow stops. Water motor gongs use available water pressure during system activation while vane-type paddle switches and pressure switches are connected to a fire alarm system. Paddle or vane-type alarms are allowed only for use in wet pipe systems because the initial force of water from dry system activation may damage the paddle flow indicator, preventing an alarm. Pressure switches are commonly used in dry systems. A signal may also be taken from the alarm check valve for a wet pipe system that opens water flow to a pressure switch. A retard adjustment is available on paddle or vane switches to prevent false activation from momentary pressure surges in the water supply system, while retard chambers are provided at pressure switches for water surge conditions. See Figures W-1 and W-2. *See also* **Alarm Check Valve; Fire Alarm System; Retard Chamber; Waterflow Alarm; Water Motor Alarm** or **Gong.**

Water Flow Test—An evaluation of water supplies and a piping distribution network to

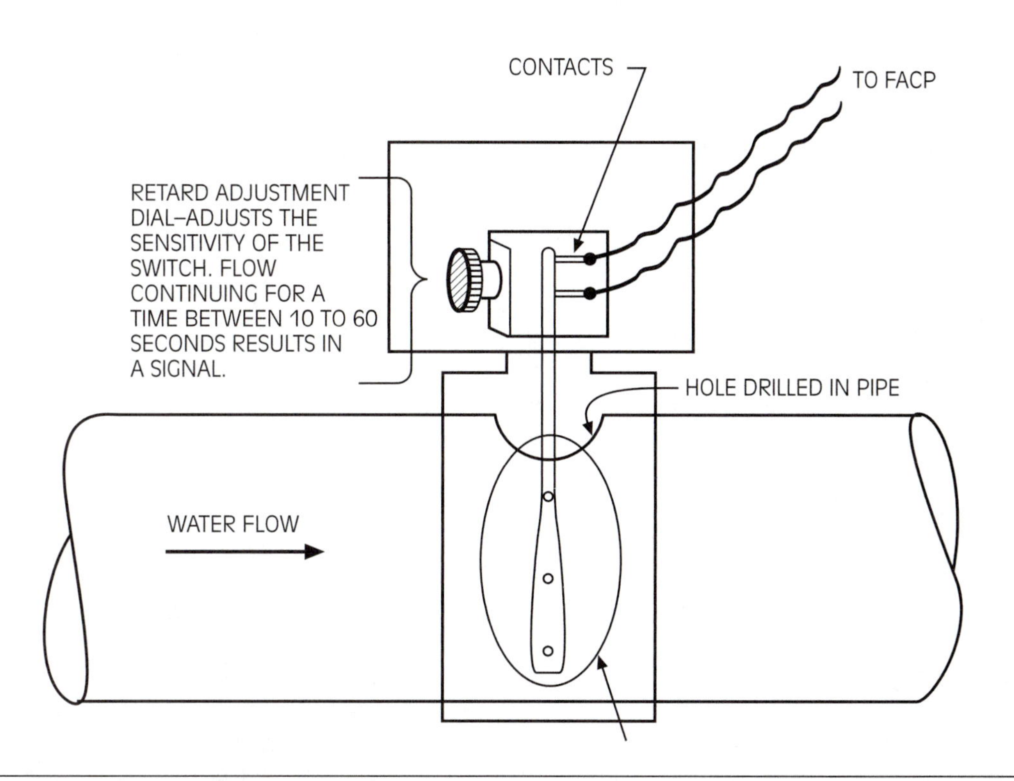

Figure W-1 Water flow alarm, vane type.

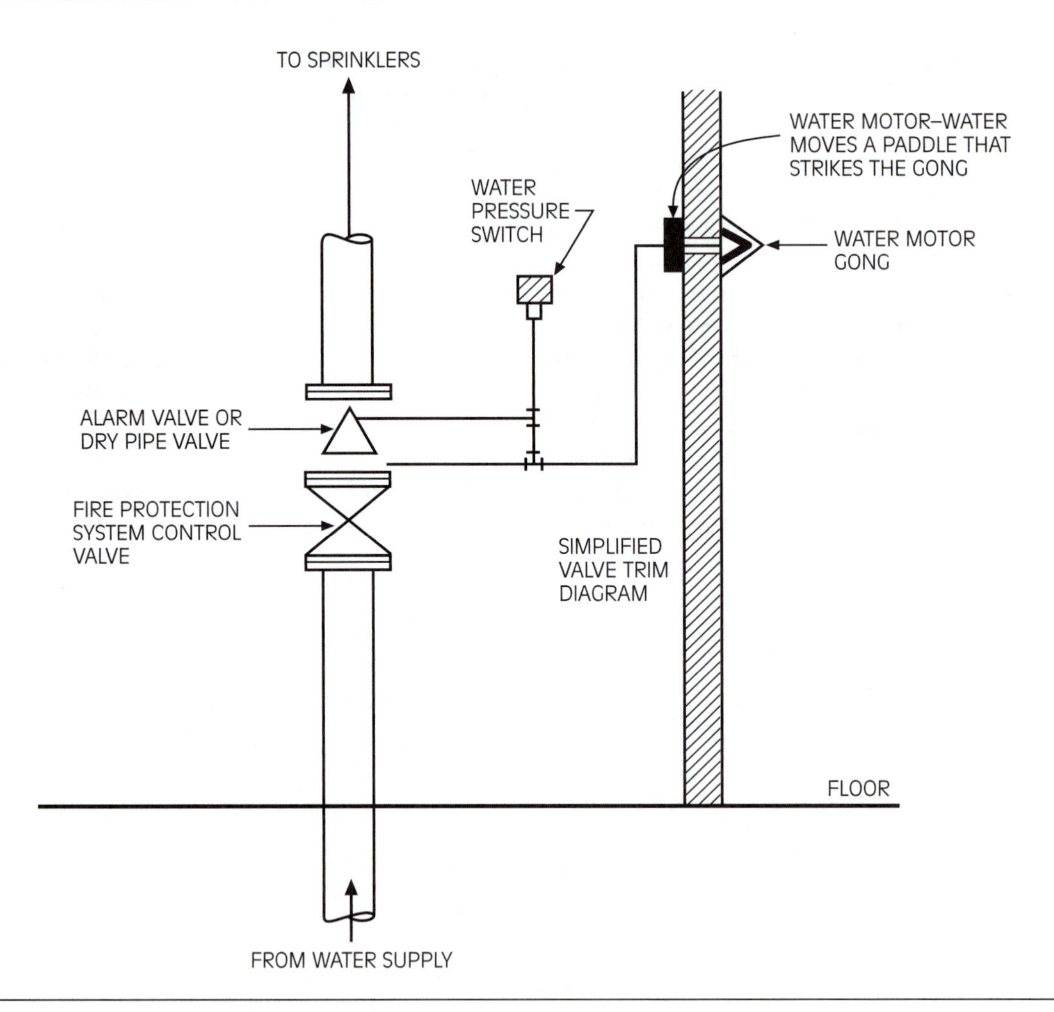

Figure W-2 Waterflow alarm, pressure switch type.

determine whether it is of sufficient capacity and pressure to provide or meet fire protection needs or requirements. Static pressure, residual pressure, hydraulic profile, and flow rates may be obtained during water flow tests.

Water Fog—A term used to describe the water droplet formation where the median diameter is less than 0.03 in. (0.075 cm) used for fire protection applications. Water spray is used to denote water droplets of a large size. Water fog is produced by the impingement of converging jets or by forcing the water through specially designed teeth that break it up into fine particles. Water fog is considered more effective than water spray for direct application to burning surfaces because, as the average water droplet size decreases, the rate of heat transfer and, therefore, the rate of conversion to steam increases. Water fog application is more susceptible to the effects of ambient conditions, wind, updrafts, etc.; therefore, water spray is commonly used to achieve depth penetration of a burning fire. It may also be called fire fog. *See also* **Nozzle, Fog; Water; Water Mist; Water Spray.**

Water Hammer—An increase or dynamic change in pressure produced as a result of the kinetic energy of the moving mass of liquid being transformed into pressure energy, which results in an excessive pressure rise. As the pressure or shock pulsations move backward and forward, they are generally accompanied by a series of rapidly succeeding noises like the rapping of a hammer against a pipe; hence the name. Water hammer may occur in a firewater system due to the sudden startup of a firewater pump, the rapid opening or closing of a valve, etc., within a pipe network made of rigid pipe materials. Water hammer causes a pipe to stretch and eventually, after numerous repeated applications, may cause it to fail. Water hammer conditions can be avoided by the prevention of rapid pressure buildup conditions in a distribution pipe network, such as by slowly opening or closing valves, etc. Mathematical analysis (surge analysis) of the possibility of water hammer can be undertaken during the design of a water supply and distribution system.

Water Mist—A fine spray of water particles at high pressure for the purpose of fire protection. The very small water droplets allow the water mist to control or extinguish fires by cooling the flame and fire plume, displacing oxygen by water vapor, and attenuating radiant heat. Water mist systems can be effective on both solid (Class A) and liquid (Class B) fuel fires. Research indicates that droplets smaller than 400 microns are essential for extinguishment of Class B fires, whereas larger drop sizes are effective for Class A combustibles that benefit from extinguishment by fuel wetting. Three classes of water mist systems have been defined according to droplet size per NFPA Standard 750, *Standard on Water Mist Fire Protection Systems.* Overall, water mist spray has droplets of 1,000 microns or less at a distance of 3.3 ft. (1 m) from the discharge nozzle.

Class 1 water mist systems are defined as 90 percent of the droplets in the water spray having less than 200 microns in size. This system represents the "finest" water mist. Many commercially available water mist nozzles produce Class 1 mists. Class 2 water mist systems are defined as having 90 percent of the droplets in the water spray less than 400 microns in size. These water mists can be generated by pressure jet nozzles, twin-fluid nozzles, and many impingement nozzles. Because of the larger water drop sizes, higher mass flow rates are easier to achieve with Class 2 sprays than with Class 1 sprays. However, the larger drops are not effective on liquid fuel fires. Ordinary combustibles are more effectively extinguished with a Class 2 water mist because considerable surface wetting occurs. A Class 3 water mist system has water droplet sizes that do not meet the criteria for Class 1 or Class 2 water mist systems but are less than 1,000 microns in size. Water mists in this class are generated by intermediate pressure, small orifice sprinklers, impingement nozzles of various sorts, and fire hose fog nozzles. High mass flow rates are possible. They are suitable for Class A combustibles, and under some circumstances provide fire control or fire extinguishment for Class B fires.

In general, Class 1 and Class 2 water mists are successful at extinguishing liquid fuel pool fires and spray fires without agitation of liquid pool surfaces. Given an appropriate geometry, Class 3 sprays are reported to have extinguished pool fires. In addition, in general, it is difficult to extinguish Class A combustibles with Class 1 sprays, which may not achieve the fuel wetting necessary to penetrate the char layer. However, Class A fires can be extinguished with Class 1 mists, particularly if the velocity is high, the burning is superficial, or enclosure effects enhance the degree of oxygen reduction. This evidence confirms that drop size distribution alone does not determine the ability of a spray to extinguish a given fire. Factors such as fuel properties, enclosure effects, spray flux density, and spray velocity (momentum) are all involved in determining whether a fire will be extinguished. The drop size distribution of a spray does not uniquely define its suitability for a given application. It is inseparable from the spray direction relative to the fire plume, its velocity, and flux density. The "momentum" of an element of spray is the product of its velocity (which includes direction as well as speed) and the mass of dispersed water droplets. Therefore, all three variables—drop size distribution, flux density, and momentum—are involved in determining the ability to extinguish a fire in a given scenario. The classification system allows a designer to distinguish between the fine and coarse end of the spectrum of sprays encompassed by the definition. *See also* **Twin Fluid Media Systems; Water Fog; Water Mist System, High Pressure; Water Mist System, Intermediate Pressure; Water Mist System, Low Pressure.**

Water Mist System, High Pressure—A water mist system where the operating pressure is 500 psi (3,445 kPa) and greater. *See also* **Water Mist.**

Water Mist System, Intermediate Pressure—A water mist system where the operating pressure is between 175 and 500 psi (1,206 and 3,445 kPa). *See also* **Water Mist.**

Water Mist System, Low Pressure—A water mist system where the operating pressure is 175 psi (1,206 kPa) or lower. *See also* **Water Mist.**

Water Motor Alarm or **Gong**—A mechanical bell alarm that is hydraulically operated by water pressure. Water is directed to a paddle wheel (pelton wheel) during activation of a fixed water-based fire protection system and rotates a shaft that turns a striker against a bell housing. The alarm will sound as long as water is flowing in the system. The bell housing is usually mounted on an external wall of the protected premises to attract the attention of watchmen or the public. It is used in conjunction with wet pipe alarm check, dry pipe, deluge, and pre-action valve fire protection systems. See Figure W-3. *See also* **Sprinkler Alarm; Waterflow Alarm; Waterflow Detector.**

Water Relay—An operation where firewater is transferred over long distances by firewater pumpers that are connected by temporary connections and spaced at intervals along a route between the water source and the point where water is needed. *See also* **Relay Operation.**

Water Screen—A series of wide angle or mist spray nozzles connected to a water supply system, which protects a specific area from the passage of heat, smoke, and billowing flames.

Water Spray—The use of water in a form having a predetermined pattern, particle size, velocity, and density discharged from specially designed nozzles or devices. Water spray is generally referred to when the mass median diameter of the droplets is more than 0.03 in. (0.075 cm) in size. Water spray fixed systems are specifically designed to provide for fire control, extinguishment, or exposure protection (cooling). Water spray fixed systems may be independent of, or supplementary to, other forms of fire protection. Water spray was not commonly employed for fire protection purposes until the 1930s and 1940s, when research highlighted the benefits of smaller water droplet size for improved heat absorption from a fire due to a higher surface

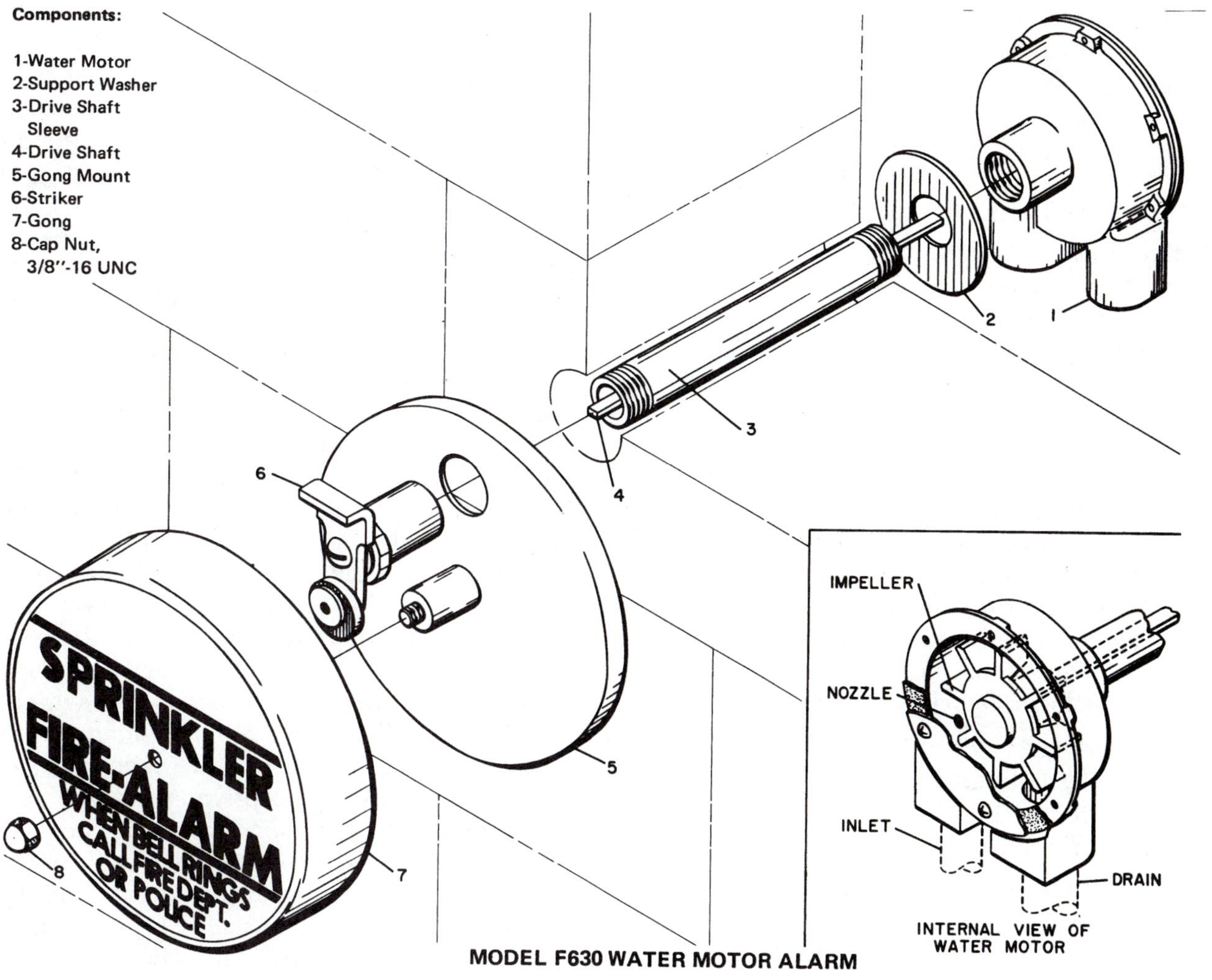

Figure W-3 Water motor alarm. *(Courtesy Grinnel Fire Protection Systems Company, Inc.)*

area availability from the water droplets. *See also* **Cooling Spray; Hose Stream; Nozzle, Fog; Water; Water Fog.**

Water Tender—Fire service term for a land-based water supply apparatus. *See also* **tanker.**

Water-Thickening Agent—A chemical additive that is provided to water to increase its viscosity. With an increase in viscosity, an improvement in fire extinguishing effectiveness is achieved. Sodium carboxymethylcellulose is a commonly used chemical compound that is used for this purpose.

Water Thief—Slang terminology used by the fire service for a firewater hose connection/valve assembly that branches one or several smaller lines from a larger fire hose, usually a 1.5 in. (3.8 cm) line from a 2.5 in. (6.4 cm) line.

Water Tower—Mobile portable standpipes of steel construction used in the late 1800s and early 1900s for the external application of fire-fighting water to upper floors of a burning building by the fire services. They are usually mounted horizontally on a trailer for traveling and raised upright for use at the fireground location. A large diameter water application nozzle is fitted at the top of the tower. The first water tower in America entered service in 1879. With the advent of modern aerial ladders and platforms fulfilling the same purpose, water towers have generally become obsolete as an individual piece of equipment. *See also* **Aerial Ladder; Aerial Platform; Snorkel; Squirt.**

Water Wastage—Firewater that does not meet its intended application, usually caused by overspray or carryover by wind effects.

Wet Barrel Fire Hydrant—*See* **Hydrant, Fire.**

Wet Chemical—A solution of water and potassium carbonate-based chemical, potassium acetate-based chemical, or a similar combination that is used as a fire extinguishing agent. Its application may cause corrosion or staining of the protected equipment if not removed. Wet chemical solutions are generally considered relatively harmless and normally have no lasting significant effects on human skin, the respiratory system, or personal clothing. *See also* **Fire Suppressant Agent.**

Wet Chemical Fire Suppression System—An automatic fire suppression system that uses a liquid agent. It is applied through a system of piping and nozzles with an expellant gas from a storage cylinder. It is usually released by automatic mechanical thermal linkage. The agent leaves a residue that is confined to the protected area that must be removed after application. Primarily applied to having cooking range hoods and ducts and associated appliances. The wet chemical agent consist of water and usually potassium carbonate or potassium acetate. *See also* **Dry Chemical Fire Extinguishing System; Pre-Engineered System.**

Wetting Agent—A wetting agent is a chemical compound that, when added to water in amounts indicated by the manufacturer, will materially reduce the water's surface tension, increase its penetrating and spreading abilities, and might also provide emulsification and foaming characteristics. Decreased surface tension disrupts the forces holding the film of water together, thereby allowing it to flow and spread uniformly over solid surfaces and to penetrate openings and recesses over which it would normally flow. Water treated in this manner not only spreads and penetrates, but displays increased absorptive speed and superior adhesion to solid surfaces. Water normally has a surface tension of 73 dynes per centimeter and wetting agents can lower it to about 25 dynes per centimeter. Leaks in piping connections and pump packing can occur that would not have occurred if the wetting agent had not been used. Visual inspection should be made during wet water operations. Wet water should be applied directly to the surface of the combustible. These agents do not increase the heat absorption capacity of water, but the greater spread and penetration of the wet water increase the efficiency of the extinguishing properties of water, as more water surface is available for heat absorption and run-off is decreased. Therefore they enhance fire control and suppression applications, especially for three dimension fires.

Wetting agents are broadly defined as being surfactants (surface acting agents). All wetting agents are concentrated and are mixed with a liquid at varying percentages (usually 1 to 2 percent). The wetting agent can be liquid or powder. The liquid into which it is mixed for firefighting purposes is water. However, the primary sales for some wetting agents are for use as a carrier for liquid fertilizers, fungicides, insecticides, and herbicides. These wetting agents can be, and are, used for firefighting purposes. They do not have additives that protect tanks, pumps, valves, and bushings, etc., so it is recommended that unused mixtures be drained out of the tank and a flush of all parts made with plain water. With all wetting agents, hard water usually requires a greater amount of additive to produce the same results. Wetting agents designed for fire department use will normally contain rust inhibitors to protect the tank, pump, piping, and valves. Generally, the mixture loses some of its rust-inhibiting characteristics if left in the tank. Wetting agents are best used as a soaking or penetrating agent for a three-dimensional burning mass such as wildland fuels, coal piles, sawdust, cotton (bales, bedding, upholstery), rags, paper, etc. These agents are used very effectively on smoldering or glowing combustibles. All of the commercially available products that fall into the preceding category will satisfactorily suppress Class A fires. *See also* **Firewater; Proportioner; Water Additive; Wet Water; Wet Water Foam.**

Wet Water—Firefighting water to which a wetting agent has been added to reduce its surface tension and increase its penetrating power into the fire environment. Wet water is useful in congested environments where normal water application may be blocked or restricted. Wet water can more easily seep into inaccessible areas. *See also* **Wetting Agent.**

Wet Water Foam—A mixture of wet water with air that forms a foam used for firefighting. Wet water foam rapidly breaks down into its liquid form at temperatures below the boiling temperature of water. *See also* **Wetting Agent.**

Wheeled Fire Extinguisher—Wheeled fire extinguishers are essentially enlarged versions of handheld portable extinguishers. They are usually mounted on a cart with two large wheels for easy maneuverability. They generally have 10 to 20 times the capacity of normal portable extinguishes. They are intended for use on fires beyond the capacity of handheld portable extinguishers or where larger fire control capacity must be undertaken by fewer individuals (aircraft operations, petrochemical plants, etc). They are usually provided with hand-directed hoses. Carbon dioxide (CO_2), dry chemical, dry powder, Halons, and an aqueous film forming foam (AFFF) pre-mixed solution wheeled fire extinguishers are commonly available. They generally range in sizes from 30 to 350 lbs. (13.6 to 159 kg) and can be self-expelling, stored pressure, pressure transfer, or regulated supply and demand types. Self-expelling types contain the extinguishing agent with a single high-pressure cylinder and use the agent's own vapor pressure to force it out of the cylinder (CO_2 types). Stored pressure units use a single pressure cylinder but contain an agent and an expelling gas. Pressure transfer types consist of separate agent and expelling gas cylinders. Upon use, the expelling gas is transferred to the agent cylinder where it fluidizes the agent and expels it from the container. The regulated supply and demand type is similar to the pressure transfer type but uses a pressure regulator to control the expelling gas pressure to the agent tank to provide stable agent flow rates.

Wheeled fire extinguishers were introduced in the 1940s to protect aircraft at military airports. They were originally referred to as wheeled engines. *See also* **Fire Extinguisher, Portable.**

Wildland Fire—An uncontrolled fire in a forest, grassland, brushland, or land cultivated for agricultural purposes. Fire danger in a wildland setting varies with weather conditions. Drought conditions, heat, and wind all participate in drying woodlands or other fuel, making it easier to ignite. After a fire has started, drought, heat, and wind will also increase its intensity. Topography also affects wildland fire, which spreads quickly uphill and slowly downhill. Dried grass, leaves, and light branches are considered flash fuels; they ignite readily, and fire spreads quickly in them, often generating enough heat to ignite heavier fuels, such as tree stumps, heavy limbs, and the matted duff of the forest floor. Such fuels, ordinarily slow to kindle, are difficult to extinguish. Green fuels, such as growing vegetation, are not considered flammable, but an intense fire can dry out leaves and needles quickly enough to allow ready ignition. Green fuels sometimes carry a special danger—evergreens, such as pine, cedar, fir, and spruce, contain flammable oils that burst into flames when heated sufficiently by the searing drafts of a forest fire. *See also* **Brushfire; Fireline; Forest Fire.**

Wired Glass—Glass that is manufactured with an embedded thin wire net in order increase its resistance to failure from fire heat conditions or hose stream applications during a fire. Wired glass is usually provided in a fire-rated barrier where visual observation is needed through the barrier for operational or aesthetic reasons. Clear glazing without embedded wires has also been developed that is fire-protection-rated. It is also able to provide a higher hourly fire rating than had been possible with wired glass. These products include fire-protection-rated glass and transparent ceramics as well as fire-resistance-rated glass. *See also* **Heat-Resistant Glass.**

Wood Frame Construction—*See* **Building Construction Types.**

Worst Case Consequence Analysis—A review of possible incident scenarios to highlight the most practical event with dire circumstances.

Zoned Application Systems—A fire protection methodology where a volume is protected by several distinct zones, each with its own detection system.

Zoned Smoke Control—A smoke control system that provides smoke exhaust for a smoke zone and pressurization of all adjacent smoke control zones, thereby providing removal of smoke from the primary area of concern and using preventive measures to avoid additional smoke infiltration to the primary area of concern. *See also* **Smoke Management System (SMS).**

Notable Fires Throughout History

The following is a chronological list of some of the most notable fires that have occurred throughout history. These fires are noteworthy for several possible reasons: a huge loss of life, significant destruction, highly unusual nature, the frequency of reoccurrence, or the aftermath of its effects, i.e., lessons learned with respect to causing a change in fire safety requirements. Other great fires that are known about but for which exact details of losses are unknown, have not been listed.

390 BC Rome

The city of Rome was burned by the Gauls under the leadership of the chieftain Brennus in 390 BC. It was a great disaster, but its effect was temporary because they rebuilt quickly. It did, however, serve as an enormous wake-up call to the Romans, who launched a campaign of aggression as a result.

47 BC Library of Alexandria, Alexandria, Egypt

The Library of Alexandria was partially or wholly destroyed on several occasions. In 47 BC, during the civil war between Julius Caesar and the followers of Pompey the Great, Caesar was besieged in Alexandria. A fire that destroyed the Egyptian fleet spread through some stores of books, about 40,000 of which were ruined. According to legend, the library at Alexandria was burned three times: in 272 by order of the Roman emperor Lucius Domitius Aurelian; again in 391 under the Roman emperor Theodosius I, who ordered that all non-Christian works be eliminated; and in 640 by Muslims under the caliph Umar I, 581–644.

64 Rome

Considered the Great Fire of ancient Rome, or Nero's fire, it raged for nine days. One third of the city was destroyed (10,000 to 12,000 buildings). The emperor Nero rebuilt the city with fire precautions that included wide public avenues, limitations in building heights, provision of fire-resistant construction, and improvements to the city water supplies to aid in firefighting.

1087 London, England

A major fire destroyed many of London's wooden houses and St. Paul's Cathedral. In the rebuilding, houses of stone and tile began to appear, and some streets were partially cleansed by introducing open sewers and conduits, but wooden houses remained the norm. At this time, London was the largest city in Europe north of the Alps.

1135 London, England

Widespread fire destroyed most of the city, including partially destroying London Bridge, which at the time was built of wood. Later in the 12th century, it was decreed that the lower parts of all city houses were to be made of stone and roofs were to be made of tile; however, it was not enforced.

1212 London, England

A fire occurred at both ends of the London Bridge, which was mostly of wooden construction with about 100 wooden houses, three to seven stories high, on top. As a result 3,000 people were trapped and killed. St. Thomas's hospital was later established to alleviate the distress inflicted by the fire. In fact, it is recorded that one of the inconveniences of London at this time was the frequent occurrence of fires.

1421 Amsterdam, Netherlands

The First Great Fire of Amsterdam occurred.

1452 Amsterdam, Netherlands

The Second Great Fire of Amsterdam occurred. Legislation was subsequently passed prohibiting the use of wood as a building material in the city.

1514 Venice, Italy

A large fire destroyed shops in the Rialto area before workers were called to protect government buildings. The city had no fire brigade.

1536 Delft, Netherlands

The town was heavily damaged by fire in this year, and later, by the explosion of a powder magazine in 1654.

1631 Boston, Massachusetts, USA

After a major fire in Boston in 1631, Boston adopted a fire code that required every householder to have 12 ft. (3.5 m) of pole with a large swab at the end and a ladder to reach the ridge of his house.

1640s Magdeburg, Germany

During the Thirty Years' War (1619–1648), Magdeburg was burned by its Lutheran defenders on the day it fell to the Catholics. In a strong wind, the largely wooden city was a furnace within minutes and was generally destroyed. About 25,000 died as a result of the fire.

1666 London, England

The Great Fire of London burned for three days and 80 percent of the city was destroyed. At the time, losses were estimated at $53 million. Nine fatalities were recorded. It was the worst fire in London's history. It destroyed a large part of the city of London, including most of the civic buildings, old St. Paul's Cathedral, 87 parish churches, the city water pumping station, and about 13,000 houses. The fire began accidentally in the house of the King's baker near London Bridge. A violent east wind blew the flames. Some houses were blown up by gunpowder to prevent the spread of the fire in an attempt to help extinguish it.

After the Great Fire of 1666, which destroyed most of the walled section of the city, the Rebuilding Act of 1667 stipulated that only brick or stone be used and building density and height be restricted. Interestingly enough, few changes were made to the municipal water supply system. Additional building ordinances in the early 1700s prohibited the use of widely projecting wooden cornices and later made the use of brick or stone mandatory. Additionally, wooden window sashes had to be built with a generous margin of brick, with 4 in. (10 cm) or more separating the wooden window frame from the surface of the wall. Fire insurance companies began to sell insurance as a result of the fire. Companies organized their own fire brigades to protect properties they insured, identifying them with distinctive fire marks. As a reaction to this fire,

Boston, New York, and Philadelphia ordered the use of brick or stone in the construction of new buildings.

1676 Boston, Massachusetts, USA

The first fire of serious consequence in America occurred when about 50 dwellings, several warehouses, and a church burned to the ground in Boston, Massachusetts. The occurrence of a sudden torrential rainfall aided in the extinguishment of the fire. The first order for a fire engine (hand-pumped) was placed from London, England.

1679 Boston, Massachusetts, USA

Eighty dwellings and 70 commercial buildings were destroyed by a fire that caused approximately 1 million dollars in damage. As a result of this fire, the first paid fire department in North America was organized. A Fire Chief and 12 firefighters were employed and America's first fire engine, ordered in 1676, was imported from London, England.

1689 Amalienburg, Denmark

The Opera House was destroyed by fire, causing a loss of approximately 350 lives. Following this catastrophe, fire regulations were introduced requiring the importation of fire engines from Holland and the use of leather fire hoses.

1698 Charleston, South Carolina, USA

The city suffered a conflagration that burned down the greater part of the town.

1711 Boston, Massachusetts, USA

A large conflagration spread over the city and destroyed 100 buildings because of inadequate means of control. The city bought three manual pumpers as a result of this fire, and divided the city into several fire districts called "firewards" under the control of officers.

1720 Rennes, France

Fire badly damaged the city, about two-thirds of which was destroyed. The fire raged for six days. It was spread and exacerbated by high winds, wooden construction, and overhanging buildings. Although they had two fire pumps, the water conduits did not function well. Soldiers and workers called to assist firefighting efforts started to loot the city. The city later modernized its fire brigade.

1729 Constantinople (Istanbul), Turkey

Six thousand fatalities and 12,000 buildings—most of the city—were destroyed by fire.

1740 Charleston, South Carolina, USA

A conflagration destroyed 334 homes, shops, and warehouses.

1750 Constantinople (Istanbul), Turkey

A conflagration destroyed 20,000 houses, with a loss estimate of $9 million.

1751 Stockholm, Sweden

One thousand homes were lost in a city conflagration.

1752 Moscow, Russia

Eighteen thousand houses were destroyed by fire.

1755, Lisbon, Portugal

Lisbon was heavily damaged by a great earthquake in 1755. The source was situated some distance off the coast. The total number of persons killed in Lisbon alone was estimated to be as high as 60,000, including those who perished by drowning and in the fire that burned for about six days following the shock.

1756 Constantinople (Istanbul), Turkey

A conflagration destroyed 15,000 buildings.

1760 Boston, Massachusetts, USA

A large conflagration destroyed 400 buildings.

1769 Brescia, Italy

A storage depot containing 100 tons of gunpower exploded, killing over 3,000 people and destroying one-sixth of the city. The suspected cause of the incident was dust in the atmosphere ignited by static.

1772 Smyrna, Asia Minor (Turkey)

A conflagration in the city destroyed 3,000 houses and 4,000 shops with an estimated property loss of $2 million.

1776 New York City, New York, USA

A fire believed to have been started by an arsonist destroyed about one-fourth of the city during the Revolutionary War against England, with property damage estimated at $10 million. Lack of firefighters and the poor condition of firefighting pumpers contributed to the inability to control the fire.

1778 Philadelphia, Pennsylvania, USA

The city suffered a "great fire," after which the Pennsylvania Executive Council urged a revival of fire companies, which had suffered during the Revolutionary War and immediately afterward.

1781 New London and Groton, Connecticut, USA

British troops commanded by Benedict Arnold largely destroyed the two cities by fire during the Revolutionary War.

1782 Constantinople (Istanbul), Turkey

A conflagration lasting three days killed 100 people and destroyed 10,000 buildings.

1784 Constantinople (Istanbul), Turkey

A conflagration destroyed 10,000 houses.

1788 New Orleans, Louisiana, USA

A fire started by a candle that fell and ignited a curtain destroyed most of the city, including a cathedral and two government buildings. Strong winds and a lack of firefighting equipment allowed the fire to spread into a conflagration. In total, 856 buildings burned down. As a result of the fire, the city made provisions for fire protection equipment and fire companies.

1805 Detroit, Michigan, USA

The city of Detroit was virtually destroyed by a fire. Property losses were estimated at $200,000 at the time of the fire. Volunteer Canadian firefighters ferried their fire engine from Windsor, Ontario to fight the Detroit fire.

1811 Newburyport, Massachusetts, USA

A conflagration destroyed 250 buildings (dwellings and shops). It was believed the fire was caused by arson.

1812 Moscow, Russia

During the Napoleonic Wars, Moscow was occupied by Napoleon's armies. Russian patriots set fire to the city soon after Napoleon's entry to prevent its use by the enemy. The city burned for five days, destroying 90 percent of it, and causing about $45 million worth of damage and the destruction of 30,000 houses. This is considered one the worst self-inflicted fires in history.

1814 Washington, DC, USA

Several government buildings, including the White House and the Capital, were set on fire by British troops commanded by General Robert Ross during the War of 1812 (between America and England).

1818 Cumberland River, Kentucky, USA

An oil well was unintentionally drilled on a tributary of the Cumberland River in an attempt to obtain salt. The drillers attempted to plug the well with sand, but the oil flowed into the river and covered its surface for a distance of 35 miles. The oil ignited and an enormous conflagration ensued, which not only destroyed trees along the banks of the river but also the salt works.

1820 Savannah, Georgia, USA

A fire destroyed 463 buildings (about one-half of the city) with $3 million in damage.

1824 Edinburgh, Scotland

A series of fires in the center of the city caused 10 fatalities and $200,000 in losses. The fire led to the modernization of the fire department and the redrafting of rules to prevent fires.

1825 New Brunswick, Canada and Maine, USA

Drought conditions led to the occurrence of a forest fire caused by an accidental fire set by lumbermen. Three million forest acres and several small towns and settlements were destroyed, and 160 people perished.

1834 London, England

The Houses of Parliament (Palace of Westminster) were destroyed by fire. The fire started when an accumulation of wooden tally sticks being burned in a furnace underneath the building went out of control.

1835 New York City, New York, USA

A fire of great destruction took place when 700 buildings across 55 acres were demolished, entailing a loss of over $20 million. Dynamite was used to destroy buildings to create a firebreak to stop the fire. As a result of this fire, a number of insurance companies that were heavily committed to city risks failed, and insurance rates were sharply increased. This was a contributing factor to the formation of Farm Mutual insurance companies (more than 50 percent of its risk is underwritten on farm property). Farmers were suspicious of catastrophic city losses and prompted the organization of local mutual insurance companies. The first factory mutual insurance companies were also organized in 1835. Insurance companies required houses be built with less risk, which meant limiting the amount of wood in houses.

1838 Charleston, South Carolina, USA

Approximately half the city of Charleston—1,158 buildings—was destroyed by fire, with an economic loss of $6 million. Four citizens were killed by gunpowder explosions used to blow up buildings to create firebreaks.

1838 London, England

London's Royal Exchange was destroyed by a fire believed to have started in Lloyd's Underwriting Offices, which were located in the building. Although attended by eight engines and 68 firefighters, fire plugs and manual pumps froze up because of cold weather conditions, resulting in ineffective firefighting efforts.

1841 London, England

A fire at the Tower of London destroyed the Grand Armory and caused £250,000 in damage.

Damages included the loss of 280,000 objects, including historically priceless armor and arms. Inadequate buckets, hand pumps, and dilapidated small engines contributed to the loss.

1842 Hamburg, Germany

The great Hamburg fire, which raged for three days, devastated one-fourth of the city and killed 100 people. Four thousand buildings, including 1,447 houses, were destroyed, with estimated damages of $21 million. A British civil engineer, William Lindley, helped organize strong measures to check it, including blowing up the town hall to form firebreaks to stop the advance of the fire. Afterward he was appointed Consulting Engineer to the burned city. He surveyed and drew up a plan for its complete rebuilding. He constructed a system of sewers, waterworks, gasworks, and public baths and washhouses, as well as extensions to the port.

1845 Pittsburgh, Pennsylvania, USA

A fire destroyed much of the city in 1845, with an estimated loss of $10 million. Eleven hundred buildings were destroyed. The fire spread resulted from high winds and a low reservoir of water for firefighting.

1845 Quebec, Canada

Two fires occurred during the year, one causing the destruction of 1,500 buildings and the second one, a month later, destroying 1,300 buildings with a loss of many lives.

1845 Canton, China

A fire in a popular theater resulted in 1,670 fatalities.

1845 New York City, New York, USA

A fire started in a sperm oil establishment and spread to adjacent commercial buildings containing explosives, and also to housing for the poor. A large explosion occurred, causing major damage. Thirty people perished, including four firefighters. Losses were estimated at the time at $10 million. Late notification of the fire to the Fire Department contributed to the losses. Builders later agreed with insurers that houses should not be built of wood, and stone (brownstone) house construction was started.

1846 St. John's, Newfoundland, Canada

Fire nearly destroyed most of the city and left 6,000 people homeless.

1846 Quebec City, Canada

A fire in the Royal Theater caused 47 fatalities and resulted in the worst large-loss fire incident in Canada up to that time.

1846 Nantucket, Massachusetts, USA

A conflagration burned 300 buildings, and as a result, destroyed the area's supremacy in the whaling industry, which it had maintained up to that time.

1848 Albany, New York, USA

A conflagration occurred in which 600 buildings, piers, and steamboats were destroyed.

1849 St. Louis, Missouri, USA

One person (a firefighter) was killed, and 425 buildings and 27 steamships were destroyed by a fire. Losses were estimated at $3.5 million. Firefighters attempted to stop the fire with a firebreak by blowing up buildings, but were unsuccessful. The fire burned itself out after it finally lacked combustibles. The city later made streets wider and structures more resistant to fire.

1849 San Francisco, California, USA

A fire caused property damage of approximately $1 million, and destroyed 50 houses.

Following this occurrence, the Town Council was directed "to take such measures as may be deemed advisable to protect the town against another calamity by organizing fire companies, and that the Town Council will supply the hooks, ladders, axes, ropes, etc., to be kept by said companies."

1850 San Francisco, California, USA

Five major fires occurred in various parts of the city (one with a property loss of $4 million). An ordinance was passed after the fire, that any person who refused to assist in extinguishing the flames or in removing goods would be fined. High winds and combustible construction were factors in all of these fires.

1850 Philadelphia, Pennsylvania, USA

Fire destroyed 18 acres and 367 buildings, killed 28, and injured 100. Property losses were estimated at $1.5 million.

1851 St. Louis, Missouri, USA

A major area of the city was destroyed by fire, including 15 blocks of homes and 23 steamboats.

1851 San Francisco, California, USA

Two major city fires occurred in this year. The first caused destruction of approximately 2,500 houses and 18 blocks in the main business district. Damage was estimated at $12 million. The second caused damage estimated at $3 million with the loss of 500 buildings. High winds and combustible construction were factors in both fires.

1852 Montreal, Canada

Approximately 1,000 houses were destroyed by a fire causing $5 million in property losses.

1852 New Orleans, Louisiana, USA

A fire in the city caused $5 million in property damage.

1853 New York City, New York, USA

The biggest clipper ship ever built, the *Great Republic,* burned in the harbor on the eve of her maiden commercial voyage. An onshore fire, originating in a bakery, caused flying embers to start a fire in the folded sails of the ship, which eventually spread into its rigging, deck, and cargo holds. Fire engines of the time did not have a fire stream that could reach high enough pressure to suppress the fires in the rigging of the vessel.

1854 Newcastle, England, UK

Eight hundred buildings were destroyed in a conflagration in the city.

1856 Santa Cruz, California, USA

The city was destroyed by fire.

1858 North Atlantic Ocean

A German steamer was destroyed by fire in the North Atlantic Ocean, with 471 fatalities.

1861 Lindsay, Ontario, Canada

A conflagration destroyed 91 city buildings. As a result of the fire, brick buildings were later constructed to replace the wooden buildings that had been consumed in the fire.

1861 Rouseville, Pennsylvania, USA

The first oil well fire on record occurred on April 17, 1861 on a farm at Oil Creek in Pennsylvania near the village of Rouseville. It caught fire shortly after the well gushed. The fire lasted for three days and resulted in 19 fatalities. The well flowed about 3,000 barrels of oil a day.

1861 Charleston, South Carolina, USA

Fire driven by hurricane force winds caused the destruction of the city's waterfront and much of the heart of the city, with damage estimated at $7 million.

1863 Santiago, Chile

Fire caused by lamps in the Church of the Compania led to the deaths of approximately 2,500 people attending a service. Lack of adequate exit facilities contributed to the enormous loss of life that has never been equaled in a single building fire.

1863 Denver, Colorado, USA

The city of Denver suffered a devastating fire.

1866 Quebec, Canada

Twenty-five hundred buildings were destroyed in a conflagration in the city.

1866 Portland, Maine, USA

A conflagration destroyed much of the central business area of the city. The fire was caused by fireworks thrown carelessly into a wood pile. The fire spread rapidly due to strong winds and lack of sufficient fire apparatus, and caused the loss of two lives and 1,500 buildings valued at $15 million.

Fifty buildings were blown up to create a fire break, but this was unsuccessful due to delay action by officials after the fire was well underway.

1868 Chicago, Illinois, USA

A block of buildings in the city was destroyed by fire with a property loss of $3 million.

1869 Cairo, Illinois, USA

Stonewall, a steamer boat, burned on the Mississippi River, causing 200 fatalities.

1870 Constantinople (Istanbul), Turkey

A conflagration destroyed 7,000 buildings, including many embassies located in the city.

1871 Chicago, Illinois, USA

Named "The Great Chicago Fire," it resulted in one-third of the city being destroyed. The fire's precise cause remains in doubt, but it did begin in the barn of Patrick and Katherine O'Leary. A city fire academy now exists on this site. It was driven by a strong southwesterly wind and destroyed the downtown area and the North Side. The three-year-old water tower was one of the few buildings in the fire's path to survive, and it remains a monument to the fire. The fire caused approximately 250 fatalities, destroyed 17,000 buildings, and was responsible for a financial loss estimated at $200 million. Sixty-four of the some 200 fire insurance companies failed as a result of the fire. This led to the reorganization of the city's fire department on a military basis and improved city building codes. The fire also started the awareness of Fire Prevention Week in the United States.

1871 Peshtigo, Wisconsin, USA

A small town in northeastern Wisconsin was the site of one of the worst fires in American history. It began as isolated peat bog fires and became a major forest fire. Winds whipped the forest fire that had been burning for several days, destroying 1,200,000 acres (500,000 hectares). The city of Peshtigo and 16 other towns were destroyed by the fire, with 1,152 known deaths. A monument commemorating those who perished is provided in the Peshtigo Fire Cemetery. It generally went unnoticed because of the famous Chicago fire that occurred on the same day. Five years after the Peshtigo disaster, and partly as a result of this fire, the US Congress set up a special commission for the protection of forest, under the Department of Agriculture.

1872 Boston, Massachusetts, USA

City fires caused 14 fatalities (including 9 firefighters), destroyed 800 buildings in the richest quarter of the city, and resulted in $85 million in losses. The fire started in the boiler room of a building. Although the city's fire department was considered one of the best in the country, the horse population required to pull the engines had suffered an epidemic, rendering them useless. The fire alarm was delayed and, coupled with the inadequate transport of engines, allowed the fire to escalate and destroy 65 acres.

1872 Tokyo, Japan

Tokyo has frequently suffered disastrous earthquakes and fires. A major conflagration occurred in 1872 when the Ginza and Maronouchi districts were devastated. They were subsequently refurbished with Western-style brick buildings.

1873 Portland, Oregon, USA

A fire originating in a furniture factory destroyed 22 city blocks of the city. Damage was estimated at $1,075,000 and resulted in the installation of a fire telegraph 2 years later.

1875 Virginia City, Nevada, USA

A conflagration destroyed most of the mining town of Virginia City, once considered the richest city in America. Combustible construction of the city buildings was a major factor in the destruction. Buildings were later rebuilt of mostly brick and stone materials.

1876 Brooklyn, New York, USA

A major fire in the Brooklyn Theater was caused by the ignition of the stage backdrop from touching a border lamp as it was lowered into place for a performance. The fire caused 295 fatalities. Panic, poor exit facilities, liberal use of combustible materials, and inadequate building design contributed to the high death toll. New York later enacted laws concerning the number and size of exits from theaters as a result of this fire.

1877 St. John, New Brunswick, Canada

After a 3-week dry period, along with strong winds at the time, one of the most prosperous cities in North America, a fire occurred which resulted in 18 fatalities and damages estimated at $27,000,000 which included 1,600 building destroyed. Lack of horses to pull newly purchased steam fire engines to the fire contributed to the destruction. New building codes and wider streets were instituted during rebuilding the city.

1881 Lower Peninsula, Michigan, USA

One hundred and thirty people died in a fire that destroyed 1,800 square acres of forest in the Lake Huron-Saginaw Bay region of Michigan.

1881 Vienna, Austria

Fire destroyed the Ring Theater and caused 850 deaths. It is considered one of the worst fire incidents in Europe.

1883 Milwaukee, Wisconsin, USA

The Newhall Hotel fire caused 71 deaths due to lack of adequate exits, inadequate fire barriers, and combustible construction. The fire started in the elevator shaft and spread to the other floors, preventing orderly evacuation. The fire department's extension ladder failed to reach the upper stories.

1887 Exeter, England

A theater fire caused 200 fatalities.

1887 Paris, France

Two hundred fatalities occurred during an opera performance (Opéra Comique) as a result of a fire, including members of the cast.

1889 Seattle, Washington, USA

A fire caused by an overturned gluepot destroyed the old downtown business district of approximately 64 acres. Estimated property losses were $10 million.

1889 Spokane, Washington, USA

Much of the city was destroyed by a fire that caused $4.8 million in property damages. The city was soon rebuilt.

1889 Lynn, Massachusetts, USA

A fire that started in the boiler room of a wooden shoe factory destroyed 80 acres of the city's central business district, causing a property loss of $5 million.

1889 Boston, Massachusetts, USA

Four people were killed and 54 buildings were destroyed with a property loss of $3.6 million.

1894 Hinckley, Minnesota, USA

A forest fire across northern Minnesota caused 418 fatalities, destroyed 12 towns and 160,000 acres of forest, and caused $2 million in property damage. Some inhabitants escaped by boarding trains that raced ahead of the fire.

1894 Chicago, Illinois, USA

A fire caused the destruction of the White City (a group of 150 buildings) of the World Exposition and caused $2 million in property damage.

1894 Santa Cruz, California, USA

As a result of an outage of the municipal water distribution system due to a burst reservoir, a fire believed to be of incendiary origin destroyed most of the downtown portion of the city. After the water supply was restored, the fire was contained and extinguished within a few hours. Afterward, improvement of the local fire department was undertaken, fire engines were purchased, and restrictions were placed on the construction of wooden buildings in the downtown area.

1896 Cripple Creek, Colorado, USA

A fire started from a kerosene lamp, destroying 30 acres of property. Later, a grease fire in a hotel kitchen ignited dynamite stored nearby and caused an explosion that devastated the remaining portion of the city.

1897 Paris, France

A charity bazaar fire resulted in 150 deaths.

1898 Salonika, Greece

One thousand people were killed when 340 barrels of gunpowder were accidentally ignited and exploded.

1900 Hoboken, New Jersey, USA

Fire at the marine docks (piers and steamships) resulted in 326 deaths and a property loss of $4.6 million.

1901 Jacksonville, Florida, USA

A conflagration occurred involving 2,368 buildings (146 blocks) valued at more than $11 million. Seven lives were lost in the fire. The fire started in a fiber factory and spread rapidly because of wooden shingled roofs. The fire was considered the biggest and most damaging fire ever to occur in a southern American city up to that time.

1902 Port-de-Paix, Haiti

The city was almost totally destroyed by fire and never regained its former prestige.

1902 Paterson, New Jersey, USA

Five hundred and twenty-five buildings were lost in a conflagration with an economic loss of $5.8 million. Fire started in railway repair shops with a wind blowing at 60 miles an hour and flying embers spread it to other buildings. Several buildings of fire-resistant construction demonstrated their performance during the fire, remaining structurally intact and helping prevent the spread of the fire.

1902 Birmingham, Alabama, USA

A church fire lead to the deaths of 115 individuals.

1903 Boyertown, Pennsylvania, USA

An opera house fire caused the deaths of 170 during a performance. This fire contributed to the formation of the NFPA's committee on Safety to Life (NFPA 101).

1903 Chicago, Illinois, USA

A fire in the Iroquois Theater killed 602 and resulted in a move for stringent fire and life safety regulations for theaters worldwide. The fire started on the stage from an arc light that ignited nearby combustible scenery. An asbestos curtain failed to close and doors at the top balcony caused a draft to develop quickly, causing the fire to carry to the balcony area. The fire lasted for only 30 minutes but managed to kill 30 percent of the occupants. The essentially new theater was of "fireproof construction" (now known to be a misnomer). This fire contributed to the formation of the NFPA's committee on Safety to Life.

1904 Baltimore, Maryland, USA

A conflagration destroyed 80 city blocks, including most of the downtown section, and resulted in a property loss of $150 million. The NFPA investigated the fire extensively, including 27 buildings labeled as "fireproof." In its report, it recommended the term *fireproof* be discarded in favor of *fire resistive* as more correctly describing the character of the structure. It showed that building fires could reach temperatures above 2,000°F (1,093°C). The Baltimore fire also highlighted the need for a national standard thread for fire hose couplings. The fire resulted in improvements in building construction standards and the development of new procedures in fire prevention. The insurance industry Municipal Grading Schedule was also developed as a result of this fire.

1904 East River, New York City, USA

An excursion steamer, *General Slocum,* burned in New York Harbor causing 1,030 fatalities. It was reported that no onboard firefighting devices were operable, personal flotation devices were substandard, and the vessel was overcrowded.

1906 San Francisco, California, USA

An earthquake caused major fires throughout the city. The fire blazed out of control for three days, destroying 4 square miles (10 square kilometers), about 514 city blocks, which included 28,000 buildings. More than 503 people perished as the city's business and industrial sections were leveled and charred, causing $350 million in damage. The building code was strengthened and the fire department was reorganized. It also directed attention nationwide to fire prevention practices required for areas subject to earthquakes.

1908 Constantinople (Istanbul), Turkey

A conflagration in the city destroyed 1,500 buildings.

1908 Collinwood, Ohio, USA

The Lakewood Grammar School fire resulted in the deaths of 176 children. The fire started in the basement and spread up the open main staircase to the first and second floors, impeding evacuation. The fire department lacked adequate equipment for firefighting and rescue operations. This fire contributed to the formation of the NFPA's committee on Safety to Life.

1908 Chelsea, Massachusetts, USA

A conflagration destroyed 3,500 buildings, burned over 492 acres, and left 17,450 homeless. Nineteen deaths were reported, and the property loss was estimated at $12 million. The fire started at a dump and spread from flying embers to poorly constructed houses and buildings. Oil tanks burst due to radiant heat, and further spread the fire. Fire mains were found to be undersized for the size of the fire and not adequately looped and interconnected.

1910 Bitterroot Mountains, Idaho, USA

Drought conditions caused forest fires to destroy three million acres of forest. Three thousand firefighters and federal army troops were used to combat the fires, and 85 lives were lost.

1911 Constantinople (Istanbul), Turkey

A conflagration in the city destroyed 2,463 buildings.

1911 New York City, New York, USA

A fire occurred in the Triangle Shirtwaist factory, a New York City sweatshop, and resulted in the deaths of 146 individuals and injuries to 70 others, mostly young immigrant women. The event touched off a national movement in the United States for safer working conditions. The fire started on the eighth floor of the Asch Building just east of Washington Square Park, and quickly spread upward to the two top floors of the building. Some workers, having no way of opening the doors that had been locked to prevent theft, leaped from windows to their deaths. Fire truck ladders, then able to reach only six stories, were of little help, and the building's overloaded fire escape collapsed. The disaster led to the creation of health and safety legislation, including the NFPA's Life Safety Code, factory fire codes, and child labor laws, and helped shape future labor laws. The New York Fire Department's Bureau of Fire Prevention was established after this fire.

1911 Bangor, Maine, USA

A conflagration resulted in two fatalities, 267 buildings destroyed, and $3.2 million in property losses.

1913 Mid Glamorgan, Wales

A colliery explosion and fire resulted in 439 deaths. The initial explosion blocked the main mine exit and started a major fire. The local fire brigade was not called until two hours after the incident, which is believed to have contributed to the tragedy. The mine ventilation system was also found to be defective and may have fed air to the fire.

1913 Hot Springs, Arkansas, USA

A fire destroyed 518 buildings.

1914 Salem, Massachusetts, USA

A fire destroyed much of the city and caused six fatalities. It was estimated that 1,600 buildings were destroyed with a reported loss of $14 million. High winds, wooden shingled roofs, and inadequate public protection contributed to the loss.

1915 Constantinople (Istanbul), Turkey

A conflagration in the city destroyed 1,400 buildings.

1916 Paris, Texas, USA

Fourteen hundred-forty buildings were destroyed by a conflagration spread to wooden-shingled buildings by high winds. The reported property loss was estimated at $11 million. This fire raised awareness of the fire hazard of wooden shingles as roofing materials and sparked the proposal of city ordinances for the prohibition of wooden shingles.

1916 Nashville, Tennessee, USA

A conflagration occurred in which one person died and 648 buildings were destroyed with a loss reported at $1.5 million. High winds and wooden-shingled roofs contributed to the loss.

1916 Augusta, Georgia, USA

Six hundred eighty-two buildings were destroyed in a fire with an estimated property loss of $4.25 million. Wooden-shingled roofs and inadequate public protection contributed to the loss.

1917 Atlanta, Georgia, USA

A fire destroyed or damaged 1,938 buildings because of wooden-shingled building construction. Property losses were estimated at $5.5 million.

1917 Halifax Harbor, Nova Scotia, Canada

Nineteen hundred and sixty people were killed and 6,000 injured in an explosion and fire from a shipping collision. A French freighter, *Mont Blanc,* packed with 5,000 lbs. of explosives and combustibles, collided with another ship in the harbor. The collision caused fuel to spill over the explosives and a fire occurred. The ship's crew did not attempt to fight the fire, and abandoned the ship. An explosive blast was felt 60 miles away. Seventy-five acres were destroyed. Property damage was estimated at $35 million and included the destruction of 1,600 buildings.

1917 Eddystone, Pennsylvania, USA

An explosion in a munitions plant killed 133 factory workers.

1918 Norman, Oklahoma, USA

Fire started as a result of faulty wiring in a linen closet in a state mental hospital and resulted in the deaths of 38 hospital patients.

1918 Cloquet, Minnesota, USA

The town of Cloquet and 25 other settlements, along with 2,000 square miles of forest, were destroyed by a forest fire that killed 559 people and caused approximately $30 million in reported damage. The city of Duluth, Minnesota was threatened by the fire.

1918 Constantinople (Istanbul), Turkey

A conflagration in the city destroyed 8,000 buildings.

1919 Mobile, Alabama, USA

A fire in a meat market destroyed 40 city blocks and 200 homes. Wooden-shingled roofs and nonstandard hoses and couplings contributed to the loss.

1919 San Juan, Puerto Rico, USA

A fire in the Mayaguez Theater caused 150 deaths.

1921 Oppau, Germany

An explosion at a chemical storage facility caused 561 deaths, and destroyed the plant building and one-third of the surrounding city. An explosion crater 390 ft. (105 meters) wide and 45 ft. (14 meters) deep was created and the blast wave was felt 50 miles (80 km) away. The explosion involved approximately 4,500 tons of a 50/50 mixture of ammonium sulfate and ammonium nitrate. Detonation was caused by blasting powder used to break up piles of the material that had caked.

1922 Astoria, Oregon, USA

A fire destroyed 30 blocks of the city during a rainstorm. Thirty-two acres of buildings worth $10 million were destroyed.

1923 Camden, South Carolina, USA

Fire and resulting panic in a two-story elementary school caused the deaths of 76. A kerosene lamp fell at a school play, causing the props to ignite. Panic to escape and limited exits caused many deaths to occur due to trampling and the crushing effect of an escaping crowd down a stairway.

1923 Berkeley, California, USA

A conflagration caused the destruction of 640 dwellings, primarily due to wooden shingled roofs and high winds. Property losses were estimated at $6 million.

1923 Tokyo, Japan

The earthquake and fire of 1923 destroyed the greater part of the city. Eight square miles of the city and the neighboring port of Yokohama were demolished by fire and approximately 150,000 people died. Seven hundred thousand buildings were destroyed, mostly dwellings, but also shrines, temples, and libraries. Strong winds contributed to the fire spread, which lasted for 36 hours.

1924 Babbs Switch, Oklahoma, USA

A fire in a school caused the deaths of 36. A candle fell from a decorative tree being used for a holiday celebration in a two-room school building. It brought the realization that fire safety is required even for very small schools.

1925 Shreveport, Louisiana, USA

A conflagration destroyed 196 buildings.

1927 Montreal, Quebec, Canada

The Laurier Palace Theater caught fire with a loss of 78 lives.

1929 Cleveland, Ohio, USA

Cleveland Hospital burned, with the loss of 125 lives. The fire was caused by the inadequate storage of nitrocellulose x-ray film in the basement of the building. Most of the fatalities were the result of the inhalation of toxic fumes from the burning x-ray film. Safety film was developed soon after this fire and laws were passed requiring its use in places of public assembly.

1929 Paisley, Scotland, UK

Seventy boys and girls were killed in a cinema hall fire due to locked exit doors and panic. A film had caught fire, releasing billowing black smoke into the corridor. In the ensuing panic, numerous children were trampled by others trying to escape.

1930 Nashua, New Hampshire, USA

Fire destroyed most of the city, leaving 2,000 homeless. Wooden shingled roofs and an inadequate water distribution system contributed to the loss. Property losses were estimated at $2 million.

1930 Columbus, Ohio, USA

Fire in an overcrowded penitentiary caused the deaths of 320 inmates. Construction activities at the prison and a strong wind spread the fire. Guards became disorganized and refused to evacuate the prisoners. Training for emergencies was inadequate. The Bureau of Prisons of the Department of Justice asked the NFPA to conduct a survey of all federal penitentiaries as a result of the fire.

1931 Pittsburgh, Pennsylvania, USA

A fire in a home for the aged caused 48 fatalities.

1934 Asbury Park, New Jersey, USA

The luxury liner vessel, *SS Morro Castle,* a US steamer going from Havana to New York, burned off the coast of New Jersey with the loss of 124 lives.

1934 Hakodate, Japan

A city conflagration destroyed the city and killed 2,018 people. Combustible building construction materials and inadequate spacing were important factors in the spread of the fire.

1934 Lansing, Michigan, USA

A fire in the Kerns Hotel resulted in 34 fatalities and 42 injuries. The fire spread rapidly, trapping occupants in their rooms. Individuals jumped into life nets and descended down fire department ladders.

1937 New London, Texas, USA

A schoolhouse explosion and fire killed 413 students, faculty, and staff. Un-odorized gas from the heating system had leaked in the building, causing an explosion and fire. This incident led to the development of state laws to protect buildings not subject to municipal ordinances or inspections.

1937 Lakehurst, New Jersey, USA

The *Hindenburg* (zeppelin) airship, filled with 7,000,000 cu. ft. (198,221 m^3) of combustible hydrogen, caught fire as it approached its mooring, killing 35 of the 97 persons aboard. Crewmen were preparing to lay out the mooring cables when they heard a "pop." They looked up to see a brilliant flash of light where the central catwalk passed through gas cell 4. The original design of the airship had specified helium gas (a nonflammable gas) for lift; however, because of political concerns, hydrogen was used. The subsequent airships' gas cells were altered and stringent fire precautions were implemented. These included having the crew wear hemp-soled shoes and anti-static asbestos coveralls without buttons or any kind of metallic surface. All matches and lighters were removed from passengers before they boarded. The smoking room was specially insulated and pressurized to prevent hydrogen from entering and it was fitted with a double door. A steward lit cigars and cigarettes from a special lighter and ensured that no fire left the room.

1939 Lagunillas, Venezuela

A fire started in an oil refinery, destroying the adjacent town and killing 500 people.

1940 Natchez, Mississippi, USA

Two hundred and seven people died due to a single exit with an inward opening door at a dance hall. The door was blocked by flames and the hall had numerous combustible decorations and overcrowded conditions. This fire emphasized the need for adequate building exits and fire safety features.

1942 Honkeiko Colliery, China

A coal mine explosion killed 1,549 workers.

1942 Boston, Massachusetts, USA

The Coconut Grove Nightclub fire caused 493 fatalities. Rapid flame-spread along the surfaces of interior finish was judged a major factor in the fire growth, along with exceeding the rated capacity for the building. Flame-spread ratings of 2,500 were found for the fabric used on the ceiling of the nightclub. As a result, ASTM E 84/NFPA 255 was developed for use by the building code authorities to regulate the use of interior finish materials. Recorded as the worst multiple-death nightclub fire in the 20th century, this fire emphasized the need for adequate building exits and fire safety features. Authorities estimated that 300 of those killed may have survived had the building exit doors swung outward.

1942 St. John's, Newfoundland, Canada

A hotel fire caused 100 fatalities and over 100 injuries. Inadequate exits, combustible decorations, and panic contributed to the loss.

1943 Hamburg, Germany

Approximately 45,000 people were killed as a result of Allied bombing (using incendiary bombs made of phosphorus) that created firestorms in the city during World War II. Four square miles of the city were destroyed, which included 16,000 buildings. Every fire engine in the city was used to control the fires.

1943 Houston, Texas, USA

Fifty-five persons were killed by fire in the Gulf Hotel.

1944 Hartford, Connecticut, USA

During a Ringling Brothers Circus act, an inept fire-eater started a fire that destroyed a crowded circus tent, killing 168 people and injuring 261. The tent material was not flame retardant. As a result of this fire, the NFPA created a committee on Places of Outdoor Assembly, which later led to the development of NFPA 102, *Standard for Assembly Seating, Tents, and Membrane Structures.*

1944 Bombay Harbor, India

As a result of an explosion of a munitions ship in the harbor of Bombay, India, over 700 people were killed and between one and two thousand were injured. The aggregate property loss was estimated at $1 billion or more at the time. The ship was carrying explosives and cotton. A fire started in the cotton hold, and eventually spread to the explosives. Nineteen other ships were also destroyed and 40 firefighters who were battling the blaze were killed.

1944 Port Chicago, California, USA

The explosion of a munitions depot adjacent to Port Chicago killed 322 people. The depot was the chief US west coast shipping point for military munitions. As a precautionary measure, the US Navy purchased the town of Port Chicago and dismantled it in 1969.

1944 Cleveland, Ohio, USA

An explosion and fire at a liquefied natural gas (LNG) plant killed 135, injured 200 to 400, and caused a large property loss. The plant was the first LNG plant in the world. A large LNG tank ruptured and spread its contents throughout the plant and to a nearby residential area.

1945 Dresden, Germany

A nighttime raid of hundreds of Allied war bombers released incendiary bombs and started a firestorm in the city of Dresden, killing 135,000 people and demolishing 80 percent of the city.

1945 Bari, Italy

A US Navy cargo ship loaded with bombs exploded and burned at Bari Harbor. Three hundred sixty people were killed and 1,730 were injured.

1945 Toyko, Japan

A nighttime raid of hundreds of Allied war bombers released incendiary bombs and started a firestorm in the city of Toyko, killing 84,000 people and demolishing an area of 15 square miles in six hours. The fire was aided by a 40 mph (65 kph) wind. It took four days to extinguish the blaze.

1946 Chicago, Illinois, USA

A fire in the LaSalle Hotel caused 61 fatalities. Contributing factors included delayed discovery, open stairways, and combustible interior finishes. Rapid flame spread along the surfaces of interior finish was judged a major factor in fire growth. It led to the adoption of standardized tests for the flame spread of interior building materials (ASTM E-84/NFPA 255). This fire emphasized the need for adequate building exits and fire safety features.

1946 Atlanta, Georgia, USA

A fire occurred in the Winecoff Hotel leading to 119 fatalities and 168 injuries. Fire issues involved delayed discovery, open stairways, and combustible interior finishes. Rapid flame spread along the surfaces of interior finish was judged a major factor in fire growth. A number of occupants jumped from windows. Most of the fatalities were from smoke inhalation. This and several similar fires at the time led to the adoption of standardized tests for the flame spread of interior building materials (ASTM E-84/NFPA 255). This fire also emphasized the need for adequate building exits and fire safety features. The State of Georgia adopted a fire safety code soon after this fire. The NFPA campaigned for the elimination of the term *fireproof* to describe building construction after this fire.

1946 Dubuque, Iowa, USA

A fire at the Canfield Hotel caused 19 fatalities. Fire issues involved delayed discovery, open stairways, and combustible interior finishes. It led to the adoption of standardized test for the flame spread of interior building materials (ASTM E-84/NFPA 255). This fire emphasized the need for adequate building exits and fire safety features.

1947 Berlin, Germany

Eighty people died in a crowded nightclub when a defective electrical system ignited the wooden building. Panic caused people to block the exits and the available fire main was apparently partially frozen. Initially, only one fire hose was operating from the British and German fire brigades that responded.

1947 Texas City, Texas, USA

The city was damaged when a chemical explosion occurred in a French cargo ship from a fire among bags of ammonium nitrate fertilizer. The explosion at the harbor triggered some 50 blasts that started widespread fires. Sixteen hours later, a ship in the next slip exploded. It is the worst industrial accident to ever occur in the United States in terms of casualities, with 561 fatalities, 3,000 injuries, and $67 million in property losses. This disaster led to improvements in the labeling for the shipping of hazardous materials.

1948 Ludwigshafen, West Germany

A railroad tank car ruptured next to a dimethyl ether plant. The resulting gas release caused an unconfined vapor cloud explosion (UVCE) resulting in 245 deaths and 2,500 injuries. Property damages were estimated at approximately $12 million.

1949 New York City, New York, USA

A truck loaded with poisonous and flammable carbon disulfide caught fire in a tunnel and exploded. Despite the destruction of 23 trucks and 500 ft. (150 m) of tunnel ceiling, no fatalities resulted. The tunnel had 84 powerful fans capable of replacing air in the tunnel every 90 seconds. Polluted air was drawn off through a duct in the roof of the tunnel with the aid of the suction fans. This undoubtedly aided in preventing loss of life from smoke inhalation.

1949 Effingham, Illinois, USA

Rapid flame spread along the surfaces of interior finish was judged a major factor in fire growth in a hospital fire, along with open stairways, corridors, laundry chutes, wooden interior construction, and lack of automatic sprinkler protection. Seventy-seven fatalities occurred. The industry later treated the fiberboard used in building construction to limit its flame spread.

1949 Chungking, China

A conflagration fire occurred at the waterfront area of the city. It was reported that 2,513 people were killed and 8,046 buildings were destroyed.

1953 Livonia, Michigan, USA

A fire in a major automotive factory had a major impact on industrial fire safety. The fire occurred in a General Motors transmission factory with an undivided floor area of 34.5 acres with only partial sprinkler protection. Three workers died and property damage was estimated at $55 million at the time. The large size of the building also hindered access and application of water by fire service personnel and equipment. The fire started from a welding operation that ignited the oil-soaked wooden floor. This fire was classified as the greatest industrial fire up to that time.

1956 Cali, Colombia

Army trucks carrying dynamite exploded, killing over 1,100 people and destroying 2,000 buildings in an area of slums, factories, and warehouses.

1957 Warrenton, Missouri, USA

A fire in a home for the aged caused 72 fatalities. Construction deficiencies, lack of an automatic sprinkler system, and several other factors contributed to the loss. This fire contributed to the improvement in nursing home fire safety regulations in a number of jurisdictions.

1957 Liverpool, England

A fire in the Windscale plutonium production reactor north of Liverpool spread smoke and combustion gases with radioactive material throughout the countryside. In 1983, the British government said that the release was the probable cause of the cancer deaths of 39 people. The fire was a result of a rapid rise in the temperature of the nuclear pile when it was under particle bombardment, during which time the graphite deformed, swelled, and released a large amount of energy.

1958 Chicago, Illinois, USA

A parochial school (Our Lady of Angels) fire caused 93 fatalities. The fire began in a trash pile at the bottom of an open stairway and spread upward through the building and horizontally through corridors, trapping the victims. Delayed notification of the fire to the fire department and icy winds contributed to the spread of the fire. This fire resulted in a major fire inspection of schools in the United States within days of the fire. Research for school fire safety (Los Angeles "School Burning" Tests, 1962) was also begun.

1958 Koniya, Japan

A conflagration in the city destroyed 1,500 buildings.

1958 Bursa, Turkey

A conflagration in the city destroyed 2,500 buildings.

1959 Arlington, Virginia, USA

A fire occurred at the Pentagon Building, headquarters of the US military, causing property damage estimated at $7 million.

1960 Guatemala City, Guatemala

A fire in a mental hospital caused 225 fatalities and injured another 300. Inadequate exit facilities, delayed fire discovery, and lack of automatic sprinkler protection contributed to the loss.

1960 Amude, Syria

A movie theater fire caused 152 fatalities (mostly children) and 355 injuries. The fire started in nitrated film next to the movie projector.

1960 Kazakhstan, USSR (Russia)

Ninety-one people were killed when a rocket exploded during fueling operations at the Baikonur Space Center.

1961 Hartford, Connecticut, USA

A fire in a hospital resulted in 16 deaths. Combustible ceiling tiles and trash and linen chutes directly open to main building corridors contributed to the loss.

1961 Sahara Desert, Algeria

The greatest gas fire in the world ignited at Gassi Touil in the Sahara Desert of Algeria. The fire was from a gas well being drilled that blew out. A pillar of flame rose 450 ft. (137 m) and burned for six months. It was extinguished by Paul "Red" Adair using 550 lbs. (250 kgs.) of dynamite to blow out the fire and cap the well.

1961 Niteroi, Brazil

A circus tent fire caused 323 fatalities.

1962 Saarbruken, Germany

A coal mine fire caused the deaths of 290 miners.

1963 Fitchville, Ohio, USA

Sixty-three lives were lost in a nursing home fire. The building had no fire protection or alarm system and was located 7 miles (11 km) from the nearest fire department. This fire contributed to the improvement in nursing home fire safety regulations in a number of jurisdictions.

1965 Watts, California, USA

Civil disturbances led to fires within the city; however, firefighters also came under attack. This was the first time firefighters had been subjected to this type of response from the public and exposed to this new occupational hazard.

1967 South China Sea, near North Vietnam

Ammunition and fuel on the US aircraft carrier *Forrestal* caught fire off the North Vietnamese coast, causing 134 deaths. The fire started as a result of a rocket fuse failure.

1967 Brussels, Belgium

A fire in a 70-year-old, large department store caused 322 deaths. Believed to have originated in a closet, the fire spread up through open stairways, escalators, elevator shafts, and an open light well. The fire alarm system was not activated (possibly damaged by the fire) and a large amount of decorations were in place in the store. It was considered the worst fire in Europe since the Ring Theater fire in Vienna in 1881.

1967 Jay, Florida, USA

Thirty-seven convicts died in a locked barracks as a result of a malicious fire set by a few of the inmates.

1967 Cape Kennedy, Florida, USA

The three-man Apollo I crew died when a flash fire swept through their space capsule during simulated launch activities. The spacecraft was arranged with the hatch locked, the power on, and an internal atmosphere of 100 percent oxygen. The crew was in space suits and performing the normal sequence of prelaunch activities. A spark inside the spacecraft ignited flammable material and instantly engulfed the closed compartment in flames. Once the hatch was opened, more than five minutes later, it was discovered that the crew had died from asphyxiation within 15–20 seconds after the fire started. The precise source of the spark and fire was never determined; nor were any individuals or specific organizations implicated in the fire. The actual cause, determined by an investigation, was due to the combination of several conditions: an oxygen-rich atmosphere; flammable interior materials such as paper, the space suits, Velcro, and other flight equipment; a vast array of exposed internal wiring, which presented many potential sources of electrical sparks; and the design and manufacture of the spacecraft. As a result of the fire, many changes were made to the design, manufacturing, testing, and checkout procedures of the vehicles and the management of the entire Apollo program.

1967 Chicago, Illinois, USA

A fire swept through a large convention hall (McCormick Place), resulting in its destruction, a loss of one person, and $10 million in property damages. The fire was believed to have been started by an electrical spark. The building had unprotected steel trusses against fire.

1968 Farmington, West Virginia, USA

Seventy-eight men died after a series of methane gas explosions and fires in a coal mine. Explosions in the mine continued for several days when it was felt the 78 men may have still been alive. However, hope was later abandoned and the mine was sealed to extinguish the fires.

1969 Golden, Colorado, USA

A fire at the US Atomic Energy Commission plant for uranium enrichment caused a property loss of $456 million. This fire led to improvement in fire safety standards for atomic energy processes.

1970 Tucson, Arizona, USA

A fire spread through the upper eight stories of an 11 story hotel. Twenty-eight died and 27 were injured. Fire officials indicated the building did not meet city safety requirements for older buildings and did not have a sprinkler system.

1970 Saint Laurent du Pont, France

A fire occurred when a match was dropped on a couch in a crowded nightclub. The building's interior walls and columns were covered with highly flammable plastic, exit signs were obscured, and exit doors were locked or blocked. By the time the local fire department arrived, the structure of the building was collapsing. One hundred forty-six fatalities occurred, mostly through inhalation of toxic fumes.

1971 Seoul, South Korea

A gas stove exploded in a coffee shop of the second floor of the Daiyunkuk Hotel. The fire raged through the two-year-old, 222-room hotel, killing 165 people. Although the original design for the hotel had been officially approved, changes had occurred that contravened safety regulations.

1972 Chelsea, Massachusetts, USA

A fire fanned by high winds destroyed 1,000 buildings in a 30-block area. Property damage was estimated at $500 million.

1972 Overland, Kansas, USA

A fire burned out of control for two days and destroyed 56 million military personnel records in a building that was used as the center for US military records.

1972 Osaka, Japan

A fire in a nightclub caused 116 fatalities. Many of the deaths were the result of individuals jumping from the roof of the seven-story building. One exit stairway was found locked and patrons were unaware of other exits. The fire originated on the third floor of the building, which was undergoing renovation, and was suspected to be the result of an electrical deficiency.

1972 Kumamoto, Japan

A department store fire resulted in 107 fatalities and 84 injuries. The 21-year-old building had no fire escapes or ladders.

1972 Hong Kong

The 83,000-ton luxury liner *Queen Elizabeth* caught fire and burned for 24 hours while in Hong Kong harbor. The vessel was undergoing modifications to be used as a floating campus. Insurance claims were estimated at $8 million.

1972 Kellogg, Idaho, USA

A fire, thought to have started from spontaneous combustion, spread in supporting timbers of a silver mine tunnel, causing 91 deaths. The spread of the fire was prevented by sealing off empty shafts. Inadequate safety training and lack of protective chemical masks also contributed to the loss of life.

1973 Staten Island, New York, USA

An explosion destroyed the world's largest liquefied gas storage tank during repairs. Forty workers were killed and damage was estimated at $31 million.

1973 Chelsea, Massachusetts, USA

A conflagration destroyed 1,000 buildings. Property damages were estimated at $500 million.

1974 Flixbourough, England

A chemical plant accidentally released 36 tons of cyclohexane because of the failure of a temporary piping bypass. The resulting vapor cloud exploded, with a force equivalent to 15 to 45 tons (13.5 to 40.8 Mg) of TNT. The explosion and fire caused 28 fatalities and total plant destruction with a property loss of $70 million. The temporary bypass was inadequately engineered and the plant layout was poor. As a result of this incident, major process industry safety regulations were adopted in the United Kingdom.

1974 Seoul, South Korea

The top two floors of the seven-story Daowang Corner Building were used as a hotel and a nightclub. A cigarette caught a hotel bed on fire, causing 88 fatalities. Sixty-four of the people who died were trapped in their rooms because the doors were locked to ensure they would pay their bill. The fire raged for three hours before it was under control.

1974 São Paulo, Brazil

Fire broke out on the 17th floor of the 25-story Joelman Building. The building had been designed with virtually no fire escapes. Highly flammable paint and plastic were used on the windows and flooring. The public fire department was also reported to be insufficient (only 20 stations were provided to the city of 8 million people), and 227 fatalities occurred.

1975 Philadelphia, Pennsylvania, USA

A fire started in a 50-year-old refinery, one of the largest on the US east coast, as a result of a storage tank explosion. Ten people were killed (including eight firefighters), and the fire lasted 24 hours. Property damage was estimated at $10 million.

1975 Pacific Ocean

A crude oil supertanker exploded while en route to Japan from Brazil, causing the hull to break apart and sink. The tanker, one of the largest vessels lost at sea, was considered one of the costliest disasters at sea at the time with insurance claims of $27 million.

1975 Beek, Netherlands

A vapor cloud explosion of propylene (5.5 tons) from an accidental release at a chemical plant caused 14 fatalities and $40 million in property damages. The government later required extensive risk analysis of new and existing process industry plants in the country after this incident.

1975 Chasnala, India

An explosion in a mine caused 431 fatalities.

1976 Lapua, Finland

An explosion in an ammunition plant caused 45 fatalities and injured 75. The incident was considered Finland's worst industrial accident to date.

1976 Kawasaki, Honshu, Japan

After a fire in 1976, the city was rebuilt using extensive fire-prevention techniques, including provision of firebreaks, parks, and roads.

1977 Moscow, Russia (USSR)

Forty-five fatalities occurred in the Rossiya Hotel, the world's largest hotel. The fire occurred in the 21-floor tower and the 12-story north wing of the hotel. The fire apparently started in an elevator motor and spread up the elevator shaft to the upper floors.

1977 Southgate, Kentucky, USA

A popular nightclub caught fire resulting in 164 deaths. Fire separations, automatic sprinklers, and adequate exits were lacking. Interior furnishings were of combustible materials. The fire started as a result of defective wiring to lights in the ceiling.

1977 Columbia, Tennessee, USA

Forty-two fatalities (33 inmates and 9 visitors) occurred in a jail when a fire started by an inmate spread toxic fumes in the facility.

1978 San Carlos de la Rapita, Spain

A tank truck carrying 1,500 cu. ft. (42 m^3) of propylene crashed into a campsite as the driver lost control on a curve, possibly due to tire failure. Two hundred eleven people were killed and several hundred injured as a result of an intense explosion and fire from the crash. A crater 60 ft. (20 m) across was left after the explosion. The crash also caused secondary explosions and fires to occur from camping cooling bottles and vehicle fuel tanks.

1979 Sargossa, Spain

The worst hotel fire in Spain's history caused 80 fatalities. The fire started as a result of spilled cooking oil, which caused an explosion of an oil tank, which allowed flames to race through the 10-story hotel.

1979 Tuticorin, India

Ninety-two persons were killed and 80 injured in a movie theater fire.

1979 Houston, Texas, USA

A fire in an apartment complex caused $20 million in property losses. The Houston City Council passed an ordinance restricting the use of wooden roof shingles for multifamily housing units after the fire.

1980 Kingston, Jamaica

A fire in a nursing home caused 157 fatalities.

1980 Las Vegas, Nevada, USA

The MGM Grand Hotel caught fire, resulting in 85 deaths and 600 injuries. It was speculated that the fire was smoldering for two hours before it was discovered. The 26-story hotel was only partially sprinklered. Although the fire started in the casino due to arcing from electrical wiring, it spread to the high-rise section where most of the victims were located. This fire led to the adoption of fire safety regulations in the state requiring retroactive fire safety improvements (sprinklers) in publicly occupied buildings. The hotel reopened later, after $26 million in repairs, including the installation of $6 million in fire protection systems.

1980 Riyadh, Saudi Arabia

Three hundred and one people aboard a plane reportedly died as the plane caught fire at the Riyadh airport.

1981 Dublin, Ireland

A popular discotheque caught on fire, resulting in 44 fatalities and 130 injuries. Plastic-covered interior finishes and furnishings contributed to the fire spread and loss. Allegations were also made that exit doors may have been locked. Fire regulations limiting the use of highly combustible materials for buildings in Ireland were later implemented as a result of this fire.

1981 Las Vegas, Nevada, USA

A Hilton Hotel fire caused eight fatalities. This fire and similar hotel fires in the state led to the adoption of fire safety regulations in the State of Nevada requiring retroactive fire safety improvements (sprinklers) in publicly occupied buildings.

1981 Java Sea

Tamponas II, an Indonesian passenger ship, caught fire and sank, resulting in 580 fatalities.

1982 Salang Pass, Afghanistan

A petrol tanker crashed head-on into a Soviet military convoy while it was in a 1.7-mile (2.5 km) tunnel through the Hindu Kush mountains. The resulting fire killed an estimated 1,000 to 2,500 people. Most victims died from inhaling smoke due to the poorly ventilated tunnel and the closing of the tunnel at both ends because of the fear it was under attack.

1983 Nasser Lake, Egypt

A passenger vessel on the Nile River caught fire, causing the vessel to sink and resulting in 357 fatalities.

1983 Madrid, Spain

A fire in a discotheque caused 83 fatalities. Most of the fatalities were the result of asphyxiation or of being trampled upon during the panic of evacuation. It was suspected the fire started from an electrical short. The discotheque was decorated with plastic curtains, wall hangings, and upholstery, which gave off toxic fumes during the fire.

Portable fire extinguishers were difficult to operate or did not work at all.

1983 Turin, Italy

A fire in a 1,074-seat movie theater caused 74 fatalities. Some rear exit doors had been locked (to prevent entry by non-paying customers), seats were covered with plastic causing dense toxic smoke, and panic ensued during the evacuation.

1984 Cubatao, São Paulo, Brazil

A 24-inch gasoline pipeline ruptured, spilling 700 tons of liquid. An illegally built shanty town near the pipeline caught on fire and at least 508 people were killed.

1984 Mexico City, Mexico

A major LPG storage area disaster occurred with multiple BLEVEs. A failure of a fill line resulted in a line break and the formation of a vapor cloud, which ignited after 20 minutes. The flame from the broken pipeline impinged on one of four spheres, causing a BLEVE to occur. Approximately 15 BLEVEs were recorded during the overall incident. A piece of one vessel weighing over 20 tons landed at a distance of 1.2 km (.74 mi.) away. Five hundred deaths and 7,000 injuries were recorded.

1986 Puerto Rico, USA

The DuPont Plaza Hotel and Casino fire in San Juan resulted in 96 fatalities and 140 injuries. The fire had been deliberately set due to labor unrest concerning recently delivered furniture stored on the property (ballroom). The government of Puerto Rico commissioned a study for the need of improvements for the fire safety code, which eventually improved the requirements for fire safety of hotels.

1986 Basel, Switzerland

A fire in the warehouse of a chemical manufacturing company resulted in one of the most serious cases of environmental pollution in Europe. Millions of gallons of water applied by firefighters during the incident became contaminated by insecticides (including dioxin), which drained into the city sewers and into the Rhine River. The contamination spread downstream causing ecological damage and affecting many forms of wildlife.

1987 London, England

A fire occurred in an inclined wooden escalator shaft and spread to the public hall above the King's Cross Subway (Underground) Station, resulting in 30 fatalities. Investigation into the incident led to the identification of the "Trench Effect" fire phenomenon.

1987 Cape Kennedy, Florida, USA

The Space Shuttle Challenger exploded and was destroyed about one minute after launch because of the failure of a sealant ring on one of its solid fuel boosters. The booster nosed into the main propellant tank of liquid hydrogen and oxygen, causing a near-explosive disruption of the entire system. Seven astronauts were killed in the disaster and a major safety review of the NASA manned space vehicle design, production, and launch program was undertaken.

1988 Lisbon, Portugal

A fire, called the worst disaster in the city's history since the disastrous earthquake in 1755, destroyed the city's shopping district.

1988 Yellowstone National Park, Wyoming, USA

In the summer and fall of 1988, wildfires raged through about one-quarter of the park, badly burning about 22,000 acres (8,910 ha). Thousands of volunteers and military personnel helped fight the fires, which officials called the worst disaster in the park's history. Dry and windy conditions contributed to the loss. Some scientists believe the fires may be a natural and necessary occurrence for the biotic community of the park.

1988 Piper Alpha Platform, North Sea

A fire and explosion on a North Sea oil and gas production platform resulted in the worst offshore facility incident. One hundred sixty-five fatalities resulted from the fire and explosions as result of a release of condensate during maintenance activities for a process pump. Most of the personnel were trapped in the accommodation module and died of smoke inhalation before the module fell into the sea after its supports failed from exposure to the high temperature fire. It resulted in the highest loss of life in the petroleum industry and the total destruction of the facility. The largest ever marine insurance loss, $836 million, was paid on this incident. This incident lead to revisions in offshore oil industry fire safety regulations worldwide, especially in the UK for the North Sea.

1988 Richmond, California, USA

A dry goods grocery warehouse was destroyed by fire. Fire protection consisted of 13 sprinkler risers. Roof vents and 21 draft curtains were provided. Storage was to within inches (centimeters) of the roof deck, and no in-rack sprinklers were installed. A fire of undetermined origin was discovered by an employee. Employees attempted to fight the fire but were driven out after the sprinkler system operated, which was soon overtaxed. The fire was finally extinguished eight days later. The damages totaled $80 million.

1989 Pasadena, Texas, USA

A Phillips Petroleum refinery fire and explosion caused 23 fatalities and $750 million in damage. Improper procedures for equipment cleaning resulted in the release of 39 tons of iso-ethylene, leading to an explosion and fire.

1989 between Ufa and Asha, USSR (Russia)

A gas pipeline rupture resulted in an explosion and fire and caused over 650 fatalities.

1990 Bronx, New York, USA

A fire caused by an arsonist spreading gasoline on a stairway at a social club resulted in 87 fatalities. Blocked fire exits contributed to the high fatalities. The individual responsible was tried for arson and murder and convicted for both crimes.

1991 Hamlet, North Carolina, USA

A fire in a food processing (chicken) plant resulted in 25 fatalities and 54 injuries. The fire's intensity, along with inoperable exits, lack of sprinklers, and a windowless facility contributed to the loss.

1991 Madison, Wisconsin, USA

A cold storage complex of five buildings was destroyed by fire. Four of the five buildings were sprinklered by two wet-pipe and two dry-pipe systems. A forklift driver heard the sound like "a torch being lit" and noticed a fire involving a battery-powered forklift. As employees discharged portable extinguishers, the fire department was notified. The roof of the warehouse collapsed within the first hour of fire department operations. Damage to the building and contents totaled over $75 million.

1991 Berkeley-Oakland, California, USA

One of the costliest wildfires fires in US history caused 26 deaths and more than $1.5 billion in damage in the residential hills above the cities of Berkeley and Oakland, California. Dry conditions and steep terrain conditions contributed to the loss of 3,469 housing units.

1991 Oil Wells, Kuwait

During the Gulf War, the retreating Iraqi Army set fire to 656 oil wells in Kuwait. Some of these oil well fires were not extinguished until nine months later. The oil well fires led to the estimated loss of 1.5 billion barrels of oil with a value of some $27 billion. Worldwide oil well firefighting experts were brought in to extinguish the fires and cap the wells.

1992 Windsor Castle, England

A fire devastated historic St. George's Hall at Windsor Castle (ancestral castle of the monarchy of the United Kingdom). The structure sustained tens of millions of dollars in damage, but no fatalities occurred. Thirty-seven appliances were brought to the scene to fight the fire.

1992 Guadalajara, Mexico

The release of combustible materials into the city sewer caused explosions and fires, resulting in 190 fatalities.

1993 Waco, Texas, USA

A religious cult compound was accidentally set on fire during an attempted siege by agents of the US government. Seventy-two deaths and destruction of the compound occurred as a result.

1993 Bangkok, Thailand

One hundred and eighty-eight workers died and 500 were injured in a factory fire that manufactured toys. A government inquiry found that faulty building design, lack of fire exits, and poor safety exits meant that workers couldn't escape in time. The main exit of the building was locked, and there were no fire alarms or extinguishers.

1995 Baku, Azerbaijan

About 300 people were killed and over 200 injured when a crowded subway train caught fire in a tunnel. Most of the fatalities were due to carbon monoxide poisoning from smoke inhalation and panic when illumination facilities failed during the fire. The cause of the fire is believed to have been the failure of electrical equipment. Combustible materials apparently composed 90 percent of the interior finish of the train.

1995 Taege, South Korea

A gas explosion in an underground railway being constructed under a busy city street caused 110 fatalities. It was believed the construction operation caused a leak in an underground gas line or ignited a leak from a line.

1996 English Channel

A fire in the Channel Tunnel (Chunnel) as a result of a train fire caused about $75 million in damages and lost revenue. Ventilation systems designed to remove or prevent smoke from entering the passenger car failed to meet expectations, forcing passengers to evacuate to a service tunnel.

1997 Hyderabad, India

An explosion and fire at a petroleum refinery killed 28 and injured 100 others. A pipeline leaked petroleum, exposing hydrocarbon storage tanks and vessels. Most of the victims died from burns suffered in the incident.

1997 Sumatra/Kalimantan, Indonesia

Up to a million hectares of forest, scrubland, and plantation burned out of control across Sumatra and Kalimantan. Land-clearing fires became uncontrolled and lasted for several months. It represented one of the most wide-ranging fires ever and presented a threat to endangered wildlife and potentially the ecosystem of the Earth.

1998 Lagos, Nigeria

A ruptured gasoline pipeline caught fire and killed 500 individuals. The victims were apparently illegally collecting the spilled gasoline when the spilled liquid ignited.

1998 Gothenburg, Sweden

Sixty-two teenagers died and 162 were injured in an overcrowded second-floor dance hall fire that swept through the facility. Blocked exits and panic contributed to the loss. Although the hall was only approved to hold 150, tickets were sold for 320 persons.

1999 Detroit, Michigan, USA

Automobile manufacturing power generation complex explosion and fire resulted in 6 fatalities and $650 million in property damages. The preliminary conclusion is that a gas built up at a boiler. The facility was 80 years old at the time of the incident. The incident resulted in the largest state fine to a company for a violation of worker safety laws.

2000 New Mexico, USA

A wildland fire named the Cerro Grande Fire of 2000, was the largest, most destructive wildfire that New Mexico has ever known. The fire swept across 47,000 forested acres in Bandelier National Monument, the Santa Fe National Forest, Los Alamos National Laboratory, Los Alamos County, and the Santa Clara and San Ildefonso Indian Reservations, causing about $1 billion in property damage. Over 400 families were left homeless, and over 100 Los Alamos Laboratory structures burned. First set as a controlled burn by the National Park Service, unprecedented winds soon whipped the fire out of control.

2000 Luoyang, China

A fire began elsewhere in a four-story shopping plaza and spread to disco/dance hall, resulting in 309 fatalities. It was reported that at least 200 people were in the disco on the top floor when the fire struck, and construction workers were working on the second and third floors. Later, the Chinese government reportedly increased efforts in fire prevention inspections for public facilities.

2001 New York City, USA

Religious extremist suicide terrorists crashed two commercial airliners into the World Trade Center twin office towers, which resulted in an ensuing fire that caused structural collapse of the building—2,666 fatalities (343 firefighter and police fatalities) and $33.4 billion in property damages occurred.

2003 West Warwick, Rhode Island, USA

A fire in a nightclub resulted in 100 fatalities. Pyrotechnics used by the stage performance ignited nearby highly combustible soundproofing materials which resulted in the rapid building destruction. A contributing factor in the high loss of life was the exits' inability to handle all the occupants in the short time available for evacuation for such a fast-growing fire. Changes to the NFPA fire codes were recommended including the need to provide sprinklers for existing nightclubs and also placed pressure on the U.S. concert and nightclub industry to adopt fire and crowd safety standards.

2003 Julian, California, USA

A wind-whipped 100,000-acre wildfire in eastern San Diego county caused 14 deaths, and destroyed more than 550 homes. The fire was apparently started when a hunter lost in the mountains near Julian lit a signal fire.

2003 San Bernadino, California, USA

Wildland fire resulted in $974 million in property damages. The fires damaged about 72,000 acres in the San Bernardino and Angeles national forests and destroyed more than 375 homes.

2004 Ryongchon, North Korea

It is believed that two trains carrying ammonium nitrate and fuel oil collided resulting in an explosion which left 161 people dead and destroyed 2,000 nearby buildings. It is thought that the collision brought down nearby electrical transmission lines which was the ignition source.

2004 Buenos Aires, Argentina

A fire at a popular nightclub resulted in 180 fatalities. Authorities probed reports that a flare launched by someone during an indoor rock concert sparked the fire. Reeling from thick smoke, crowds surged toward the doors, but exits were reportedly locked. Although the club had a reported capacity of 1,500, authorities said investigators were looking into accounts that 4,000 people were in the club at the time of the fire. The Buenos Aires mayor fired many city inspectors and ordered hundreds of discos and nightclubs to close temporarily amid rigorous inspections.

2005 Texas City, Texas, USA

15 people were killed and at least 100 others injured when an isomerization unit exploded at a major oil refinery. Preliminary indications from company reports indicate process safety management concerns including the proximity of occupied temporary buildings to process vessels.

2005 Fuxin City, China

Two hundred and three miners were killed and 22 others injured in a gas explosion that occurred after an earthquake.

US Firefighter Fatalities, 1977–2003

Year	Fatalities	Year	Fatalities
1977	152	1992	75
1978	172	1993	79
1979	125	1994	104
1980	138	1995	97
1981	136	1996	96
1982	127	1997	98
1983	113	1998	91
1984	119	1999	112
1985	128	2000	103
1986	120	2001	439 (includes 340 from World Trade Center incident)
1987	131	2002	97
1988	136	2003	105
1989	118		
1990	107		
1991	108		

Source: National Fire Protection Association, Quincy, MA 2005.

Fire Safety Organizations

American Fire Sprinkler Association (AFSA)
9696 Skillman St.
Suite 300
Dallas, Texas 75243-8264
Tel: (214) 349-5965, Fax: (214) 343-8898
Email: afsainfo@firesprinkler.org
http://www.firesprinkler.org

Automatic Fire Alarm Association (AFAA)
841 Eagle Claw Court
P.O. Box 951807
Lake Mary, FL 32795-1807
Tel: (407) 322-6288, Fax: (407) 322-7488
Email: ruppies@iglou.com
http://www.afaa.org

Congressional Fire Service Caucus
1233 Longworth
House Office Building
Washington, DC 20515

Congressional Fire Services Institute (CFSI)
Railway Express Building
900 Second Street, N.E.
Suite 303
Washington, DC 20002
Tel: (202) 371-1277, Fax: (202) 682-3473
Email: cfsi@cfsi.org
http://www.cfsi.org

Factory Mutual Global
1301 Atwood Ave.
P.O. Box 7500
Johnston, Rhode Island 02919
Tel: (401) 275-3000, Fax: (401) 275-3027
www.fmglobal.com

Federal Emergency Management Association (FEMA)
500 C. Street S.W.
Room 820
Washington, DC, 20472
Tel: (202) 566-1600
http://www.fema.gov

Fire and Emergency Manufacturers and Services Association, Inc. (FEMSA)
P.O. Box 147
Lynnfield, Massachusetts 01940-0147
Tel: (781)334-2771, Fax: (781)334-2771
Email: info@femsa.org
http://www.femsa.org

Fire and Emergency Services Higher Education (FESHE)
C/o National Fire Academy
16825 South Seton Avenue
Emmitsburg, MD 21727
http://www.usfa.fema.gov

Fire Protection Association

Melrose Avenue

Borehamwood, Herfordshire WD6 2BJ, UK

Tel: 0181 207 2345, Fax: 0181 207 6305

Email: info@lpc.co.uk

http://www.lpc.co.uk

Fire Suppression Systems Association (FSSA)

5024-R Campbell Road

Baltimore, MD 21236

Tel: (410) 931-8100, Fax: (410) 931-8111

Email: fssahq@aol.com

http://www.podi.com/fssa

Higher Education Prog (EMI) (FEMA)

C/o Emergency Management Institute/NETC

Emmitsburg, Maryland

http://www.training.fema.gov/emiweb/edu

Institute of Fire Engineers (IFE)

148 New Walk

Leicester LE1 7QB, UK

Tel: 01 16255 3654, Fax: 01 16 247 1231

http://www.dspace.dial.pipex.com/firesafety/ife.htm

Institute of Fire Engineers (IFE), American Branch

P.O. Box 131145

Birmingham, AL 35213-6145

Tel: (205) 599-7976, Fax: (205) 599-0590

Email: fire@incendiis.com

http://www.ife-usa.org

Insurance Services Office (ISO), Inc.

545 Washington Blvd.,

Jersey City

New Jersey 07310-1686

Tel: (800) 888-4476, Fax: (201)748-1472

http://www.iso.com

International Association of Arson Investigators, Inc. (IAAI)

300 S. Broadway, Suite 100

St. Louis, MO 63102-2808

Tel: (314) 621-1966, Fax: (314) 621-5125

Email: IAAIHQ@aol.com

http://www.fire.investigators.org

International Association of Black Professional Fire Fighters (IABPFF)

1020 North Taylor Avenue

St. Louis, Missouri 63113

Tel: (786) 229-6914, Fax: (305) 249-5230

Email: execdir411@hotmail.com

http://www.iabff.org

International Association of Fire Chiefs (IAFC)

4025 Fair Ridge Drive

Fairfax, VA 22033-2868

Tel: (703) 273-0911, Fax: (703) 273-9363

Email: IAFCOLS@connectinc.com

http://www.ichiefs.org

International Association of Fire Fighters (IAFF)

Department of Public Relations and Communications

1750 New York Avenue NW

Washington, DC 20006-5395

Tel: (202) 824-1588

http://www.iaff.org

International Fire Buff Associates, Inc. (IFBA)
7509 Chesapeake Avenue
Baltimore, MD 21219
http://www.ifba.org/cffb.htm

International Fire Marshals Association (IFMA)
P.O. Box 9101
One Batterymarch Park
Quincy, MA 02269-9101
http://www.nfpa.org

International Fire Service Accreditation Congress (IFSAC)
Oklahoma State University
Stillwater, Oklahoma
http://www.ifsac.org

International Fire Service Training Association (IFSTA)
Fire Protection Publications
Oklahoma State University
930 N. Willis
Stillwater, OK 74078-8045
Tel: (800) 654-4055, Fax: (405) 744-8204
http://www.fireprograms.okstate.edu

International Society of Fire Service Instructors (ISFSI)
P.O. Box 2320
Stafford, VA 22554
Tel: (800) 435-0005, (540) 657-9375
Fax: (540) 657-9471
Email: webmaster@isfsi.org
http://www.isfsi.org

Loss Prevention Council (LPC)
Melrose Avenue
Borehamwood, Hertfordshire WD6 2BJ, UK
Tel: 0181 207 2345, Fax: 0181 207 6305
Email: info@lpc.co.uk

National Arson Prevention Initiative (NAPI)
Federal Emergency Management Agency (FEMA)/United States Fire Administration (USFA)
Tel: (888) 603-3100
http://www.usfa.fema.gov/napi

National Association of Fire Equipment Distributors (NAFED)
One East Wacker Drive, Suite 3600
Chicago, IL 60601-4267
Tel: (312) 923-8500, Fax: (312) 923-8509
Email: nafed@raybournechicago.com
http://www.nafed.org

National Association of State Fire Marshals (NASFM)
1319 F. Street NW
Washington, DC 20004
Tel: (202) 737-1226, Fax: (202) 393-1296
Email: staff@firemarshals.org
http://www.firemarshals.org

National Burglar and Fire Alarm Association (NBFAA)
Email: staff@alarm.org
http://www.alarm.org

National Fire Academy (NFA)
16825 South Seton Avenue
Emmitsburg, MD 21727
Tel: (301) 447-1000
http://www.usfa.fema.gov/nfa

National Fire Information Council
P.O. Box 23221
Lansing, MI 48909
Tel: (517) 655-5355

National Fire Protection Association (NFPA)
P.O. Box 9101
One Batterymarch Park
Quincy, MA 02269-9101
Tel: (617) 770-3000, Fax: (617) 770-0700
Email: Library@NFPA.org
http://www.nfpa.org

National Fire Sprinkler Association, Inc. (NFSA)
Robin Hill Corporate Park
Route 22, Box 1000
Patterson, NY 12563
Tel: (914) 878-4200, Fax: (914) 878-4215
Email: info@nfsa.org
http://www.nfsa.org

National Interagency Fire Center (NIFC)
3833 S. Development Ave.
Boise, Idaho 83705
Tel: (208) 387-5512, Fax: (208) 387-5386
http://www.nifc.gov

National Institute of Standards and Technology (NIST)
Fire Safety Division
Gaithersburg, MD 20899-0001
Tel: (301) 975-2000, Fax: (301) 975-4052
http://www.bfrl.nist.gov

National Volunteer Fire Council (NVFC)
Tel: (888) 175-6832
Email: nvfcoffice@nvfc.org
http://www.nvfc.org

National Wildfire Coordinating Group (NWCG)
National Interagency Fire Center
3905 Vista Ave.
Boise, ID 83705

Society of Fire Protection Engineers (SFPE)
Suite 1225 West
7315 Wisconsin Ave.
Bethesda, MD 20814
Tel: (301) 718-2910, Fax: (301) 718-2242
Email: SFPEHQTRS@sfpe.org
http://www.sfpe.org

Underwriter's Laboratories (UL) Inc.
333 Pfingsten Road
Northbrook, IL 60062
Tel: (800)595-9844, (847) 272-8800,
Fax: (847) 272-8129
http://www.ul.com

United States Fire Administration (USFA)
16825 South Seton Avenue
Emmitsburg, MD 21727
Tel: (301) 447-1000
http://www.usfa.fema.gov

Fire Service Job Descriptions

Firefighter

General Job Description

Responsible for controlling, extinguishing, and preventing fires and protection of life and property through firefighting activities. Duties may also include but not limited to inspections, training, and maintaining equipment and quarters.

Typical Tasks

Responds to fire alarms, connects hose, holds nozzle, and directs water streams. Forces entry of premises for firefighting, rescue, ventilating, and salvage operations. Uses portable fire extinguishers, bars, hooks, lines, and other equipment. Safely remove individuals from danger and administers first aid to injured persons. Performs salvage operations such as throwing covers, sweeping water, and removing debris. Participates in preplanning studies, fire drills, and attends training classes in firefighting, first aid, and related subjects. May be required to conduct tours of station for the public and participate in public fire safety campaigns.

Knowledge, Skills, and Abilities

Knowledge of fire prevention and state and city regulations as applied to firefighting and prevention. Knowledge of firefighting equipment and its intended uses. Ability to react quickly and calmly in an emergency situation and to determine the proper course of action. Ability to learn to operate a variety of firefighting equipment. Knowledge of building types, construction, and expected fire behavior. Ability to wear heavy protective clothing for extended periods of time. Prepare and submit reports.

Education and Experience

Usually graduation from high school, preference for 2-year college education.

Fire Apparatus Driver or Operator

General Job Description

Drive and operate fire apparatus and equipment, participate in fire suppression, prevention, and rescue activities in protecting life, property, and the environment.

Typical Tasks

Drive and operate fire apparatus and equipment. Operate engine pumping equipment to ensure proper water flow and pressure. Operate a variety of equipment related to fire suppression, rescue, and hazardous materials emergency activities. Respond to firefighting situations including laying hose lines, pulling working lines, holding the nozzle to direct the stream of water on the fire, placing, raising, lowering, and climbing ladders, and assisting in overhaul and salvage operations; participate in continuous training in fire suppression, prevention, and inspection. Inspect commercial, residential, and other occupancies for fire hazards and compliance with fire prevention codes and ordinances. Conduct fire prevention inspection and education programs. Inspect and perform routine maintenance on rescue equipment, fire apparatus, hydrants, hoses, and other support equipment.

Knowledge, Skills, and Abilities

The operating and mechanical principles of fire apparatus and equipment. Field calculations of hydraulics for the proper and effective operations of equipment. Operation of firefighting equipment. The physical layout of the municipality including street location, water distribution system, and major traffic and fire hazards. Rules, regulations, and operational procedures of the Fire Department. Firefighting methods and techniques; basic life-support medical procedures. Hazardous material first responder operational level methods and techniques.

Education and Experience

Usually a combination of education and experience equivalent to graduation from high school, preference for 2-year college education, and experience in fire suppression and prevention.

Fire Captain or Assistant Chief

General Job Description

The Fire Captain or Assistant Chief is a supervisory, administrative, firefighting, and safety education position. This position is typically second-in-command in the Fire Division, usually working under the Fire Chief. Responsible for managing personnel, planning, directing, coordinating, and evaluating all fire service activities, fire suppression training, and safety education programs for their assigned shift.

Typical Tasks

Responds to emergencies and assumes command of the fire scene unless relieved by a ranking officer; determines cause of fire, directs personnel and equipment on the emergency scene. Ensures full readiness of assigned shift; makes inspections and evaluations; ensures coordination and cooperation with other shifts, departments, and agencies. Initiates, approves, and coordinates personnel assignments and disciplinary actions within area of responsibility; maintains a proper balance of fire crew assignments on shift and investigates reported violations of rules and regulations; writes reports on findings. Prescribes actions based on findings of facts. Performs major staff administrative assignments, such as preparing operational plans to comply with state, federal, and district mandates; evaluates policy implementation and designs new policies or modifies existing policies for approval of Fire Chief.

Knowledge, Skills, and Abilities

Thorough knowledge of principles and practices of fire control, suppression, and prevention. Knowledge of fire service codes, rules, and regulations. Knowledge of operation and maintenance of fire service apparatus and equipment. Supervisory ability and skills. Ability to undertake supervision during emergency conditions and fire ground operations.

Education and Experience

Usually combination of experience and education equivalent to graduation from an accredited college or university with a major in fire science or related field, and experience in fire ground operations.

Fire Training Officer

General Job Description

The Fire Training Officer is responsible for managing, planning, directing, coordinating, and evaluating all fire service education programs.

Typical Tasks

Assess baseline skills of fire division personnel and evaluate present drill procedures. Plan, schedule, and conduct training drills. Create and maintain a training system showing all mandatory, specialized, optional, and miscellaneous training. Develop and maintain a system which identifies the training available and who should participate in various offerings. Make specialized training recommendations such as arson investigation, emergency medicine, etc. Assist in the preparation and maintenance of a training budget request. Respond to alarms to observe safety and operational practices for use in assessing training needs. Select, requisition, and distribute study texts.

Knowledge, Skills, and Abilities

Thorough knowledge of principles and practices of fire control, suppression, and prevention. Knowledge of fire service codes, rules, and regulations. Knowledge of operation and maintenance of fire service apparatus and equipment. Knowledge and practice of educational methods of teaching. Ability to evaluate effectiveness of training programs.

Education and Experience

Usually a combination of education and experience equivalent to graduation from high school, preference for college education, and extensive experience in fire suppression and prevention.

Fire Chief

General Job Description

Performs protective service and administrative work supervising the activities of suppression forces on an assigned shift within an assigned district.

Typical Tasks

Plans and directs the work of the fire companies. Responds to alarms and directs firefighting operations of all fire companies at the scene. Ensures that fire equipment, stations, and personnel are adequately able to the support the community through upkeep, response patterns, and assigned personnel. Develops fire protection arrangements with nearby districts and industries. Performs job performance, training reports, personnel functions, manning levels. Administers training programs and prepares administrative reports as required. Surveys buildings, grounds, and equipment to estimate needs of department. Prepares departmental budget and monitors finances. Confers with officials and community groups and conducts public relations campaigns to present need for changes in laws and policies and to encourage fire prevention. May investigate causes of fires and inspect buildings for fire hazards. May control issue of occupancy permits and similar licenses.

Knowledge, Skills, and Abilities

Thorough knowledge of principles and practices of fire control, suppression, and prevention. Knowledge of fire service codes, rules, and regulations. Knowledge of operation and maintenance of fire service apparatus and equipment. Supervisory ability and skills. Ability to undertake supervision during emergency conditions and fire ground operations. Administrative and training skills.

Education and Experience

Usually combination of experience and education equivalent to graduation from an accredited college or university with a major in fire science or related field and experience in fire ground operations in a supervisory capacity.

Fire Service Apparatus

Figure 1 Pumper.

Figure 2 Aerial ladder truck.

Figure 3 Ariel platform truck.

Figure 4 Rescue truck.

Figure 5 Aircraft rescue firefighting truck.

Figure 6 Mobile water supply apparatus.

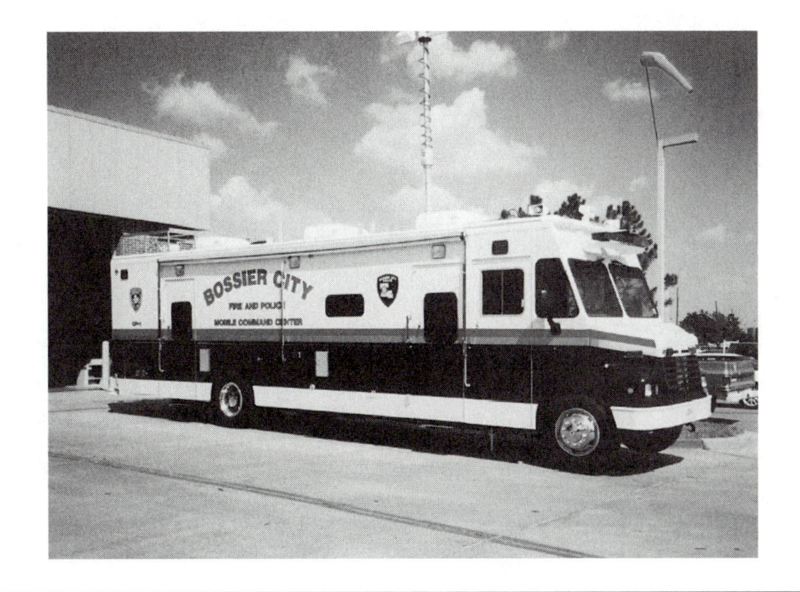

Figure 7 Mobile command post truck.

Code- and Standard-Making Organizations Containing Fire Safety Requirements

Organization	Code, Standards, or Publications
American National Standards Institute (ANSI) 1819 L Street NW, 6th Floor Washington, DC 20036 Tel: (202) 293-8020, Fax: (202) 293-9287 Email: info@ansi.org www.ansi.org	American National Standards developed or adopted from other organizations
American Petroleum Institute (API) 1220 L Street NW Washington, DC 20005-4070 Tel: (202) 682-8000 www.api.org	Standards and Recommended Practices
Consumer Product Safety Commission 4330 East-West Highway Bethesda, MD 20814-4408 Tel: (301) 504-7923 www.cpsc.gov	Voluntary Standards for consumer Products

(continues)

Organization	Code, Standards, or Publications
FM Global 1301 Atwood Avenue P.O. Box 7500 Johnson, RI 02919 Tel: (401) 275-3000, Fax: (401) 275-3029 www.fmglobal.com	Material and Equipment Testing Standards
International Code Council 5203 Leesburg Pike, Suite 600 Falls Church, VA 22041 Tel: (888) 422-7233, Fax: (703) 379-1546 www.iccsafe.org	International Building Code, International Fire Code
International Standards Organization (ISO) Central Secretariat, 1, rue de Varembé Case postale 56, CH-1211 Geneva 20, Switzerland Tel: (41) 22 749 01 11, Fax: (41) 22 733 34 30 www.iso.org	
National Fire Protection Association (NFPA) One Batterymarch Park Quincy, MA 02169-7471 Tel: (617) 770-3000, Fax: (617) 770-0700 www.nfpa.org	National Fire Codes and Standards Fire Protection Handbook
Occupational Safety and Health Administration (OSHA) Department of Labor 200 Constitution Avenue NW Washington, DC 20210 www.osha.gov	Code of Federal Regulations

Organization	Code, Standards, or Publications
SMACNA Sheet Metal and Air Conditioning Contractors National Association (SMACNA) 4201 Lafayette Center Drive Chantilly, VA 20151-1209 Tel: (703) 803-2980, Fax: (703) 803-3732 www.smacna.org	Voluntary Standards ASHRAE Handbook
Underwriter's Laboratories (UL) 333 Pfingsten Rd. Northbrook, IL 60062-2096 Tel: (847) 272-8800, Fax: (847) 272-8129 www.ul.com	Material and Equipment Testing Standards

Fire-Resistance Testing Standards (US)

Building Materials

Organization/ Specification	Name of Test	Sample	Property Measured
ASTM E-69	Combustible properties of treated wood by the crib test		
ASTM E-84	Surface burning of building materials	Building materials	Flame spread index, Smoke developed
ASTM E-108 Building Codes UBC 32-7 UL-790	Fire rating of roof coverings	Coatings, shingle shake, insulation, etc.	Spread of flame, intermittent flame, burning brand, flying brand
ASTM E-136	Behavior of materials in vertical tube furnace	Building materials	Combustibility or non-combustibility of building materials
ASTM E-160	Combustible properties of treated wood by the crib test		
ASTM E-162	Surface flammability of materials using a radiant heat source	Sheet laminates, tiles, fabrics, liquids, films	Flame spread index, visual characteristics
ASTM E-648 NFPA 253	Critical radiant flux of floor covering systems	Floor covering systems	Critical radiant flux at flameout

(continues)

Organization/ Specification	Name of Test	Sample	Property Measured
ASTM E-662	Specific optical density of smoke generated by solid materials		
CPSC HH-I-515D, HH-I-521F, HH-I-1030B 16CFR 1209.6	Critical radiant flux of attic insulation	Exposed attic floor insulation	Critical radiant flux at flameout
NIST NBSIR-82-2532	Combustion product toxicity	All materials	Inhalation toxicity
NY State, Dept. of State 15,1120	Modified Pittsburgh Test	All materials	Inhalation toxicity

Paints Aerosols and Liquids

Organization/ Specification	Name of Test	Sample	Property Measured
ASTM D-56, D-92, D-93, D1310	Flash point	Liquids	Flash point
ASTM D-3243 D-3278	Flash point-setaflash	Liquids, aviation turbine fuels	Flash point
ASTM D-1360	Fire retardancy of paint	Paint	Fire retardancy of coating or coating system on wood
FHSA ASTM-API 16 CFR 500.43	Flash point (tag open cup)	Aerosols	Flash point

Organization/ Specification	Name of Test	Sample	Property Measured
FHSA CSMA 16 CFR 500.45	Flame projection	Aerosols	Flame projection
CSMA Aerosol Guide	Drum test	Aerosols	
NIST NBSIR-82-2532	Combustion product toxicity	All materials	Inhalation toxicity
NY State, Dept. of State 15,1120	Modified Pittsburgh test	All materials	Inhalation toxicity

Textiles

Organization/ Specification	Name of Test	Sample	Property Measured
AATCC-33-1962	Flammability of clothing textiles	Fabrics	Time for flame spread, ease of ignition, flame intensity
ASTM D2859	Surface flammability of carpets and rugs	Carpets and rugs	Area of flame spread (greatest diameter)
Canvas Products Association Intl CPAI-84	Camping tents	Tenting materials	Area of flame spread, afterflame, char length
Canvas Products Association Intl. CPAI-75	Sleeping bags	Sleeping bags	Burning rate
City of Boston	Fire Code Section 11.2	Treated fabrics	Afterflame, afterglow
City of Boston	Fire Code Section 11.3	Inherent fire-resistant fabrics	Afterflame, afterglow

(continues)

Organization/ Specification	Name of Test	Sample	Property Measured
City of New York 290-40-SR	Board of Standards and Appeals	Fabrics	Afterflame, afterglow
CPSC CS-191-53 16 CFR 1610.4	Flammability of clothing textiles	Fabrics	Time for flame spread, ease of ignition, flame intensity
CPSC MAFT	Wearing apparel	Clothing and fabrics	Ignition time, maximum heat transfer rate
CPSC FF 3-71 16 CFR 1615.4	Children's sleepwear	Sleepwear sizes 0–6X	Residual flame time, char length
CPSC FF 3-71 16 CFR 1615.4	Children's sleepwear	Sleepwear sizes 7–14	Char length
CPSC FF 4-72 16 CFR 1632.4	Mattresses	Mattresses and mattress pads	Char length
CPSC FF 1-70 16 CFR 1630.4	Surface flammability of carpets and rugs	Carpets and rugs	Area of flame spread (greatest diameter)
CPSC FF 2-70 16 CFR 1631.4	Surface flammability of carpets and rugs	Carpets and rugs	Area of flame spread (greatest diameter)
CPSC PFF 6-76	Upholstery materials	Fabrics and furniture	Char length, substrate ignition
Depart of Transportation, State of California FMVSS 302	Flammability of interior materials: cars, trucks, multipurpose passenger vehicles, buses	Interior materials	Time of flame spread

Organization/ Specification	Name of Test	Sample	Property Measured
Federal Aviation Authority Par. 25.853	FAA regulations for compartment interiors	Compartment materials	Time of flame spread, afterflame, burn length
Federal Specification Defense Dept. DDD-C-95	Surface flammability of carpets and rugs	Carpets and rugs	Area of flame spread (greatest diameter)
Federal Test Method Std. FTMS 191 Method 5903	Flame resistance of cloth: vertical	Fabrics	Afterflame, afterglow, char length
Federal Test Method Std. FTMS 191 Method 5906	Burning rate of cloth: horizontal	Fabrics	Time of flame spread
Federal Test Method Std. FTMS 191 Method 5908	Burning rate of cloth: 45° angle	Fabrics	Ease of ignition, rate of burning
Federal Test Method Std. FTMS 191 Method 5910	Burning rate of cloth: 30° angle	Fabrics	Rate of burning
Federal Test Method Std. FTMS 191 Method 5920	Mackey	Cloth, related materials	Tendency of material to undergo self-heating at moderate temperature
NFPA 701 small scale	Fire tests for flame-resistant textiles and films	Fabrics and films	Afterflame, afterglow, char length
NFPA 701 large scale	Fire tests for flame-resistant textiles and films	Fabrics and films	Afterflame, afterglow, char length
NIST NBSIR-82-2532	Combustion product toxicity	All materials	Inhalation toxicity

(continues)

Organization/ Specification	Name of Test	Sample	Property Measured
NY State, Dept. of State 15.1120	Modified Pittsburgh test	All materials	Inhalation toxicity
State of California Par. 1273.1	Fire Code Title 19	Film, synthetic and coated fabrics	Afterflame, afterglow
State of California Par. 1237.1 small scale	Fire Code Title 19	Fabrics	Afterflame, afterglow
State of California Par. 1237.3 large scale	Fire Code Title 19	Fabrics	afterflame, char length
US Forestry Department 5100.3	Modified USTC-comparative inhalation		Inhalation toxicity

Plastics

Organization/ Specification	Name of Test	Sample	Property Measured
ASTM D-568	Flammability of plastics 0.050" and under	Plastic sheets and film	Non-burning, self-extinguishing, burning rate, visual characteristics
ASTM D-635	Rate of burning (self-supporting plastics)	Rigid plastics	Burning rate, visual characteristics
ASTM D-757	Incandescence resistance (rigid plastics)	Rigid plastics	Burning rate, visual characteristics

Organization/ Specification	Name of Test	Sample	Property Measured
ASTM D-1929, Procedure B	Ignition properties of plastics	Plastic sheets and films, thermoplastic pellets	Flash ignition temperature, self-ignition temperature, visual characteristics
ASTM D-2843	Smoke density from the burning of plastics	Plastic material	Percent of light absorption
Bureau of Ships NObs 84814 MIL-M-14g	Flammability and toxicity	Generally melamine plastic; any material	Flash ignition, self-ignition, composition and toxicity of gases evolved
CPSC CS 192-53 16 CFR 1611.4 ASTM D-1433	Flammability of plastic film	Plastic films, coated fabrics	Ignition time, rate of burning
Federal Test Method Std. FTMS 406 Method 2023	Flame resistance of plastics	Plastics difficult to ignite	Ignition time, burning time, flame travel
NIST NBSIR-82-2532	Combustion product toxicity	All materials	Inhalation toxicity
NY State, Dept. of State 15,1120	Modified Pittsburgh test	All materials	Inhalation toxicity

(continues)

Other

Organization/ Specification	Name of Test	Sample	Property Measured
ASTM D-2863	Oxygen index flammability test	Plastics, rigid papers, and cellular film	Oxygen index value
ASTM D-240 D-2015	Heat of combustion	Any combustible material	Heating chemical value
Bureau of Ships UL-44 IEEE P-45, P-383	Flammability of insulated wire	Insulated electrical conductors	Burn rate
CPSC 16 CFR 1500.44	Hazardous substances labeling act (toys)	Rigid and pliable solids	Burn rate
NIST ASTM E-662 NFPA 258	Smoke density/glow/ flame	All types of materials	Specific optical density
US. Forestry Dept. 5100.3	Modified US testing company	Fire tent laminate	Inhalation toxicity

Fire-Resistance Ratings

Some of the most common fire-resistance ratings used in construction and industry are identified in the following tables. The fire ratings stated are based on a response to a prescribed standard "test" fire, which is used as a basis of comparisons. Actual fires will present a more or less fire exposure than the standard fire, dependent on the site fuel loading, ventilation rate, and the enclosure thermal properties. "A," "B," and "C" ratings were originally defined by the Safety of Life at Sea (SOLAS) regulations. Hydrocarbon fire exposures for pool and jet fires have also recently evolved.

Fire Barriers (NFPA 255)

The average temperature rise of any set of thermocouples for each class of element protected is more than 250°F (121°C) above the initial temperature; or the temperature rise of any one thermocouple of the set for each class of element protected is more than 325°F (163°C) above the initial temperature.

Where required by the conditions of acceptance, a duplicate specimen shall be subjected to a fire exposure test for a period equal to one-half of that indicated as the resistance period in the fire endurance test, but not for more than one hour. Immediately after, the specimen shall be subjected to the impact, erosion, and cooling effects of a hose stream directed first at the middle and then at all parts of the exposed face, with changes in direction made slowly.

Exception: The hose stream test shall not be required in the case of construction having a resistance period, as specified in the fire endurance test, of less than one hour.

A Barriers (SOLAS or Title 46, Code of Federal Regulations, 72.05–72.10)

A 0 Cellulosic Fire, 60-minute barrier against flame/heat passage, no temperature insulation

A 15 Cellulosic Fire, 60-minute barrier against flame/heat passage, 15-minute temperature insulation

A 30 Cellulosic Fire, 60-minute barrier against flame/heat passage, 30-minute temperature insulation

A 60 Cellulosic Fire, 60-minute barrier against flame/heat passage, 60-minute temperature insulation

Class A divisions are those divisions formed by decks and bulkheads that comply with the following:

1. They are constructed of steel or material of equivalent properties.
2. They are suitably stiffened.
3. They are constructed to prevent the passage of smoke and flame for a one-hour standard fire test.
4. They are insulated with approved non-combustible materials so that the average

temperature of the unexposed side will not rise more than 356°F (180°C) above the original temperature within the time listed (A60: 60 minutes; A30: 30 minutes; A15: 15 minutes; A0: 0 minutes).

B Barriers (SOLAS or Title 46, Code of Federal Regulations, 72.05–72.10)

B 0 Cellulosic Fire, 30-minute barrier against flame/heat passage, no temperature insulation

B 15 Cellulosic Fire, 30-minute barrier against flame/heat passage, 15-minute temperature insulation

Class A divisions are those divisions formed by decks and bulkheads that comply with the following:

1. They are constructed to prevent the passage of flame for 30 minutes for a standard fire test.
2. They have an insulation layer such that the average temperature on the unexposed side will not rise more than 282°F (139°C) above the original temperature, nor will the temperature at any one point, including any joint, rise more than 437°F (225°C) above the original temperature (B15: 15 minutes; B0: 0 minutes).
3. They are of noncombustible construction.

C Barriers (SOLAS or Title 46, Code of Federal Regulations, 72.05–72.10)

C Noncombustible Construction

Class C barriers are of noncombustible materials and are not rated to provide any smoke, flame, or temperature passage restrictions.

H Barriers (UL 1709)

An exposure rating to a hydrocarbon (petroleum) fire is given an H rating. Typical ratings are:

H 0 Hydrocarbon Fire, 120-minute barrier against flame/heat passage, no temperature insulation

H 60 Hydrocarbon Fire, 120-minute barrier against flame/heat passage, 60-minute temperature insulation

H 120 Hydrocarbon Fire, 120-minute barrier against flame/heat passage, 120-minute temperature insulation

H 240 Hydrocarbon Fire, 120-minute barrier against flame/heat passage, 240-minute temperature insulation

J Ratings

Jet fire exposure or impingement ("J" ratings) are specified by some vendors or property owners for resistance to hydrocarbon jet fire exposures. Currently, no standardized test or test specification has been adopted by an industry or governmental body. Some recognized fire testing and experimental laboratories (SINTEF, Shell Research, etc.) have conducted extensive research on jet fire exposures and have proposed a test standard based on these studies (Ref. Offshore Technology Report OTO 93028, *Interim Jet Fire Test Procedure for Determining the Effectiveness of Passive Fire Protection Materials*).

Fire Doors (NFPA 252)

0.3 hours (20 minutes), Cellulosic fire

0.5 hours (30 minutes), Cellulosic fire

0.75 hours (45 minutes), Cellulosic fire

1.0 hours (60 minutes), Cellulosic fire

1.5 hours (90 minutes), Cellulosic fire

3.0 hours (180 minutes), Cellulosic fire

Over 3.0 hours (in hourly increments), Cellulosic fire

Except for 20-minute rated door assemblies where it is optional, immediately following the fire endurance test, the door test assembly shall be subjected to the impact, erosion, and cooling effects of a hose stream. Temperature rises are listed at 250°F, 450°F, and 650°F (121°C, 232°C, and 343°C); absence of a temperature rating indicates a rise of over 650°F (343°C) on the unexposed surface of the door after 30 minutes of testing.

Fire Windows (NFPA 257)

Fire ratings of windows were normally limited to the failure of wired glass at approximately 1,600°F (870°C); however, advances in glazing technology have increased the available fire resistance ratings of window assemblies.

0.3 hours (20 minutes), Cellulosic fire

0.5 hours (30 minutes), Cellulosic fire

0.75 hours (45 minutes), Cellulosic fire

Higher ratings available based on the application of other fire resistance standards fire tests (NFPA 255).

1.0 hours (60 minutes), Cellulosic fire

1.5 hours (90 minutes), Cellulosic fire

3.0 hours (180 minutes), Cellulosic fire

Over 3.0 hours (in hourly increments), Cellulosic fire

Within two minutes following the fire endurance test, the fire-exposed side of the fire window assembly is subjected to the impact, erosion, and cooling effects of a standard hose stream.

Fire Dampers (UL Std. 555)

0.3 hours (20 minutes), Cellulosic fire

0.75 hours (45 minutes), Cellulosic fire

1.0 hours (60 minutes), Cellulosic fire

1.5 hours (90 minutes), Cellulosic fire

Smoke Dampers (UL Std. 555S)

Smoke dampers are specified on the leakage class, maximum pressure, maximum velocity, installation mode (horizontal or vertical), and degradation test temperature of the fire.

Roof Coverings (NFPA 256)

Class A: flame spread less than 6 feet (1.82 meters)

Class B: flame spread less than 8 feet (2.44 meters)

Class C: flame spread less than 13 feet (3.96 meters)

For all classes of roof coverings, there is to be no significant lateral spread in the flame spread, no flying brands or particles are to continue to flame or glow after reaching the floor, no flaming is to be produced on the underside of the deck of the test sample, and the roof deck should not be exposed.

Fusible Links

Fusible links are available in temperature ratings of 125°F to 500°F (51.6°C to 260°C) and in various load ratings.

College/University Programs in Fire Safety/Science Administration and Management

University/Program (Resident Programs)	Degrees
Southern Illinois University Fire Science Management	BS
Western Illinois University Fire Science Management	BS
Allegheny University Emergency and Public Safety Services	BS, MS
Anna Maria College Emergency Planning and Response	MS
Arizona State University Emergency Management and Fire Science	BAS
George Washington University Crisis, Disaster Emergency Management	MS, PhD
Hope International University Management with focus in Crisis and Emergencies	MS
John Jay College Protection Management	MS
Jacksonville State University Public Administration with concentration in Emergency Management	MS
Oklahoma State University Fire and Emergency Management	MS
Saint Joseph's University Public Safety	MS

(continues)

University/Program	Degrees
California State University Fire Science	BS
Columbia Southern University Fire Science	BS
Lake Superior State University Fire Science	BS
University of New Haven Fire Science	BS
Eastern Kentucky University Fire & Safety Engineering Technology	AS
Elizabethtown Community and Technical College Fire Science	AS
University Center Gaylord Criminal Justice and Fire Science	BS
Henenpin Technical College Fire Science Technology	AAS
Houston Community College System Fire Science and Safety Technology	AS
Kansas City Community College Fire Science	AAS
Lakeland Community College Fire Science Technology	AS
Montgomery County Community College Fire Science Technology	AS
Passaic County Community College Fire Science Technology	AAS
Pima Community College Fire Science	AAS
Red Rocks Community College Fire Science	AAS
Seminole Community College Fire Science	AS
Southwest Tennessee Community College Fire Science	AS

University/Program	Degrees
Truckee Meadows Community College	
Fire Science Technology	AS
San Antonio College	
Fire Science	AS
University of Cincinnati	
Fire and Safety Engineering Technology	BS
Volunteer State Community College	
Fire Science Technology	AAS

University Programs in Fire Safety Engineering

Location/Program	Degrees
United States	
Interdisciplinary Graduate Program in Fire Safety Science Engineering Science University of California, Berkeley	MS, PhD
Department of Fire Protection Engineering University of Maryland http://www.enfp.umd.edu	BS, MS
Fire Protection and Safety Engineering Technology Oklahoma State University	BS
Center for Fire Safety Studies Worcester Polytechnic Institute http://www.wpi.edu	MS, PhD
Canada	
Department of Chemical Engineering University of British Columbia	MS
Fire Safety Center University of New Brunswick	BS, MS, PhD
United Kingdom	
Carleton University Fire Safety Engineering	MS, PhD

(continues)

Acronyms

The following acronyms are commonly used within the fire protection profession. In some cases, the acronym may have several meanings and its exact definition should be interpreted within the context in which it is or may be used.

ACV	Alarm Check Valve
ADD	Actual Delivered Density
AFAA	Automatic Fire Alarm Association
AFD	Aspirating Fire Detection
AFFF	Aqueous Film Forming Foam
AFP	Active Fire Protection
AFSA	American Fire Sprinkler Association
AHJ	Authority Having Jurisdiction
AIA	American Insurance Association
AISC	American Institute of Steel Construction
AIT	Autoignition Temperature
ALARP	As Low As Reasonably Practical
ALV	Alarm Valve
ANFO	Ammonia Nitrate Fuel Oil
ANSI	American National Standards Institute
AP	Annunciator Panel (fire alarm)
APAC	Asian Pacific Fire Safety Association
API	American Petroleum Institute
AR-AFFF	Alcohol Resistant-Aqueous Film Forming Foam
ARFF	Aircraft Rescue Fire Fighting
ARV	Air Release Valve
AS	Automatic Sprinklered *or* Associate of Science
ASD	Aspirating Smoke Detection *or* Automatic Smoke Detection
ASHRAE	American Society of Heating, Refrigeration, and Air Conditioning Engineers, Inc.
ASME	American Society of Mechanical Engineers
ASSE	American Society of Safety Engineers
ASTM	American Society for Testing and Materials
ATC	Alcohol-Type Concentrates
ATS	Automatic Transfer Switch
AWWA	American Water Works Association
BLEVE	Boiling Liquid Expanding Vapor Explosion

BMS	Burner Management System
BNFPC	BOCA National Fire Prevention Code
BNICE	Biological, Nuclear, Incendiary, Chemical, Explosives
BOCA	Building Officials and Code Administrators
BOP	Blowout Preventor (valve)
BS	Bachelor of Science
BSD	Beam Smoke Detector
Btu	British Thermal Unit
CAD	Computer Aided Dispatch
CAFS	Compressed Air Foam System
CAFSS	Clean Agent Fire Suppression System
CBRNE	Chemical, Biological, Radiological, Nuclear, Explosive
CCl_4	Carbon Tetrachloride
CFD	Computational Fluid Dynamics
CFR	Code of Federal Regulations
CFSI	Congressional Fire Services Institute
CFT	Cool-Flame Reaction Threshold
CISD	Critical Incident Stress Debriefing
CMA	Chemical Manufacturers Association
CO	Carbon Monoxide
CO_2	Carbon Dioxide
COPE	Construction Characteristics, Occupancy, Protection, and Exposures
CPSC	Consumer Product Safety Commission
CV	Check Valve
CVD	Combustible Vapor Dispersion
DC	Dry Chemical
DD	Duct Detector
DH	Double Outlet, Fire Hydrant
DOT	Department of Transportation (USA)
DP	Dry Pipe *or* Dry Pendant Sprinkler
DPV	Dry Pipe Valve
EDITH	Exit Drills in the Home
EIT	Enclosure Integrity Test *or* Engineer-In-Training
ELO	Extra Large Orifice (sprinkler)
EMI	Emergency Management Institute
EMS	Emergency Medical Services
EOLR	End of Line Resister
EPA	Environmental Protection Agency
ERG	Emergency Response Guidebook
ESFR	Early Suppression Fast Response (sprinkler)
FA	Fire Alarm
FACP	Fire Alarm Control Panel
FAMA	Fire Apparatus Manufacturer's Association
FAST	Firefighter Assist and Search Team
FD	Fire Department, Fire District, Fire Damper, *or* Fire Detection

FDC	Fire Department Connection
FDIC	Fire Department Instructors Conference
FE	Fire Endurance, Fire Escape, *or* Fire Extinguisher
FEHM	Fire and Explosion Hazard Management
FEMA	Federal Emergency Management Agency (USA)
FEMSA	Fire Equipment Manufacturers and Service Association
FESHE	Fire and Emergency Services Higher Education Project
FFFP	Film Forming Fluoroprotein Foam
FH	Fire Hydrant *or* Fire Hose
FHA	Fire Hazards Analysis *or* Assessment
FHR	Fire Hose Reel
FHZ	Fire Hazardous Zone
FI	Formal Interpretation
FIA	Factory Insurance Association
FM	Factory Mutual
FMANA	Fire Marshal's Association of North America
FMEA	Failure Mode and Effects Analysis
FMRC	Factory Mutual Research Corporation
FP	Fire Pump *or* Fluoroprotein (foam)
FPE	Fire Protection Engineer
FPH	Fire Protection Handbook
FPI	Fire Propagation Index
FPR	Fire Protection Rating
FRA	Firewater Reliability Analysis *or* Fire Risk Analysis
FRC	Fire Retardant Coveralls
FS	Flow Switch (water)
FSA	Formal Safety Assessment
FSCS	Firefighters' Smoke-Control Station
FSI	Flame Spread Index
FSR	Flame Spread Rating
FSSA	Fire Suppression Systems Association
FSSD	Fire Safe Shutdown
FTA	Fault Tree Analysis
FW	Firewater
FWP	Firewater Pump
FWS	Firewater System
GN_2	Gaseous Nitrogen
gpm	gallons per minute
HAD	Heat Actuated Device
HALON	Halogenated Hydrocarbon
HAZMAT	Hazardous Materials
HAZOP	Hazard and Operability Study
HAZWOPER	Hazardous Waste Operations and Emergency Response
HBPFM	High Backpressure Foam Maker
HC	Hose Cabinet *or* Hose Connection
HCP	Halon Control Panel
HD	Heat Detector
HFT	Hydraulic Forcible Entry Tools
HMIS	Hazardous Materials Identification System

HMR	Hazardous Material Regulations
HPR	Highly Protected Risk
HRR	Heat Release Rate
HSI	Hydraulics Standards Institute (USA)
HSSD	High Sensitivity Smoke Detection
HVAC	Heating, Ventilation, and Air Conditioning
IAAI	International Association of Arson Investigators
IABPFF	International Association of Black Professional Fire Fighters
IAFC	International Association of Fire Chiefs
IAFF	International Association of Fire Fighters
IAFSS	International Association of Fire Safety Science
IAP	Incident Action Plan
ICBO	International Conference of Building Officials
ICS	Incident Command System
IDC	Initiating Device Circuit
IEEE	Institute of Electrical and Electronics Engineers, Inc.
IFBA	International Fire Buff Associates, Inc.
IFSAC	International Fire Service Accreditation Congress
IFSTA	International Fire Service Training Association
ILBP	In-Line Balanced Proportioner
IME	Institute for Makers of Explosives
IR	Infrared
IRI	Industrial Risk Insurers
IS	Intrinsically Safe
ISFSI	International Society of Fire Service Instructors
ISO	Insurance Services Office *or* International Organization for Standardization
JCNFSO	Joint Council of National Fire Service Organizations
kPa	kiloPascals
LDH	Large Diameter Hose
LEL	Lower Explosive Limit
LEPC	Local Energy Planning Committee
LFL	Lower Flammable Limit
LHD	Linear Heat Detection
LNG	Liquefied Natural Gas
LO	Large Orifice (sprinkler)
LPG	Liquefied Petroleum Gas
LSC	Life Safety Code (NFPA 101)
LSD	Linear Smoke Detection
MABAS	Mutual Aid Box Alarm System
MAC	Manual Activation Callpoint (fire alarm)
MDH	Medium Diameter Hose
MEC	Minimum Explosible Concentration
MFL	Maximum Foreseeable Loss
MIC	Minimum Ignition Current
MIE	Minimum Ignition Energy
MORT	Management Oversight and Risk Tree

MPL	Maximum Possible Loss
MPS	Manual Pull Station (fire alarm)
MSDS	Material Safety Data Sheet
MTBF	Mean Time Between Failure
NAPI	National Arson Prevention Initiative (USA)
NAFED	National Association of Fire Equipment Distributors (USA)
NAFI	National Association of Fire Investigators
NASFM	National Association of State Fire Marshals (USA)
NBFAA	National Burglar and Fire Alarm Association, Inc. (USA)
NBFU	National Board of Fire Underwriters
NEC	National Electrical Code (NFPA 70)
NFA	National Fire Academy (USA)
NFAC	National Fire Alarm Code (NFPA 72)
NFC	National Fire Codes
NFDC	National Fire Data Center (USA)
NFDRS	National Fire Danger Rating System (USA)
NFIC	National Fire Information Council (USA)
NFIRS	National Fire Incident Reporting System (USA)
NFPA	National Fire Protection Association
NFSA	National Fire Sprinkler Association (USA)
NHT	National Hose Thread
NICET	National Institute for Certification in Engineering Technologies (USA)
NIFC	National Interagency Fire Center (USA)
NIMS	National Incident Management System
NIST	National Institute of Standards and Technology (USA)
NLE	Normal Loss Expectancy
NPLFA	Nonpower-Limited Fire Alarm (circuit or cables)
NPQB	National Professional Qualifications Board
NPSH	Net Positive Suction Head
NPSHa	Net Positive Suction Head Available
NPSHr	Net Positive Suction Head Required
NRP	National Response Plan
NRS	Nonrising Stem (gate valve)
NRTL	Nationally Recognized Testing Laboratory
NS	Nonsprinklered
NSC	National Safety Council (USA)
NST	National Standard Thread
NVFC	National Volunteer Fire Council (USA)
NWCG	National Wildfire Coordinating Group (USA)
ODP	Ozone Depletion Potential

OI	Oxygen Index
OIA	Oil Insurance Association
OIL	Oil Insurance Limited (Bermuda)
OSFM	Office of State Fire Marshal
OSHA	Occupational Safety and Health Act (USA)
OS & Y	Outside Stem and Yoke (valve)
PAV	Pre-Action Valve
PASS	Personal Alert Safety Systems
PCA	Practical Critical Fire Area
PDP	Pump Discharge Pressure
PDS	Point of Demand-Supply
PE	Professional Engineer
PFP	Passive Fire Protection
PHA	Process Hazard Analysis
PIO	Public Information Officer
PIV	Post Indicator Valve
PLFA	Power-Limited Fire Alarm (circuits or cables)
PML	Probable Maximum Loss
ppm	parts per million
PPV	Positive Pressure Ventilation
PRA	Probabilistic Risk Assessment
PS	Pressure Switch *or* Pull Station (fire alarm, manual)
psi	pounds per square inch
PSM	Process Safety Management
QH	Quadruple Outlet, Fire Hydrant
QOD	Quick Opening Device
QR	Quick-Response Sprinkler
QRA	Quantitative Risk Analysis
QRES	Quick-Response Early Suppression (sprinkler)
RDD	Required Delivered Density
RIC	Rapid Intervention Crew
RIT	Rapid Intervention Team
RIV	Rapid Intervention Vehicle
ROR	Rate of Rise
RP	Recommended Practice
RPP	Radiant Protective Performance
RPZ	Reduced Pressure Zone (water)
RTI	Response Time Index
RTT	Pre-Flame Reaction Threshold
RV	Riser Valve (water) or Relief Valve
RVP	Reid Vapor Pressure
SBC	Standard Building Code
SBCCI	Southern Building Code Congress International, Inc.
SCBA	Self-Contained Breathing Apparatus
SD	Smoke Detector *or* Smoke Damper
SDH	Small Diameter Hose
SDI	Smoke Density Index *or* Smoke Damage Index
SEP	Surface Emissive Power
SFM	State Fire Marshal
SFPC	Standard Fire Prevention Code (USA)
SFPE	Society of Fire Protection Engineers (USA)

SI	Système International d'Unités (metric system)
SINTEF	Norges Branntekniske Laboratorium (Norwegian Fire Research Laboratory)
SIP	Shelter-In-Place
SMACCNA	Sheet Metal and Air Conditioning Contractors National Association, Inc.
SMS	Smoke Management System
SOLAS	Safety of Life at Sea
SPF	Single Point Failure
SSP	Standard Spray Pendant (sprinkler)
SSU	Standard Spray Upright (sprinkler)
STTC	Standard Time-Temperature Curve
SV	Smoke Vent
TCE	Test Connection
TCFA	Theoretical Critical Fire Area
TH	Triple Outlet, Fire Hydrant *or* Test Header
TNT	Trinitrotoluene
TPP	Thermal Protective Performance
TS	Tamper Switch
UBC	Uniform Building Code
UEL	Upper Explosive Limit
UFC	Uniform Fire Code
UFL	Upper Flammable Limit
UL	Underwriter's Laboratories, Inc.
ULC	Underwriter's Laboratories, Inc., Canada
US	United States
USFA	United States Fire Administration
UV	Ultraviolet
UVCE	Unconfined Vapor Cloud Explosion
VCE	Vapor Cloud Explosion
VSS	Vented Suppressive Shield
WC	Wet Chemical
WCCE	Worst Case Creditable Event
WH	Wall Hydrant
WMD	Weapons of Mass Destruction
WOM	Water Oscillating Monitor
WPIV	Wall Post Indicator Valve
ZV	Zone Valve (sprinkler)
1OO2	One Out Of Two
2OO2	Two Out Of Two
2OO3	Two Out Of Three

Bibliography

American Iron and Steel Institute, *Fire Protection Through Modern Building Codes,* Fourth Edition, American Iron and Steel Institute, New York, NY, 1971.

American Petroleum Institute (API), RP 14G, *Recommended Practice for Fire Prevention and Control on Open-Type Offshore Production Platforms,* Second Edition, API, Washington, DC, 1986.

American Petroleum Institute (API), RP 521, *Guide for Pressure Relieving and Depressuring Systems,* Fourth Edition, API, Washington, DC, 1997.

American Petroleum Institute (API), RP 752, *Management of Hazards Associated with Location of Process Plant Buildings,* First Edition, API, Washington, DC, 1995.

American Petroleum Institute (API), RP 2001, *Fire Protection in Refineries,* Sixth Edition, API, Washington, DC, 1984.

American Petroleum Institute (API), Publication 2021, *Guide for Fighting Fires In and Around Flammable and Combustible Liquid Atmospheric Petroleum Storage Tanks,* Third Edition, API, Washington, DC, 1991.

American Petroleum Institute (API), Publication 2028, *Flame Arresters in Piping Systems,* Second Edition, API, Washington, DC, 1991.

American Petroleum Institute (API), RP 2218, *Fireproofing Practices in Petroleum and Petrochemical Processing Plants,* First Edition, API, Washington, DC, 1988.

American Society for Testing Materials (ASTM), C1055, *Standard Guide for Heated Systems Surface Conditions that Produce Contact Burn Injuries,* ASTM, West Conshohocken, PA, 1997.

Americana Review, *Fire Fighting of Long Ago,* Americana Review, Scotia, NY, 1981.

Bare, William K., *Introduction to Fire Science and Fire Protection,* John Wiley & Sons, New York, NY, 1978.

Blackstone, G. V., *A History of the British Fire Service,* Fire Protection Association/Loss Prevention Council, London, UK, 1996.

Braidwood, John S., *Fire: Its Prevention & Extinguishing,* Second Edition, H. & J. Pillians & Wilson, Edinburgh, Scotland, 1913.

Brannigan, Francis L., *Building Construction for the Fire Service,* Third Edition, National Fire Protection Association (NFPA), Quincy, MA, 1992.

Brown, A. A. and Davis, Kenneth P., *Forest Fire Control and Use,* Third Edition, McGraw-Hill, New York, NY, 1980.

Bryan, John L., *Automatic Sprinkler & Standpipe Systems,* National Fire Protection Association (NFPA), Quincy, MA, 1976.

Bryan, John L., *Fire Suppression and Detection Systems,* Third Edition, Macmillan Publishing Company, Los Angeles, CA, 1993.

Buck, Charles C. and Schroeder, Mark J., *Fire Weather, A Guide for Application of Meteorological Information to Forest Fire Control Operations,* Forest Service, US Department of Agriculture, Agriculture Handbook 360, US Government Printing Office, Washington, DC, 1970.

Bugbee, Percy, *Men Against Fire: The Story of the National Fire Protection Association 1896–1971,* Fidelity Press, Boston, MA, 1971.

Bugbee, Percy, *Principles of Fire Protection,* National Fire Protection Association (NFPA), Quincy, MA, 1982.

Burklin, Ralph W., *Fire Terms: A Guide to Their Meaning and Use,* National Fire Protection Association (NFPA), Boston, MA, 1980.

Calderone, John A., *The History of Fire Engines,* Barnes and Noble Books/ Brompton Books, Greenwich, CT, 1997.

Cannon, Donald J., General Editor, *Heritage of Flames: The Illustrated History of Early American Firefighting,* Doubleday Publishers, Garden City, NY, 1977.

Center for Chemical Process Safety (CCPS), *Guidelines for Evaluating Process Plant Buildings for External Explosions and Fires,* American Institute of Chemical Engineers, New York, NY, 1996.

Conway, W. Fred, *Firefighting Lore,* Fire Buff House Publishers, New Albany, IN, 1993.

Cornell, James, *The Great International Disaster Book,* Third Edition, Charles Scribner's Sons, New York, NY, 1982.

Costello, Augustine E., *Our Firemen: A History of the New York Fire Departments, Volunteer and Paid, 1609 to 1887,* Knickerbocker Press, New York, NY, 1997.

Cote, Arthur E., Editor, *Fire Protection Handbook,* Eighteenth Edition, National Fire Protection Association, Quincy, MA, 1997.

Cote, Arthur E., *Handbook of Industrial Fire Protection,* Third Edition, National Fire Protection Association, Quincy, MA, 1987.

Crosby, E. U., *Hand-Book of the Underwriters' Bureau of New England,* The Standard Publishing Company, Boston, MA, 1896.

Davletshina, Tatyana, *Industrial Fire Safety Guidebook,* Noyes Publications, Westwood, NJ, 1998.

Diamantes, David, Fire Prevention Inspection and Code Enforcement, Thomson Delmar Learning, Clifton Park, NY 2007

Dinneno, P. J., Editor, *SFPE Handbook of Fire Protection Engineering,* Second Edition, National Fire Protection Association, Boston, MA, 1995.

Ditzel, Paul, *Fire Alarm! The Story of the Red Alarm Box on the Corner,* Fire Buff House Publishers, New Albany, IN, 1990.

Ditzel, Paul, *Fire Engines, Firefighters: The Men, Equipment, and Machines from Colonial Times to the Present,* Bonanza, New York, NY, 1984.

Ditzel, Paul, *Fireboats: A Complete History of the Development of Fireboats in America,* Fire Buff House, New Albany, IN, 1989.

Dunshee, Kennth H., *Enjine! – Enjine! A Story of Fire Protection,* Home Insurance Company, New York, NY, 1939.

Drysdale, Dougal D., *An Introduction to Fire Dynamics,* John Wiley & Sons, New York, NY, 1985.

Earnest, Ernest P., *The Volunteer Fire Company Past and Present,* Stein & Day Publishers, Briarcliff Manor, NY, 1979.

Editors of Country Beautiful, *Great Fires of America,* Country Beautiful Corporation, Waukesha, WI, 1973.

Factory Mutual Engineering Corporation, *Handbook of Industrial Loss Prevention,* Second Edition, McGraw-Hill, New York, NY, 1967.

Freitag, Joseph K., *Fire Prevention and Fire Protection as Applied to Building Construction: A Handbook of Theory and Practice,* Second Edition, John Wiley & Sons, New York, NY, 1921.

Gagnon, Robert M., *Design of Water-Based Fire Protection Systems,* Thomson Delmar Learning, Clifton Park, NY, 1997.

Gagnon, Robert M., *Design of Special Hazard & Fire Alarm Systems,* Thomson Delmar Learning, Clifton Park, NY, 1998.

Gagnon, R. M., Heckler, K. A, et al., *Department of Fire Protection Engineering 40th* Anniversary History Book, 1956–1996, Jostens Publishing, Minneapolis, MN, 1996.

Garrison, Webb B., *Disasters That Made History,* Abingdon Press, New York, NY, 1973.

Goodenough, Simon, *Fire! The Story of the Fire Engine,* Chartwell Books, Secaucus, NJ, 1978.

Goodman, M. W., MD, *Inventing the American Fire Engine: An Illustrated History of Patented Ideas for Fire Pumpers,* Fire Buff House Publishers, New Albany, IN, 1994.

Goudsblom, Johan, *Fire and Civilization,* Loss Prevention Council, London, UK, 1996.

Griswold, J., *The Fire-Insurance Agents Text-Book: An Annotated Dictionary of the Terms and Technical Phrases in Use Among Fire Underwriters,* Smith, Montreal, Canada, 1888.

Hass, Ed, *Fire Equipment,* PRC Publishing, Ltd., London, UK, 1998.

Hickey, Harry E., *Hydraulics for Fire Protection,* National Fire Protection Association, Quincy, MA, 1980.

Hoover, Stephen R., *Fire Protection for Industry,* Van Nostrand Reinhold, New York, NY, 1991.

Ingram, Arthur, *A History of Fire-Fighting and Equipment,* Chartwell Books, Secaucus, NJ, 1978.

International Code Council, *2003 ICC Performance Code for Buildings and Facilities™,* ICC, Country Club Hills, IL, 2003.

International Fire Service Training Association (IFSTA), *Fire Service Orientation and Terminology,* Third Edition, Oklahoma State University, Fire Protection Publications, Stillwater, OK, 1993.

Institute of Makers of Explosives, *Glossary of Commercial Explosives Industry Terms,* Safety Library Publications No. 12, Institute of Makers of Explosives, Washington DC, 1985.

Jacobs, T. A., *The Fire Engine,* Brompton Books Corp., Greenwich, CT, 1993.

Kearne, Paul W., *The Ward LaFrance Book of Conflagrations,* Ward LaFrance, Emira, NY, 1944.

Kimball, Warren Y., *Fire Department Terminology,* Fourth Edition, National Fire Protection Association, Boston, MA, 1970.

Klinoff, Robert, Introduction to Fire Protection 3E, Thomson Delmar Learning, Clifton Park, NY, 2007

Kuvhinoff, B. W. et al., Editors, *Fire Sciences Dictionary,* John Wiley & Sons, New York, NY, 1977.

Layman, Lloyd, *Attacking and Extinguishing Interior Fires,* Third Edition, National Fire Protection Association, Boston, MA, 1958.

Lloyd's, *Lloyd's List, 1734–1984, 250th Anniversary Special Supplement,* Lloyd's of London Press Ltd., London, UK, 1984.

Mahoney, Eugene, *Introduction to Fire Apparatus and Equipment,* Second Edition, Fire Engineering, 1986.

Minton, A. S. et al., Editors, *Dictionary of Fire Technology,* Chartry Publications Limited, London, UK, 1952.

Morris, John V., *Fires and Firefighters,* Bramhall House, New York, NY, 1955.

Morrison, Ellen E., *Guardian of the Forest: A History of Smokey Bear and the Cooperative Forest Fire Prevention Program,* Morielle Press, Alexandria, VA, 1991.

Moulton, Robert S., Editor, *Crosby-Fiske-Forster Handbook of Fire Protection,* Ninth Edition, National Fire Protection Association, Boston, MA, 1941.

NFPA, *Fire Pump Handbook,* National Fire Protection Association, Quincy, MA, 1998.

NFPA, *NFPA Inspection Manual,* Seventh Edition, National Fire Protection Association, Quincy, MA, 1994.

Nolan, Dennis P., *Fire Fighting Pumping Systems at Industrial Facilities,* Noyes Publications, Westwood, NJ, 1998.

Nolan, Dennis P., *Handbook of Fire and Explosion Protection Engineering Principles for Oil, Gas, Chemical, and Related Facilities,* Noyes Publications, Westwood, NJ, 1996.

Quintiere, James G., *Principles of Fire Behavior,* Thomson Delmar Learning, Clifton Park, NY, 1998.

Perry, Donald, *Wildland Firefighting, Fire Behavior, Tactics & Command,* Second Edition, Fire Publications, Inc., 1990.

Petrovich, Wayne P., *A Fire Investigator's Handbook: Technical Skills for Entering, Documenting, and Testifying in a Fire Scene Investigation,* Charles C. Thomas, Springfield, IL, 1998.

Pyne, Stephen J., *Fire in America: A Cultural History of Wildland and Rural Fire,* University of Washington Press, Seattle, WA, 1997.

Rafferty, Ken, *Fire Control, Flammable Liquids and Gases: California Fire Training Program,* California State Department of Education, Bureau of Industrial Education, Sacramento, CA, 1964.

Roetter, Charles, *Fire Is Their Enemy: Fire-Fighting Through the Ages,* Sydney Angus and Robertson, London, UK, 1962.

Russel, Henry, *Fire Catastrophes, Fire Disasters: The Truth Behind the Tragedies,* Greenwich Editions, London, UK, 1998.

Smith, Dennis, *Dennis Smith's History of Firefighting in America: 300 Years of Courage,* The Dial Press, New York, NY, 1978.

Soloman, Robert E., Editor, *Automatic Sprinkler Systems Handbook,* Sixth Edition, National Fire Protection Association, Quincy, MA, 1994.

Stephens, Hugh W., *The Texas City Disaster, 1947,* University of Texas Austin, TX, 1997.

Stephens, Peter J., *The Story of Fire Fighting,* Harvey House, Irvington-on-Hudson, NY, 1966.

Tamarin, Alfred, *Fire Fighting in America,* The Macmillan Company, New York, NY, 1971.

Technomic Publishing, *Handbook of Fire Retardant Coatings and Fire Testing Services,* Technomic Publishing Company, Lancaster, PA, 1990.

Thiery, Pierre, *Fireproofing: Chemistry, Technology, and Applications,* Elsevier Science, London, UK, 1970.

Thompson, Norman J., *Fire Behavior and Sprinklers,* National Fire Protection Association, Boston, MA, 1974.

Thomson Delmar Learning, *Firefighters Handbook, Essentials of Firefighting and Emergency Response,* Second Edition, Clifton Park, NY, 2004.

Tuve, Richard L., *Principles of Fire Protection Chemistry,* National Fire Protection Association, Boston, MA, 1976.

Underwriter's Laboratories, UL 555S, *Leakage Rated Dampers for Use in Smoke Control Systems,* Third Edition, Underwriter's Laboratories, 1996.

Vanderveen, B. H., Editor, *Fire-fighting Vehicles, 1840–1950,* Frederick Warne & Co., New York, NY, 1972.

Wass, Harold S. Jr., *Sprinkler Hydraulics,* IRM Insurance, White Plains, NY, 1983.

Williams, Earnest W., *Fire Fighting: A Complete Treatise on all Subjects Pertaining to Fire Fighting, Fire Prevention, Fire Protection, Tools, Appliances, etc.,* Nicholas-Ellis, Lynn, MA, 1942.